CONSERVATION BIOLOGY AND BIODIVERSITY

CRITICAL RESEARCH TECHNIQUES IN ANIMAL AND HABITAT ECOLOGY

EXAMPLES FROM INDIA

Conservation Biology and Biodiversity

Additional books and e-books in this series can be found
on Nova's website under the Series tab.

CONSERVATION BIOLOGY AND BIODIVERSITY

CRITICAL RESEARCH TECHNIQUES IN ANIMAL AND HABITAT ECOLOGY

EXAMPLES FROM INDIA

KAUSHALENDRA KUMAR JHA
AND
MICHAEL O'NEAL CAMPBELL
EDITORS

nova
science publishers
New York

Library of Congress Cataloging-in-Publication Data

ISBN: 978-1-53619-720-4

Published by Nova Science Publishers, Inc. † New York

CONTENTS

PREFACE

This book covers selected topics on research methods in modern ecology, focussing on animal ecology, landcover assessment and habitat change. This is plainly a vast area, due to the multiplicity of research foci and the great range of research techniques available. Two main clusters of research methods have emerged: laboratory studies and field techniques. Both are now totally reliant on computer hardware and software and associated technology. Recent developments in ecological research methods largely concern the adoption of enhanced computer software techniques and the syntheses of these into pre-existing research methods, the better to record, analyze and explain the behavioral patterns of the studied animals and/or plants, and their ecologies. Chemical analyses, including studies of animal dietary and foraging patterns are also important, given reductions in food resources, environmental degradation and animal intrusions into human life spaces. This volume focusses on specific examples that typify some of the recent developments in research methods. In this regard, the research methods are spread over several fields of research and the interests of several researchers. Animal and integrated ecology in these specialisms is also considered broadly, to include habitat analysis, animal behavior, landcover, habitat and plant ecological change and even human/animal relations, and international case studies along these lines. Chapters on plant ecology and genetics are included, to indicate that animal ecology is inherently linked to vegetation assessment and plant dynamics, as part of the animal species interactive environment and animal bodies and related genetic studies run through any studies of animal ecology.

Technological and methodological advancements in the applied sciences may appear to be the main drivers in the development of animal ecology, but the latter discipline, based on people's understanding of animals remains the main focus and guide for research method development. Remote sensing and geographical information systems are emerging as relevant, cutting-edge research methods, at small, medium and large-scale levels. Animal behavioral studies at the more detailed, individual subject level are also complementing this trend. To better explore animal behaviors and the relations between these and the environmental context, the key developments concern closer, less intrusive and observable recording of animal life activities. This includes more critical human observation of animal behavior, and more detailed study of animal environments, including plant ecology. From a technological perspective, developments include more accurate positioning systems, more sensitive tracking systems, the removal of obstacles to clearer observation and species identification, such as darkness and poor lighting, dense vegetation and coarse image resolution and more comparative studies across different local contexts and global ecosystems. More detailed ecological studies and shared analyses of the research results may serve as guides for the development of more sophisticated research tools. The syntheses of these new developments with older research methods such as field foot surveys, coarse resolution satellite imaging and aerial photography, georeferencing with hand-held devices and eyeball observation of animals in the field may ensure the continuance of constructive research engagements.

The current book examines these issues through with some carefully selected and relevant case studies, each occupying one chapter. The first chapter, by Professor Kaushalendra Kumar Jha, a member of most distinguished technical Indian Forest Service, titled *Forest Management in India: A Paradigm Shift and the Use of Geoinformatics*, looks at how the management of forests in India developed historically with the introduction of new ideas and GIS systems. The view and exploitation of the Indian forests changed, with the focus shifting from a resource derivation model to a conservation model that acknowledged the forest's status as an animal habitat and a vital component of a wider ecosystem. From this model emerged the concept of sustainable development. The chapter argues that India has a very long history of forest management, with the period of scientific forestry lasting more than 150 years. The historical timeline includes the ancient period of the Gupta and Maurya empires and the Vedic period. Research materials include ancient religious scriptures and writings, the medieval records and some legal institutions of modern India. During this period, the exploitation model dominated, hence the forests slowly declined from levels with sufficient supplies for local communities to levels of scarcity, mostly through timber exploitation. Conservation mindsets emerged with the development of new forest policies, acts and rules, and international conventions. This led to a change from production to conservation forestry, with the inclusion of local, indigenous people and their needs for food security and livelihoods. Forest sustainability was defined within the principles of sustainable forest management (SFM). SFM and geoinformatics were integrated into a constructive National Working Plan Code. The chapter concludes that there are positive results.

The second chapter, titled *Biogeography as an Ecological Science and Methodology* by Michael O'Neal Campbell, examines biogeography in its relationship with ecology as a research methodology, and its adaptation to changes in ecological sciences. The ideas presented in this chapter examine issues relevant to the recent developments in ecological contexts such as India. Biogeography is noted as a spatial science, with almost universal application to biological topics, including ecology and the applications of this science. This chapter refers to the earlier chapters as relevant to the dynamics between biogeography and ecology, as most of the examined topics may have spatial variation. The key issue of disciplinary duplication is mentioned, as is the evidence of the unique contribution of both biogeography and ecology to the larger project of environmental change and dynamics. It is concluded that currently, most ecological topics are becoming more spatialized, hence the tools of biogeography may be increasingly relevant to ecology, especially spatially oriented technologies like geomatics. The chapter concludes that ecology and biogeography have a joint future, for the increasingly complex environmental issues of the 21st century.

The third chapter, by Piara Lal Loi and Professor Kaushalendra Kumar Jha, who are authorities of animal and habitat ecology, is titled *Landuse-Landcover Dynamics Analysis Using Remote Sensing and GIS: A Case of Central Himalayan Terai, India.* This chapter looks at the role of remote sensing and GIS in the assessment of natural resource development, clustering around the central concept of change detection, a key investigation at the ecosystem level. This application is tested in a case study of the underdeveloped Terai districts within the Central Himalayan region of India, selected for studying landuse and landcover (LULC) dynamics and their driving factors.

In this study, the LULC was classified into 5 classes (forest, agricultural land, waterbodies, wasteland and built-up areas) and 15 subclasses using LISS I and LISS III images of 1989 and 2002, respectively. Over 13 years, the forest land, wasteland and waterbodies lost area to the expanding agricultural landcover, largely due to urbanization, infrastructural development, plantation/regeneration/orchard creation and management, and tree losses due to overexploitation and the wood-based industry. Monsoon occurrence and livelihood security actions also played a role. The findings indicated change-detection based planning is useful for the wider issues of Forest Working Plans for the sustainable management of forest, water and wildlife resources. Considering the high biodiversity of India's animal species, these considerations are of vital significance.

The fourth chapter by Professor Kaushalendra Kumar Jha, titled *Geoinformatics, Climate Change, Habitat Dynamics and a Case of Vultures in Central India*, looks at vulture ecology and related climatic variables. The central thesis is that the imperilled situation of vultures is highly dependent on habitat issues, and the analysis of this situation may be better understood by employing the theoretical concepts of geoinformatics, with applications to the parameters of climate and habitat change. Geoinformatics enables the categorization of environments according to vulture habitat suitability, providing an informed background for positive conservation planning, especially if the studies go forward into more advanced techniques. The justification for this type of analysis is that such environmental variables, are intrinsic for the evaluation of species conservation possibilities. The chapter focusses on habitat (niche) modelling, species distribution models such as heuristic and statistical models and predictive law optimization and machine learning and how these may be calibrated for accurate measurement. This is linked to habitat projection modelling and climate models based on different scenarios of emission. It is inferred that GIS, ensemble modelling for Species Distribution Models (SDMs) and Global Circulation Models (GCMs) are a constructive approach.

The fifth chapter, titled *A Symbiosis of Habitat Conservation and Eco-Tourism Development: A Case Study of the Valmiki Tiger Reserve, India* by Debashis Debnath and Jaynath Ram focuses on two important ecosystem-related human activity systems: habitat conservation and eco-tourism. The first of these comprises the processes of preservation, protection and restoration of biodiversity, natural resources, ecosystem or environment for sustainable development. The second is a conservation tool representing an economic alternative to more invasive and extractive activities and facilitating more public awareness of environmental issues. The entwining of these two systems contributes effective environmental services for sustainable development. This assessment tests the effectiveness of the integrated approach, using an empirical-based study in the Valmiki tiger reserve in Bihar, India. This study used a participatory approach including sociological methods such as random sampling among villagers and tourists. The results of the study indicate the influences of process of globalization, and the convergence of the integration of the social and environmental processes. These include the issues of awareness, education, knowledge, and Naxalite (social unrest) complexities in the studied areas. This also indicated the scopes of habitat conservation within the social issues of ethos, religious values and ecotourism activities, and the balance between livelihoods system and ecological development.

The sixth chapter, by legal expert Arvind Kumar Jha, titled *Technology in Good Governance: A case of FRA and Geoinformatics*, looks at forest policy and the role of geoinformatics in the development of effective forest conservation. In discussing the role of geomatics in forest assessment, the history of forest policy must be elaborated at least on those points that affected the spatialization of forest structure and issues. The chapter notes that the policies governing Indian forests have evolved from the 1952 policy that focussed on revenue generation, to the 1988 policy that moved towards environmental stability, maintenance of ecological balance and derivation, rather than only economic development. The latter policy also supported the rights and concessions granted to forest, rural inhabitants to retrieve forest products for their own gain.

Another policy was the Scheduled Tribes and Other Traditional Forest Dwellers (Recognition of Forest Rights) Act, 2006 (Forest Rights Act or FRA) which was legislated by the Ministry of Tribal Affairs (MoTA), which combined the recognition of forest rights for specific peoples with participatory conservation actions for forest health and maintenance of the ecological balance. Problems included the need for effective management as other stakeholders still sought to exploit the forest. Hence, the implementation of the FRA required effective monitoring and geoinformatics applications were developed for decision making in the state of Maharashtra in India. The assessment of this development is the main topic of this chapter.

The seventh chapter, by Professor Kaushalendra Kumar Jha, titled *Similar Yet Different Rice-Farming Ecosystems: UNESCO Associated Cases of Two Culturally Rich Asian Nations* looks at the biophysical and cultural issues that surround the ecology of a historical case study of the *Subak* landscape, which has been recognized as a cultural heritage site by UNESCO in Indonesia. A second case study is of the *Apatani* cultural landscape of India, which is listed as a proposed cultural heritage site. The two sites differ in the form of cultural faith, the crops cultivated, the landscape topography, natural resources, and other possible characteristics. Common attributes include terraced rice cultivation, using historic indigenous irrigation systems. There are biophysical and cultural parameters that affect the ecology of the two contexts. These parameters include the people's culture, farming and irrigation systems, agriculture model and biodiversity, environmental problems, and other issues. These issues are evaluated for the sustainability of the ecosystems.

The eighth chapter, *Thermal Drones and Large Mammalian Carnivores* by Michael Campbell examines thermal infrared cameras mounted on drones, which have taken the capabilities of drones in field research a step further, enabling detection and recording of ecological issues in the night and total darkness. The issue of measuring temperature differentiation between an animal's body and the environment is examined. The chapter examines the application of thermal infrared image and video captures to large mammalian carnivores (LMCs), which are among the most important species that engage with people. These species may challenge people in their life spaces and even threaten human life, but they are generally difficult to study, due to their rarity, elusiveness, aggressive abilities, and large ranges. They are also important due their charismatic status in the human social media and the arguably polarized political context that surrounds their conservation and possible extirpation.

The evidence presented indicates that thermal infrared drones are a crucial addition to the methodological tools for LMC research, but there are limitations that may restrict their utility. These include the remoteness and irregularity of LMC movements and habitats, and the limited capacity of thermal infrared drones to penetrate the thickly vegetated landcover of some LMC habitats. There are also legal restrictions like those limiting other drone-usages, such as privacy issues linked to human presence or land ownership. Drones must also be closely monitored, but LMC presence can pose a physical threat to drone operators, whether in natural or urban landscapes. Evidence indicates that these restrictions are surmountable, however, the political will may be lacking, indifferent or even hostile. Case studies from North America and India are cited that point to increasing rather than weakening of legal restrictions on drone use. Further evidence points to minimal impacts on LMCs, but in more research is needed on this topic. It is concluded that thermal infrared drones are vital research tools, especially for the study of nocturnal LMCs.

The ninth chapter, titled *Rehabilitation of Mine Spoils: A Review on Problems and Strategies with Cases of Indian Mines* by Abhay Kumar Singh and Kaushalendra Kumar Jha examines the environmental impacts of mining, especially the topic of land restoration after the cessation of the mining activity. It is argued that such restoration is a major concern for ecologists, and a crucial challenge in several countries, with problems including the reversal of soil strata, instability of slopes, and poor quality of left-over soils, especially with the proximity of the mining areas with forest ecosystems and human habitation. The case study is of the bauxite mining industry in India. The research is based on abstract book consultation, and online and hardcopy literature searches, examining solutions such as legal provisions, and rehabilitation problems, such as plantation versus mining spoil, mine-spoil impoverishment, erosion and water conservation, and mine-spoil toxicity, top-soil replacement, soil amendments, seeding and planting, and species selection and revegetation. A narrower case study focuses on bauxite mine rehabilitation in Koraput District, Odisha, India. It is argued that a classical restoration model, such as the Bradshaw model would be difficult to implement here, because the original ecosystem of the area would have to be recreated. An incomplete restoration or 'rehabilitation' is assessed. It is argued that the extraction method (and the refilling of the extracted area) for bauxite, using the trench method is easier than the opencast mining of iron ore or coal, as the gradient of the filled trenches is usually gentler in bauxite mines, making it easier to rehabilitate with plantation development. The chapter concludes by assessing the application of the classical Bradshaw model of rehabilitation for the stabilization of disturbed areas, and the social and environmental implications externalities.

The tenth chapter, titled *What Could be the Present and Future Habitats of Residents and Wintering Vultures for Conservation in the Indian Subcontinent* by Radhika Jha and Kaushalendra Kumar Jha, looks at the Indian subcontinent as one of the three strongholds of Old-World vultures, and the recent declines in the numbers of resident and migrant. It aims to predict the potential habitat expanse of vulture species in the Indian subcontinent, using knowledge about these species gained from cross-continental studies. The species distribution model, MaxEnt was used, incorporating the inputs of bioenvironmental covariates and species occurrences, and presence data of nine vulture species derived from published literature and citizen science observations. CCSM4 (Community climate system model) Global circulation model along with lowest (RCP2.6) and highest (RCP8.5) emission scenarios were selected for predictions. The results indicated that models developed were robust enough (0.704-0.729; residents and 0.812-0.826; migrants) and were replicable for conservation. Across all the categories and temporal scenarios, the most important variables influencing the prediction were bio1 (Mean annual temperature), bio12 (Annual precipitation), bio15 (Precipitation seasonality and LULC (Land use land cover). The suitable habitat declined relative to the unsuitable habitat. Migratory vultures had a smaller area suitable landcover, though overlapping, as compared to resident ones. The predictions indicated habitat dynamism, with losses of suitable habitats and gains in unsuitable areas of the habitat. There was mostly a decline of area suitable for resident and increases in areas suitable for migratory vultures. The chapter suggests conservation methods to ameliorate these issues, especially related to local conditions.

Kaushalendra Kumar Jha and Michael O'Neal Campbell

INTRODUCTION

This current book edited by two specialists in forestry and physiography focuses on the latest techniques adopted in animal and habitat ecology research and management. It is based on research and reviews of specific and general topics covering the habitats as well as the species of importance in selected case studies, and the overall general scenarios. Different chapters of the book are written by the academic and field experts bringing in their acquired skills, insight and research findings. It is daunting to cover such a large field of animals and habitats in just one volume, however, a few topics have been selected on concurrent and relevant subjects, such as paradigm shifts in forest management, impacts of climatic changes on animals and their habitats, the use of the latest technology for endangered species management, the intricacies of traditional farming ecosystems, and the importance of integrated scholarship etc.

For example, the management of wild animals and habitats in India, like that of many countries is as old as forest history itself, spanning over almost three millennia, though in different forms due to its ownership to different regimes of human organization. The era of sufficiency was followed by an insufficient amount of goods and services available to the users, which affected the management methodologies designed for effective and efficient extraction of resources. The initial need-based structured management was displaced by the scientific management methods focused on material productivity. In the latter half of this phase ecological principles of management entered the scene and conservation took the centre stage to optimise the productivity as well as save the biodiversity and its habitat.

Several technological developments have occurred internationally, due to the needs of conservation. Some of these are highlighted in this book. These include the impacts of the paradigm shifts in habitat management, manifested through legal changes and supported by geoinformatics, for more certain resource sustainability, the correction of historical injustice and improvements in good governance. Geographical Information Systems, and the related techniques of remote sensing, allow more accurate assessments. Species distribution modelling enables evaluations habitat suitability and the impacts of habitat alteration and the requirements for the improvement of animal conservation. Live-stream real-time geographical positioning data and the effectiveness of remotely controlled drones, flying wings and robotic 'prey' as tools could help assess the rehabilitation of falconry species. In some cases, the application of social science techniques and continued practicing of heritage knowledge systems are vital for resource management, even without the supports of modern technology. Other contributions come from studies of the relationship between ecotourism practices and habitat conservation, and relations with human and natural capital development. Management using such integrated research results, may reduce the gaps in existing scholarship and achieve the objective of protecting ecosystems.

It is believed that the contents presented through the chapters of this book will enhance scientific understanding of different, relevant topics on research methods on ecology and the contribution to conservation. Applied research findings could be replicable in the field to achieve the objective of ecological sustainability. The facts and findings presented, and the strategies recommended in different chapters will help the policymakers and decision implementers, scientists, resource managers, research scholars and other stakeholders to attain sustainable animal conservation and habitat management.

Kaushalendra Kumar Jha

In: Critical Research Techniques in Animal … ISBN: 978-1-53619-720-4
Editors: Kaushalendra Kumar Jha et al.

Chapter 1

FOREST MANAGEMENT IN INDIA: A PARADIGM SHIFT AND THE USE OF GEOINFORMATICS

Kaushalendra Kumar Jha*
Indian Institute of Forest Management, Bhopal, India
Department of Forest, Environment and Climate Change, Uttar Pradesh, Lucknow, India

ABSTRACT

India has an extensive history of forest management spanning more than 150 years of scientific forestry in the modern period, further extending in the past up to the ancient period of the Gupta and Maurya empires and even the Vedic period. This article examines the religious scriptures and treatises, medieval records, and legal institutions of ancient and modern India. It was revealed that ancient to modern forest management systems passed through sufficiency to scarcity of resources due to high internal consumption and export of timber products. Sequential formulation of forest policies, promulgation of acts and rules, signing of international conventions etc. oversaw the changes in forest management. The control over forest resources was initiated, then tightened and then eased off by the shift from production to conservation forestry. Simultaneously, indigenous tribes and forest dwellers were relieved of historical injustices in order to provide opportunities of food security and livelihoods. The objective was to ensure the conservation of forest resources. To achieve sustainability, principles of sustainable forest management (SFM) were adopted. The methodology of forest management has been improved by bringing in changes in the Working Plan preparation by introducing SFM and geoinformatics through National Working Plan Code. Such steps appear to be a positive move in meeting people's aspiration and the nation's commitment towards humanity.

* Corresponding Author's E-mail: jhakk1959@gmail.com.

Introduction

India, being committed to International Conventions and people's aspirations, has taken several steps in natural resource management during the past 2-3 decades, especially in the bio-green sector. This development has resulted in an important paradigm shift in forestry management. The main sequence of change continued to be from abundant resources and associated management to limited and then to scarce resources and related management change. The Adoption of National Criteria and Indicators for sustainable forest management (1998), promulgation of Biodiversity Act (2002) and Forest Rights Act (2006), the implementation of National Working Plan Code (2014) among other legislative actions are some of the examples of Forest Policy actions that brought in modern scientific management in forestry and socio-political change in management perception. A major new development was the use of geoinformatics and other advanced technologies for the data generation and facilitation of informed decision making in order to achieve the objectives. This development was however based on a long history of continual adaptation to the needs of forestry development. Literature sources include ancient scriptures dating to over 4000 years. There are three phases: ancient India (pre-external influence period), medieval India (period under influence of close north western neighbors) and modern India (from western influence till date). These divisions are debated, but adequate.

Ancient Period

There is evidence from this period that people were acutely aware of forest structures. Although there is little information on the processes of forest utilization and conservation of forest resources during this period, the Hindu scriptures like "*Vedas,*" "*Puranas,*" "*Upnishadas*" etc. embody abundantly the description of large trees, small plants and wild-animals and their relationship with forest inhabitants. This was evidently a relationship based on the close harmony with nature and the conservation of vital natural resources. In Hindu Vedic society destruction of vegetation and forests was considered to be a great sin. Thakur (1997) reviewed the scriptures and gleaned from them that nature had the potential to control climate, increase fertility and improve human life as described in the '*Rigveda*'. Trees were considered the abode of various gods and goddesses in '*Atharva Veda*'. '*Yajur Veda*' emphasized that the human relationship with nature and animals was one of mutual respect and kindness. Trees were personified as god in '*Narsimha Puran*'. In '*Skandpuran*' Peepal (*Ficus religiosa*) was considered supreme to all other trees. '*Varahapuran*' advocated regular plantations by human beings to achieve post-life abode in heaven. '*Matsyapuran*' equated the planting of a tree with the begetting of sons. '*Vishnu Dharmasutra'* explained that a planted tree turns into one's son in the next generation. '*Skand Puran*' provided a long list of prohibited trees from felling. In

'*Yagyavalkya Smriti*', the cutting of trees and forests was described as a punishable offence and a penalty of twenty to eighty *pana* (old coin) was also prescribed.

However, Kumar (2008) explains the concept of utilization and conservation of resources quoting a hymn of '*Atharva Veda*' in his review that goes thus "Whatever I dig out from you, O mother Earth! May that have quick regeneration again; May we not damage thy vital habitat and heart." He further elaborates that Vedic traditions affirm completeness of a village only if it preserved certain categories of forests e.g., '*Mahavan*' (production forest yielding fodder, timber, roots and herbs, besides maintaining soil fertility air and water quality), '*Shrivan*' (equivalent to modern days plantations for revenue and agroforests to meet societal needs) and '*Tapovan*' (the home of sages). During the Vedic age each village was also responsible for maintaining the forests in its own territory which was done through the *Panchayat* (a committee of five elders).

Around 3500 BP, the Aryans studied several plants and came to know about their poisonous and non-poisonous characteristics. This initiated the rudimentary elements of the later science of toxicology. Thus, the use of forest resources entered social life through a new discipline. Vedic philosophies formed the basis of the Ayurvedic tradition (*Ayurveda* a subsidiary of *Atharva Veda*), which is considered to be one of the oldest known systems of medicine. Two sets of texts formed the foundation of Ayurvedic medicine, the *Susruta Samhita* and the *Charaka Samhita*. While the former provided important surgical and anatomical information of the understanding of anatomy by Indians in the 6th century BCE (Loukas et al., 2010), the latter prescribed medicinal formulations for various ailments.

During the post-Vedic period two well established kingdoms, Maurya and Gupta, developed some ideas of forest management and multipurpose forest resources were used for various purposes. *Arthashashtra*, an ancient Indian treatise on statecraft, economic policy and military strategy, describes the forestry affairs during the Mauryan period. The state used to appoint Kupya Adhyaksha (Superintendent of Forests), responsible for maintaining health of forests; protecting forests for wild animals especially elephants, lions, tigers etc.; producing timber of important species like teak (*Tectona grandis,* L.f.), palmyra (*Borassus flabellifer,* L.), mimosa (Genus *Mimosa*, L.), sissu (*Dalbergia sissoo*, Roxb.), kauki (*Manilkara kauki*, (L.) Dubard), sirisha (*Albizia lebbeck*, (L.) Benth.), catechu (*Senegalia catechu*, (L.f.) P.J.H.Hurter & Mabb.), latifolia (*Dalbergia latifolia*, Roxb), arjuna (*Terminalia arjuna* (Roxb.) Wight & Arn.), tilaka (*Symplocos racemosa*, Roxb), tinisa, sal (*Shorea robusta*, Roth), pine (Genus *Pinus*, L.), somavalka (*Acacia leucophloea* (Roxb.) Willd), dhava (*Anogeissus latifolia*, (Roxb. ex DC.) Wall. ex Guill. & Perr.), birch (Betula, L.); and minor forest products e.g., bamboo (*Bambusoideae* spp, Luerss.), hemp (*Cannabis sativa*, L.), balbaja (*Eleusine indica*, (L.) Gaertn.) (used for ropes), munja (*Saccharum munja*, Roxb.), fodder, firewood, bulbous roots and fruits for medicine, flowers (Olivelle, 2013) through their forest guards (*Vanapala*).

One of the important duties of *Vanapala* was to catch elephants from their forest habitat for royal use. Killing of wild animals, particularly elephants, was a punishable offence. Forests located at the boundaries of the state were well protected from the neighbouring kingdoms. The forests were classified as: (i) Reserved forests – for the King (ii) Reserved forests – for the State (iii) Forests donated – to *Brahmins*, and (iv) Forests – for public use. One of the greatest Mauryan kings, Emperor Ashoka took pride in the roads which were constructed by his administration and were lined with shade-giving trees and wells. He also provisioned the game laws into *Dhammaniyam* (laws of piety).

The forest was considered as the source of societal welfare and revenue. Therefore, most of the good practices were carried forward to the Gupta period. Additionally, improvements were made by introducing new arrangements. *Gaulmikas* were appointed as a new class of officers with additional responsibilities. Forest paths were created dividing the forest into ranges, beats and blocks as smaller units of administration. Hunting was banned in general except the king who could kill only wild animals. Birds were not to be killed. The King initiated the planting of trees which could be used for various purposes. As a welfare measure, recluses and tribal groups were allowed to dwell in the forest.

Medieval Period

After the end of the Gupta empire, the country reverted to small states which perpetually fought amongst themselves. Virtually every war took a toll on the forest. The loser in wars, took refuge in previously undisturbed forests and cleared them to create new abodes for themselves. The political instability caused by local conflicts was aggravated by repeated Muslim invasions from the north west (Lal, 1989). Their continued efforts led to the establishment of the Mughal empire in India after the Delhi Sultanate. During this period of instability forest management took a back seat. In fact, it remained confined to hunting except for a few notable acts of renowned rulers like Jalaluddin Akbar, Shershah Suri etc.

In the *Sunnud* (a charter) of Emperor Akbar it was directed that - “On both sides of the canal down to Hissar, trees of every description, both for shade and blossom be planted, so as to make it like the canal under the tree of Paradise; and that the sweet flavour of the rare fruits may reach the mouth of every one, and that from these luxuries a voice may go forth to travelers calling them to rest in the cities where their every want will be supplied” (Journal of Asiatic Society of Bengal, 1846). During the time of Shershah Suri (1486 - 1545) a tree plantation was developed along the Delhi-Patna Highway. The Emperors who were not threatened by destabilization focussed on hunting and the establishment of wonder gardens. Some of them were involved in imposing the exotic religion on to indigenous people. Engaged totally in political and administrative affairs, they ignored public welfare through forestry. As a result, well-founded forestry institutions under the Mauryans and Guptas collapsed. Unregulated extraction of forest wealth was on the rise without any conservation measures, perhaps laying the foundation of anthropological degradation of forests of India.

Modern India

During ancient India, royal decrees were used to manage resources for the benefit of both the state and the general population. At that time, the population of India was less than that of the medieval and later, modern India. As suggested by authentic records formal management started in the later Vedic period which included extraction of resources from the forest. The literature is almost silent about regeneration of resources, an obligatory component of management, indicating perhaps that it was lacking. Given the uncertain political conditions in medieval India, it is assumed that this further deteriorated. However, modern India, run by the Presidencies and Princely states and later under British control, saw forest management affairs obtain more attention, but the scientific management focussed on 'Sustained Yield' rather than 'Sustainable Management'. The most valuable wood "Teak" was exported for the British navy, in addition to another valuable timber, "Sal." Forests from the northern plains and the southern peninsula were exploited, with various management schemes being employed. The British had, meanwhile, initiated control over the forest by promulgating various acts under a well-directed forest policy (FP). However, two current photographs of forests with historical significance are given in Figure 1 a & b.

Independent India not only continued with the British legacy, it further reinforced control over forest through new policies and more defining acts. Nevertheless, the forest sustainability was not achieved through these actions and institutions. Therefore, policy changes were initiated that focussed more on the social and ecological aspects of environmental change and degradation impacts. In the process, monopolistic control over forest resources declined, stakeholders were included in policy actions and international developments were constructed. Geoinformatics was adopted as a research, data collection and decision-making tool to aid management and monitoring of the resources. The succeeding sections of this chapter examined the processes and events that enabled and constrained these changes. These include the aims and objectives and a few important provisions of different instruments.

FOREST MANAGEMENT AND GUIDING PRINCIPLES

Forest management in modern India (1855-1947-Current) was guided by forest policies and governed by different acts. The management orientation during this period is well demarcated into two phases: production forestry (pre-1988) and conservation forestry (post 1988). The former is state oriented while the latter is people welfare centric. Socio-political phenomena in the form of legal institutions related to forest management are depicted in graphical form in Figure 2 and detailed in the following paragraphs.

Photo taken by author during field visit in 2018.

Figure 1a. Alicutta forest on the Pench river (now Pench National Park, Madhya Pradesh). Two foresters met here Mr Brandis and Mr Pearson (ca. 1862) serving under the British Crown in India. They discussed the administration of forests in India and submitted a "Policy of future working of forests" which aided forestry in India.

Photo taken by author during field visit in 2018.

Figure 1b. One of the oldest forest reserves "Bori" on Bori Tawa river. This forest was known to be managed as the first Working Plan implemented in Central Province. There was also first attempt at organised forest fire protection in this reserve (ca. 1867). Currently, forming part of Satpura National Park, Madhya Pradesh.

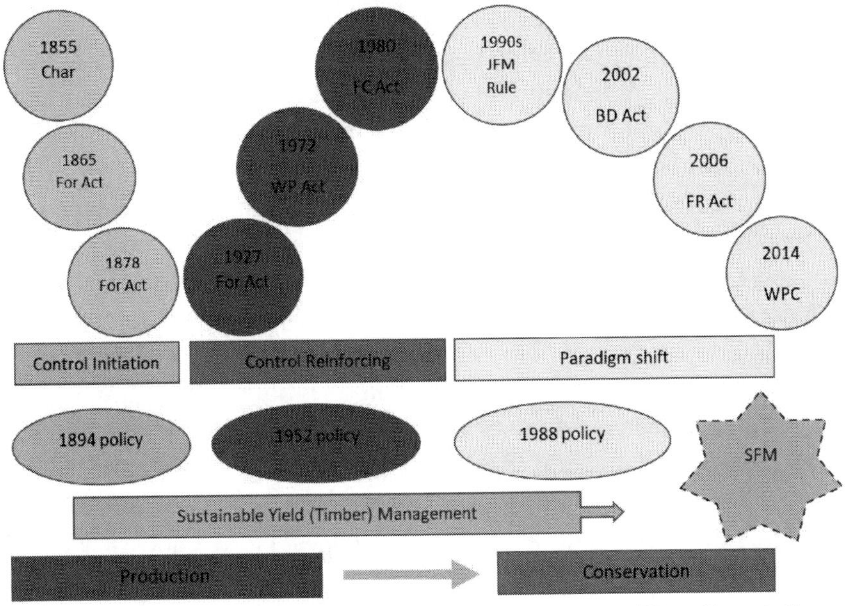

Char = Forest Charter, For = Forest, WP = Wildlife Protection, FC = Forest Conservation, JFM = Joint Forest Management, BD = Biodiversity, WPC = Working Plan Code.

Figure 2. Acts, Rules and Policies of the Phases of Forest Management (Production to Conservation).

The succession of almost four Forest Policies (1894, 1952, 1988 and a Draft 2018) in around 125 years indicates the changing priorities of the country with regard to the forestry sector. The first forest policy which was derived from the report on agriculture improvement in British India, suggested by the agriculture scientist Volcker in 1893 had, naturally, inherited its secondary importance to agriculture. The main objective of forest management was to promote the general wellbeing of the country. The influence of forest over the physical and climatic conditions of the country was quite well recognized. However, realization of maximum revenue from commercial areas was given priority. Permanent cultivation was given preference over forestry; the policy did not make any provision for management of private forests (Negi, 1986; Balaji, 2002).

However, the focus remained on continued production of timber without due emphasis on non-commercial species and NTFPs, wildlife management, catchment area treatment, forestry research, education and training in FP 1894. Therefore, these missing areas were covered, to some extent, in the succeeding FP 1952 which was in consonance with independent India's needs (Balaji, 2002). The six areas identified were to evolve a system of balanced and complementary landuse, to check denudation of mountainous regions; erosion along rivers; sand-invasion of coastal and desert area, establish tree-lands, to increase supply of grazing material and small wood, to ensure sustained supply of timber for defence, communication and industry and to realise maximum revenue.

Over the years, forests in the country suffered serious depletion due to relentless pressures arising from the ever-increasing demand for fuel-wood, fodder and timber; inadequacy of protection measures; diversion of forest lands to non-forest uses without ensuring compensatory afforestation and essential environmental safeguards; and the tendency to look upon forests as revenue earning resources. This led to the revision of FP 1952 and thus FP 1988 came into existence. The new policy emphasized the maintenance of environmental stability through preservation and restoration of depleted forests, soil erosion control and denudation in catchment areas of rivers, lakes etc. This preserved the remaining natural forests, conserving the biodiversity and increasing substantially the forest/tree cover that supplied the requirements of fuel-wood, fodder, minor forest produce and small timber, and increasing the productivity of forests to meet essential national needs. The Draft Policy 2018, while maintaining the conservation requirement and other needs emphasized in the prevailing policy, integrates the vision of sustainable forest management by incorporating elements of ecosystem security, climate change mitigation and adaptation, forest hydrology, participatory forest management, urban forestry, robust monitoring and evaluation frameworks, and establishes mechanisms to oversee multi-stakeholder convergence in forest management, while building on our rich cultural heritage of co-existence and relying on our rich and diverse forest resources.

Forest Management and Governing Acts

The Charter of Indian Forestry (1855) was intended to restrict and regulate the unchecked exploitation of forests by private individuals. This was followed by the Indian Forest Act (1865) which emphasized the protection of forests (fire and grazing) and the judicious use of timber. The prevalent uses were conceded as rights, justifying the creation of government reserves. This act indicated the start of objective-oriented forestry, with greater control by the state over forest resources that had earlier been open to public use. Thus, the seed of control initiation over the forests was sown. However, this continued for a few years and evolved into a new Indian Forest Act (1878). Here forests were categorized into Reserve, Protected and Village Forests. The first one was exclusively for the government, while in the second local use was allowed. The third one was created for fulfilling the subsistence needs of villagers. This act was an extension of the government policy of establishing control over forests (Ghildiyal, 2000). This initiation was further strengthened in the Indian Forest Act, 1927 (IFA 1927), which was enacted to consolidate the law relating to forests, the transit of forest products and the duty levied on timber and other forest-produce. The emphasis was on the protection of reserve forest managed by the state government. Some procedural and provisional amendments were added to IFA 1927, related to management issues.

The Forest Conservation Act 1980, protected forests from use and set out central government control over forest use, due to the consideration and permission of the central government. This seems to be tightening control over forests. A complementary act, the Wildlife (Protection) Act (1972) was enacted to ensure management of wildlife, an inseparable component of forestry, by providing for the protection of wild animals, birds and plants and for matters connected therewith or ancillary or incidental thereto. Consequently, in the interests of wildlife conservation, National Parks and Sanctuaries were created and rights of the people were reduced to different orders.

Forest Management and Paradigm Shift

Conventions, Acts and Rules

Sustainable Forest Management

Sustainable Forest Management (SFM) is defined as "the management of forest areas to maintain and enhance the economic social and environmental value of all types of forests, for the benefit of present and future generations." The three components of SFM, economic, social and environmental were mostly present in different proportions in Indian history. Though the information is staggered and incomplete, especially in the medieval period, it indicates a conscious effort of moving towards sustainability (Figure 3). However, formal SFM came into being in recent times (e.g., the Rio Earth Summit) when it was considered imperative to counteract the anticipated negative impact of greenhouse gas emissions and global warming. Agenda 21 was agreed upon by 179 nations at the United Nations Conference on Environment and Development (UNCED) in 1992 in Rio de Janeiro and laid out the guiding principles for sustainable development in the twenty-first century (UN 1992). In forestry, the guiding principles for sustainable development were derived and applied within the concept of sustainable forest management. As a result of the Montreal Process 1994, Criteria and indicators (C&I) provided a definition of SFM as well as a means to measure progress towards that goal in temperate and boreal forests. This was a major step forward and contributed to the paradigm shift away from sustained yield to a far broader and more holistic view of forest sustainability (Linser et al., 2018).

Forest management for a sustained timber yield is generally the objective of Sustainable Forest Management. The mindset of managing forests for production of wood does not address the wider issues of ecological and social functions of forests, with which timber production may only incidentally be compatible or may even be in conflict. Moreover, SFM is not to be confined to areas where a sustained yield of forest products and services is, at least in principle, achievable. It also calls for drawing attention to huge areas of degraded and disappearing forests (Prasad, 2001; Prasad & Kotwal, 2001). Dry zone forests which provide sustenance to millions of forest dependent people are fast degrading, depriving communities of optimum benefits from products and services. Such ecosystems are very fragile and need immediate action. Community participation in this region has been institutionalized, and there has been no mechanism to monitor the progress towards different dimensions of sustainability. Communities generally perceive improvements in forest conditions. But in the absence of an acceptable tool, they are unable to quantify the gains (Prasad, 2001). As the conditions in these tropical forests are different from boreal and temperate forests, C&I must be investigated differently.

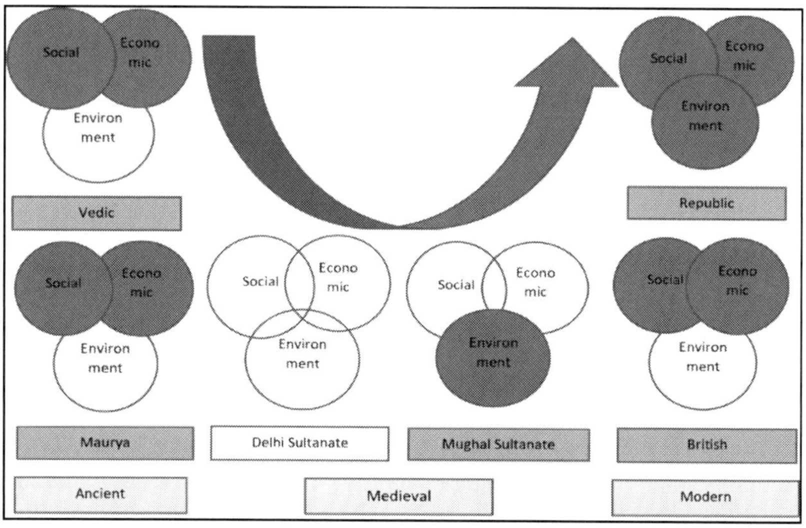

Figure 3. Conceptual Depiction of SFM in India Through History. The presence of the green components (social, economic, environmental) in the representative era of a period are based in accessed records and mention of relevant activities. White circles indicate their absence. Green arrows indicate the status and path of SFM in Indian history.

National Criteria and Indicators (1998)

In dry zone Asia, the regional initiative for the development of "National Level Criteria and Indicators" for SFM was initiated jointly by the Food and Agriculture Organization (FAO), United Nation's Environment Program (UNEP), International Tropical Timber Organization (ITTO), United States Forest Service (USFS) and Indian Institute of Forest Management (IIFM). However, with varying agro-climatic conditions and rich cultural and biological diversity, India initiated its action to evolve C & I for SFM and named it as the Bhopal-India Process (Prasad et al., 1999). Bhopal-India Process developed 8 criteria and 43 related indicators (Table 1) along the lines of the Montreal Process. SFM is now presented as the multipurpose management of forests, not just sustainable timber yield management (Prasad & Kotwal, 2001). Strong emphasis is given to environmental resources, as presented in Tables 1a,1b and 1c below. This is another paradigm shift from a focus on sustained timber yield (through strictly regulated forestry) to sustainable forest management (partly through Community Forestry or Joint Forest Management), encompassing environmental, economic and social dimensions (Rawat et al., 2008).

Table 1a. National SFM Criteria and Indicators and relevant mapping for indicator monitoring

SFM Criteria	SFM Indicators	Mapping layers
1. Maintain/ increase in the extent of forest and tree cover	1.Area/type, natural forest/tree plantations) 2. Forest diverted for non-forestry purposes 3.Forest area under encroachment 4.Area of dense, open and scrub forests 5.Tree cover outside forest area	i. Nurseries*; ii. Forest cover* iii. Forest type/stock map* iv. Plantation of last 15/20 years v. Encroachment* vi. Management areas and working circles vii. Diverted non-forestry land viii. Tree outside forest
2. Biodiversity Maintenance, Conservation Enhancement	1.Number of animal and plant species 2.Number and status of threatened species 3.Status of locally significant species 4.Species prone to overexploitation 5.Non-destructive wood/non-wood produce	i. Fire locations in the last 5 years ii. Threatened species presence iii. Endemic species presence

* indicates mandatory layers as per NWPC 2014, Indicators: source Rawat et al. (2008).

Table 1b. National SFM Criteria and Indicators and relevant mapping for indicator monitoring

3. Maintenance and Enhancement of Forest Health and Vitality	1. Status of natural regeneration 2. Incidence of forest fire 3. Extent of livestock grazing 4. Forest area open for grazing 5. Number of livestock grazing in forest 6. Occurrence of weeds in forest area and weed type 7. Incidence of pest and diseases	i. Fire affected/ Burnt area ii. Fire sensitive spots iii. Natural regeneration iv. Grazing area v. Invasive species presence vi. Pest and disease prone area
4. Conservation and maintenance of soil and water resources	1. Area under watershed treatment 2. Area prone to soil erosion 3. Area under ravine, saline, alkaline soils, deserts 4. Soil fertility/site quality 5. Stream flow duration of 6. Groundwater in the vicinity of the forest areas	i. Stream network* ii. Soil type iii. Altitude zones* iv. Slope v. Aspect vi. Mining areas vii. Grid layer of rainfall viii. Grid layer of temperature ix. Erosion liability x. Groundwater depth
5. Maintenance and enhancement of forest resource productivity	1. Growing stock of wood 2. Increment in volume of identified species of wood 3. Efforts towards enhancement of forest productivity 4. Technological inputs 5. Area under hi-tech plantations 6. Area of seed production, clonal seed orchards, etc.	i. Forest research plots ii. Seed production areas /Plus trees/CSOs iii. Timber species distribution iv. High-tech plantation

Table 1b. (Continued)

6. Optimization of forest resource utilization	1. Recorded removal of wood 2. Recorded collection of non-wood forest produce 3. Efforts towards reduction of wastage 4. Aggregate/per capita consumption of forest produce 5. Direct employment in related industries 6. Forest-based income of forest-dependent people 7. Demand and supply of all forest produce 8. Import and export of forest produce	i. Forest depots ii. Location of saw mills iii. Timber production area
7. Maintain/ enhancement of social, economic, cultural and spiritual benefits	1. JFMCs, protected areas 2. Local people/gender participation 3. Use of indigenous technical knowledge: 4. Level/provision of concessions and privileges 5. Cultural/sacred protected landscapes 6. Type and area of landscape 7. Number of visitors	i. Village locations* ii. Road/rail network* iii. Rest houses. iv. JFMCs*. v. Eco-tourism spots vi. Sacred groves vii. Sacred places
8. Adequacy of Policy, Legal and Institutional framework	1. Existence of policy and legal framework 2. Number of forest-related offences 3. Level of investment in research and development 4. Human resource capacity-building efforts 5. Forest resource accounting 6. Forestry contribution to Gross Domestic Product. 7. Budgetary allocations to the forestry sector 8. Monitoring and evaluation mechanisms 9. Status of information dissemination and utilization	i. Forest check gates* ii. Police station/chowkis iii. Offices/office locations* iv. Division/Range/RF/PF boundary* v. Conservation/protected areas* vi. Beat boundary/Block* vii. Compartment boundary* viii. Satellite imagery* ix. Wildlife corridors

* Indicates mandatory layers as per NWPC 2014, Indicators: source Rawat et al. (2008).
JFMC refers to Joint Forest Management Committees.

Joint Forest Management & JFM Rules 1990s

Joint Forest Management (JFM) is a participatory approach in forest management as envisaged in the National Forest Policy, 1988 for creating a massive people's movement for the protection and conservation of forest resources through democratic governance and project implementation. JFM rules were framed in 1998 by different states like Assam, Kerala etc. within the provision of IFA 1927 following Government of India guidelines. Other states also initiated JFM implementation around this period. State governments passed their enabling resolutions based on local conditions involving composition of village institutions, the nature of forest areas to be taken up under JFM, rights and responsibilities of the partners and usufructs sharing arrangements. The nomenclature of Joint Forest Management Committees varies in different states. For instance, it is called *Van Panchayat* in Uttranchal and *Vana Samrakshana Samiti* (VSS) in Andhra Pradesh. The democratic constitution of JFM committees with due representation of weaker sections of the society and women strengthens the democratic decision-making process. As of now, there are more than 1,060,482 JFM committees across 28 States and UT of Andaman & Nicobar Islands which are managing 22.01 million ha of forests in the country. The concept of JFM was a positive step towards decentralized governance and forest management, with the potential of empowering the community and increasing the livelihood security of impoverished forest dependent communities (Bhattacharya et al., 2010). Joint Forest Management have had some problems but have been successful in empowering the people socio-politically (Figure 4).

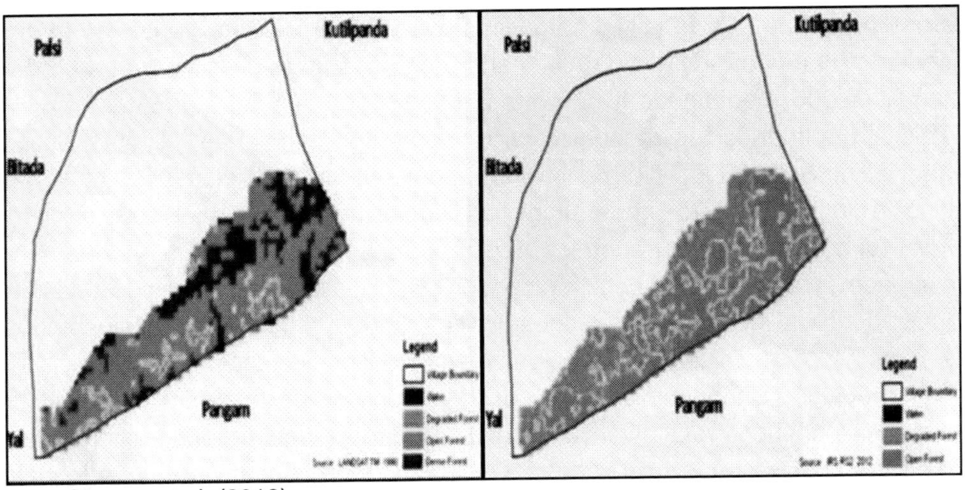

Source: Sharma et al. (2018).

Figure 4. Forest Management by JFMC, Koliawada Village, Narmada Division, Gujrat. The condition of forest may be noted at the time of JFMC formation (1990) on left and after 22 years of management (2012) on right. Area under dense forest has decreased (dark green) while area of degraded forest has increased (orange).

Biodiversity Act (2002)

India is rich in biological diversity and associated traditional and contemporary knowledge systems. It has 45,000 species of plants of which 18% are endemic. Exceptional diversity exists in the case of ferns with 900 species and orchids with 1,082 species. It is also home to 77,000 faunal species (Kothari, 1994; and Kothari et al., 1999 in Balaji, 2002; Balaji & Rai, 1998). On the other hand, it is also a party to the United Nations Convention on Biological Diversity which reaffirms the sovereign rights of the States over their biological resources. The main objective of the convention was the conservation of biological diversity, sustainable use of its components and fair and equitable sharing of the benefits arising out of the utilization of genetic resources. Therefore, the Biodiversity Act, 2002 has been promulgated to provide for the conservation of biological diversity, sustainable use of its components and fair and equitable sharing of the benefits arising out of the use of biological resources, knowledge and for matters connected therewith or incidental thereto.

Scheduled Tribes and Other Traditional Forest Dwellers (Recognition of Forest Rights) Act 2006

India's forests are home to hundreds of millions of people, including many Scheduled Tribes, who live in or near the forest areas of the country. Nearly 250 million people live in and around forests in India, of which the estimated indigenous *Adivasi* or tribal population stands at about 100 million. Their rights were regulated by different forest related acts, such as IFA 1927, WPA 1972 etc. For various reasons, there used to be a high order of dissatisfaction and therefore, conflicts between the government and people. Finally, their claims were recognised and regulated through Scheduled Tribes and Other Traditional Forest Dwellers (Recognition of Forest Rights) Act, 2006, also popularly known as Forest Rights Act 2006 (FRA, 2006).

This Act is intended to recognize forest rights and occupation of forest land by forest dwelling Scheduled Tribes and other traditional forest dwellers who have been residing in such forests for generations but whose rights could not be recorded. It provides for a framework for recording the forest rights vested in them and the nature of evidence required for such recognition with respect to forest land. Their recognised rights included the responsibilities and authority for sustainable use, conservation of biodiversity and maintenance of ecological balance. This strengthened the conservation regime of the forests while ensuring livelihood and food security of the forest dwelling Scheduled Tribes and other traditional forest dwellers. As the forest rights on ancestral lands and their habitat were not adequately recognised in the consolidation of state forests during the colonial period as well as in independent India, the result was historical injustice to the forest dwelling Scheduled Tribes and other traditional forest dwellers who are integral to the very survival and sustainability of the forest ecosystem. It was necessary to address their long-standing insecurity of tenurial and access rights, including those who were forced to relocate their dwelling due to state development interventions.

Singh (2005, in Saravanan, 2018) anticipated the main effect as being the destruction of forests and not the betterment of tribal living standards. However, the provisions of FRA 2006 when seen with IFA 1927 explain the unmistakable shift from strict control over resources by the government to release the control in favour of the tribals and forest dwellers. FRA 2006 explicitly allowed tribals to use forest resources for their livelihood, hitherto restricted by IFA 1927 and preceding Acts. Activities such as fresh clearing for cultivation and other purposes in notified areas, cattle trespass and grazing, felling of trees and damaging of any tree besides hunting activities in the reserve forests were offences and punishable either by imprisonment or penalty or both. Additionally, damage to property had to be compensated. Contrastingly, FRA 2006 endowed the tribals with the rights to live in the forest, to cultivate for their livelihood, to collect minor forest produce, to graze cattle, to convert leases or grants (*pattas*) to titles, to convert forest villages into revenue villages, to settle in their old habitations and unsurveyed villages, to access community rights over intellectual property and traditional knowledge related to forest biodiversity and cultural diversity, to manage the community forest resources and to enjoy any customary/traditional practice, excluding hunting (Saravanan, 2018).

Mainstreaming REDD+

Reducing Emissions from Deforestation and Forest Degradation (REDD+) is the United Nations programme initiated in 2005 to mitigate climate change through enhanced forest management in developing countries. Parties to the UNFCCC have agreed to consider policy approaches and positive incentives on issues relating to REDD and the role of conservation, sustainable management of forests and enhancement of forest carbon stocks in developing countries. The National REDD+ Policy aims to provide a roadmap for building comprehensive strategies for implementing REDD+ projects and programmes effectively in the country in the context of international development in this sector. The policy along with the strategies will lead to REDD+ readiness in the country and enable India to gain from international REDD+ mechanism for its pro conservation policies and efforts in the future and at the same time create financial incentives for local communities which are at the forefront of conservation of forests. The National Mission for a Green India (GIM) is one of the eight Missions under the National Action Plan on Climate Change (NAPCC). GIM acknowledges the influences that the forestry sector has on environmental amelioration through climate mitigation, food security, water security, biodiversity conservation and livelihood security of forest dependent communities. Major objectives of GIM are (i) Increased forest/tree cover on 5 m ha of forest/non-forest lands and improved quality of forest cover on another 5 m ha (a total of 10 m ha) (ii) Improved ecosystem services including biodiversity, hydrological services and carbon sequestration as a result of treatment of 10 m ha. (iii) Increased forest-based livelihood income for 3 million forest dependent households (iv) Enhanced annual CO_2 sequestration of 50-60 million tonnes by the year 2020. GIM puts the "greening" in the context of climate change adaptation and mitigation, meant to enhance ecosystem services like carbon sequestration and storage (in forests and other ecosystems), hydrological services and biodiversity; along with provisioning services like fuel, fodder, small timber and NTFPs (http://naeb.nic.in).

Management Tool Upgradation

A Working Plan is a document or an instrument/tool for forest working in the country for scientific management of forests. It is very useful in evaluating the status of forests and bioresources of an administrative unit (forest division), assessing the impact of past management practices and deciding about future management interventions. This document is periodically, generally ten years, updated and revised to keep pace with emerging trends and obligations (NWPC 2014).

There has been a paradigm shift in the objectives of forest management, which has become more people centric and oriented to provide goods and services from forests on a sustained basis, with an emphasis on ecological services and harvest of usufructs as well. The working plan should be in consonance with general planning, which is village based. Therefore, the working plan should encompass the village as a unit and realign the compartments accordingly. Proper guidelines for sustainable use of community forests; extraction, processing, market and trade of minor forest produce, etc. may be provided under separate working circles. Forest community rights related to community forest resources, minor forest produce, grazing grounds, water bodies, etc. recognized under the provision of the Forest Rights Act, 2006, can be exercised within the framework of sustainable use. On the other hand, management of forests adversely affected by mining, industries, urbanisation and other nonforest activities is also a great challenge for which special planning initiatives are needed (NWPC 2014).

As recently as April 2014, it has become obligatory to follow National Working Plan Code (NWPC) 2014 for working plan preparation due to management regime changes and availability of very efficient technology for resource assessment and evaluation. However, all the states are not on the same pace as desired. Gujrat, for example, has introduced SFM related chapters and followed use of geoinformatics (WP of Narmada Division 2017-18 to 2026-2027) while Andaman and Nicobar Islands are still on the older lines (WP of Mayabandar Division 2017-18 to 2026-27). Uttar Pradesh, once upon a time the leading state in forest management and research, despite having experienced the advantage of new technology, is surprisingly, lagging. Contents presented in box 1 give a fair idea of the win-and-lose situation in many of the states not catching up the obligation under NWPC 2014.

National Working Plan Code 2014

The country has a long history of scientific management of forests through Working Plan (WP), to achieve the objectives set for different needs and purposes like colonial, industrial, local and environmental, during different socio-political phases. The WP traces its origin in Munro in 1837 and Brandis in1856, two well-known foresters, who started specialised work in the forest related to tree felling on the basis of pre-decided and well-drawn principles and plans, instead of indiscriminate felling for profit. WP as a document kept evolving in terms of silvicultural need and took formal shape after the formulation of guidelines to prepare working plans by another forester Darcy in 1891. During the first 50 years of new India WP preparation rested with the provincial governments based on codes as per regional requirements.

However, in order to fill in the gaps between policy making (primarily national role) and implementation by the provincial government it was decided to have a uniform working plan code for the country bringing into existence the National working plan code (NWPC) 2004. This was the abridged compilation of prevailing practices in different states. Keeping pace with a number of developments in the forestry sector, it was decided to modify this code on the following lines: (i) Incorporation of criteria and indicators for Sustainable Forest Management. (ii) Address new concepts like climate change, carbon sequestration, etc. (iii) Application of modern technology like GIS, GPS etc. (iv) Address the issue of forest certification through C&I for SFM. (v) Focused management of NTFPs in view of their high economic and livelihood importance. (vi) To adopt uniform, approved formats for development areas in all states (Singh et al., 2014). Thus, came NWPC 2014 in existence. As a result additional chapters were introduced almost identical to the National Criteria of Sustainable forest Management, like maintenance/ increase in the extent of forest and tree cover, maintenance and enhancement of forest health and vitality, conservation and maintenance of soil and water resources, maintenance and enhancement of forest resource productivity, optimization of forest resource utilization, maintenance and enhancement of social, economic, cultural and spiritual benefits, adequacy of policy and legal and institutional framework. The corresponding indicators were included in sub-chapters. The current NWPC also emphasized the use of technological advancements like remote sensing, GIS, GPS, computational and analytical systems available for real time monitoring of forest dynamics.

Geoinformatics

Geographic Information System (GIS) is analogous to a compound of three independent elements like, geographic data, computer hardware- software, and personnel designed to collect, store, manage, query, analyze and present a large volume of spatial data and associated attributes in order to allow its users to make better decisions, to improve productivity, to save time, money and man power (Altan et al., 1991). In general, the term data corresponds to discrete, recorded facts about phenomena from which people gain information about the real world. In the context of geography, data may be considered, as data about geographic features such as roads, buildings, rivers etc, and phenomena, such as weather forecast, earthquakes, etc. A very effective graphic information of geographic features is provided by conventional maps, but they are not very useful for supporting data retrieval. Data retrieval on these maps is usually done by visual search, a process that is neither efficient nor reliable (Figure 5). Spatial analysis is even more difficult. Even for simple applications such as calculating the total area of the clearing in a forest, the amount of manual work involved is sometimes prohibitive.

McKendry and Eastman (2000), in an internet document, reviewed GIS applications in forestry and categorized them in two groups: i. resource inventory and monitoring and ii. analysis, modeling, and forecasting to support decision making. In fact, these are the two stages of development in a fully operational GIS based forest management (Crain & MacDonald, 1983; Jordan & Erdle, 1990). These reviewers use two scientific terms, resource assessment and resource management, to discuss applications of GIS in forestry.

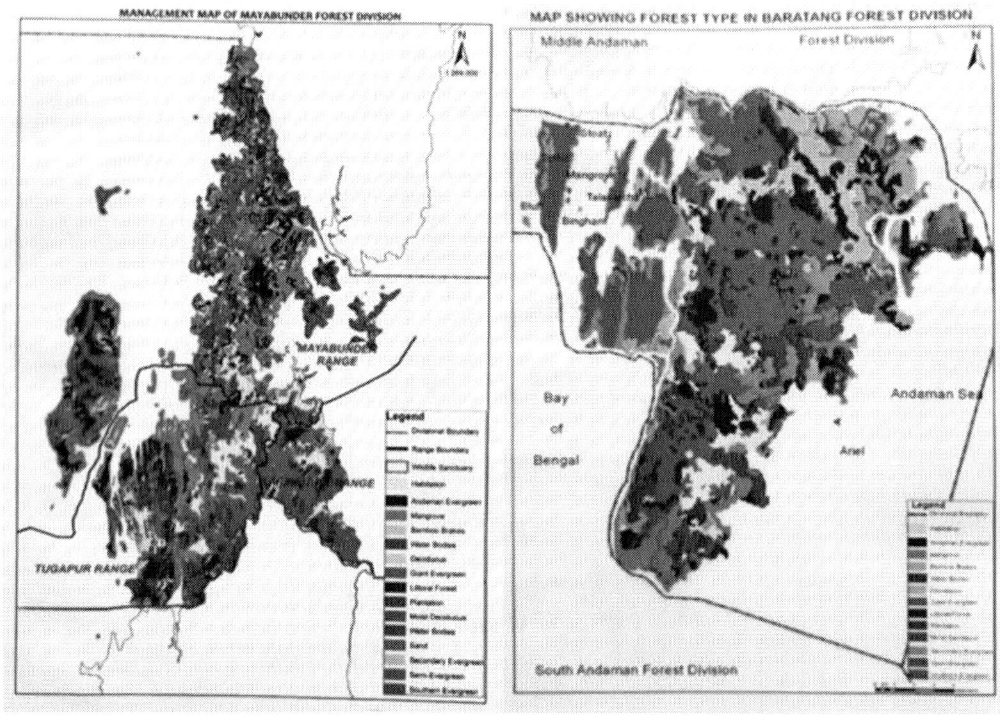

Source: Mallick (2018).

Figure 5. Management map of Mayabandar Division (left), and Forest Types map of Baratang Division (right), A&N Island. Presentation of 2D map looks clumsy due to several features (colours) making data retrieval so difficult.

Resource Assessment

Activities related to resource assessment include: 1) inventorying forest resources available for harvest, fuel, food, recreation, or conservation purposes, along with related data such as topography, soils, roads, and hydrology, 2) monitoring changes that occur to these resources over time, and 3) evaluating potential land productivity for forest types given certain biophysical and climatic factors. It is in forest resource assessment that other technologies related to GIS, remote sensing, and global positioning systems, make direct and substantial contributions. Resource assessment aspects of data collection for preparation of WP are grouped SFM criteria wise in Indian context (NPWC 2014) as below:

i. *Maintenance, Conservation and Enhancement of Biodiversity:* Forest composition and distribution, plant species diversity, status of biodiversity conservation of forests, status of species prone to over exploitation, conservation of genetic resources, fauna and their habitats, threats and challenges to wildlife, protection and management of fauna.

Box 1: Slow Adoption of Geoinformatics: Not a Win-Win Case

The following text is personal opinion of the respondent and not in any manner the state's viewpoint. However, it reflects the impact of adoption of GIS techniques due to early appreciation of the advantages of new technology, now a compulsion under WPC 2014. Despite a lapse of a considerable number of years many states have not adopted WPC 2014 arguably due to its impractical nature. While doing qualitative research on the subject, a highly experienced forester who wrote WP during both pre and post entry of GIS technology in the forest department, drew attention towards significant points mentioned below based on comparison of his first WP tenure and the second one after a twenty-year gap.

Earlier WP used to be complete in itself on Division (administrative unit of forest department) and other boundary accounts when seen in isolation. GIS helped improve the anomaly in coterminous alignment with neighbouring divisions when seen at Circle (upper level in administrative hierarchy consisting of many divisions) level and later at State level too. Currently, boundary accuracy has improved enormously.

Growing stock estimation was also a tough challenge due to several compelling reasons resulting into either over-estimation or in some cases under-estimation as well. GIS implementation has brought in unbiased and machine accuracy resulting in fallacy removal significantly.

Silvicultural and felling operations at Working Circle (operational unit in a division based on plan prescription) level has improved due to clarity in confounding and overlapping areas of different operations, like improvement, protection, plantation etc. on similar area.

Reporting quality of ecologically important events like fire incidence, impact of drought or long waterlogging etc. has improved from underreporting towards exactness. Cases of illicit tree felling, encroachment on forest lands mostly for crop raising and habitation etc. are also rid of underreporting.

Soil erosion, especially riverbank erosion and resulting loss of bio-wealth, is getting much larger attention due to visibility of unmistakable change from previous years.

Carrying capacity estimation of the forest used to be vague due to insufficient sampling and other reasons. Similarly, the density class of the forest used to be unprecise. Currently, the reliability of estimation has increased due to complete coverage of the division in one RS image.

Flow of errors which crept in the past and their magnification in successive plans has got very effectively rectified.

Most striking and current use of geoinformatics in FRA claims, about which the author found, many foresters were astonished, asking the question, "why was this technique not adopted in all states?" Being political in nature, it is not desirable to seek an answer to this question but pertinent to quote an example here. Many untenable claims by a few gullible and some treacherous forest dwellers under FRA 2006 were rejected with ease and comfort, providing the unquestionable evidences generated using GIS. For details Jha and Chaudhry (2018) could be consulted.

ii. *Maintenance and Enhancement of Forest Health and Vitality:* Status of regeneration, area affected by forest fires, area damaged by natural calamities, area protected from grazing, lopping practices, area infested by invasive weed species in forests, incidents of pest and diseases, forest degradation and its drivers.

iii. *Conservation and maintenance of soil and water resources:* Assessment of excess runoff from discharge zone and conservation measures for soil, groundwater, and soil moisture. Area treated under soil and water conservation measures, duration of water flow in the selected seasonal streams, wetlands in forest areas, water level in the wells in the vicinity (up to 5km) of the forest area, status of aquifers.

iv. *Maintenance and enhancement of forest resource productivity:* Growing stock of wood/ bamboo, increment in volume of identified timber species, efforts towards enhancement of forest productivity through quality plantation activities, carbon stock, carbon sequestration and mitigation.

v. *Optimization of forest resource utilization:* Recorded removal of timber, fuel wood, bamboo/ rattans, and locally important NTFPs including MAPs, demand and supply of timber and important non-timber forest produce, removal of fodder, valuation of the products.

vi. *Maintenance and enhancement of social, economic, cultural and spiritual benefits:* Number of JFM committees and area protected by them, status of empowerment of JFMCs, labour welfare, use of indigenous knowledge, extent of cultural/ sacred groves, social customs, status of compliance of Forest Right Act (FRA), other rights and concessions, ecotourism areas and activities, etc.

Resource Management

Collecting forest inventory data and monitoring changes is critical to forest management activities. Yet, GIS can build on these activities by incorporating models to guide timber harvesting, silviculture and fire management activities, or predict fuel wood and other resource supplies. Other priorities, such as providing for wildlife habitat, ensuring recreation opportunities and minimizing visual impact of harvesting, are also growing in importance (McKendry & Eastman, 2000, Wulder & Franklin, 2007). Some important forest management issues are discussed as below:

i. *Timber harvesting:* The main objective of timber harvesting is continuous supply of wood to consumers which is generally based on economic considerations. Basic harvesting models could be improved by adding spatial and other inputs like, locations of forest stands, timber inventory data, ecologically sensitive areas, key attributes of the terrain, topography, forest type, soil classification, management compartments, roads, streams etc. Spatial forest modelling using GIS can substantially enhance the planning of harvesting strategies (Upadhyay, 2009). Spatial model like, GISFORMAN facilitates the manager to see the economic, environmental and social impacts of the proposed harvest and take decisions for better planning (Jordan & Baskent, 1991). Other GIS based model SHG could be more useful as it can incorporate ecological inputs like, the year of stand origin, productivity of the site, area, relative stocking factor, and silvicultural treatment class etc. (Moore & Lockwood, 1990).

ii. *Fuelwood supply:* The availability of fuel wood supplies for local use is an important forest management issue in many parts of the world, especially in developing countries. GIS can contribute to assessments of fuel wood supply and demand and offers the potential to predict future needs. There are two useful studies on fuelwood management modelling in Nepal (Schreier et al., 1991) and Sudan (Olsson, 1985) which was done using GIS.

iii. *Forest fire management:* Fire has been one of the major hazards in protecting the forest or managing the timber or fuelwood resources. It is an ecological process that governs the composition, distribution, and successional dynamics of vegetation in the landscape (Johnson, 1992). Landscape fire management activities include fire prevention, wildlife control, prescribed burning, and post-fire recovery actions. It is essential to collect spatial data of the landscape like, vegetation types, topography, natural features, etc. in order to predict forest fire occurrence, location, direction, rate of spread and intensity etc. Forest managers have used GIS for fuel mapping, weather condition mapping, and fire danger rating (Holder et al., 1990). It has also been used to more accurately, quickly and thoroughly present, analyze and interpret annual and historical records of hazards (Vansickle, 1989).

iv. *Afforestation:* Plantation work within the forests or outside as reforestation or afforestation activities is the major task of a forester in Asia since China and India top in the world. Millions of saplings are planted every year, but the success is not seen in due proportion. Probably this lacks perfect planning, at least use of modern technology. GIS can be used for modeling area suitability by considering spatial data like, area expanse, soil type, topography, biotic pressure, climatic conditions, plant species, water points, fire breaks etc.

v. *Wildlife habitat suitability determination:* Species distribution modeling or ecological niche modeling are frequently used to determine the suitable habitat of plant or animal species (Kumar & Stohlgren, 2009; Hernandez-Baz et al., 2016; Keya et al., 2016; Abolmaali, et al., 2018). This is very useful in the conservation or rehabilitation program of endemic and endangered species in the light of continued loss of biodiversity. The algorithmic models use geospatial data of presence or absence of the target species, climatic factors (rainfall, temperature) and environmental determinants like, vegetation, waterbodies, soil type, elevation, slope etc. The model predicts the habitat index which determines suitability based on a combination of determinants and the presence of species in a particular locality. Future habitat prediction has also been made possible by using future climatic conditions in the model. Thus, the possible impact of climate change on habitat expanse in future can be predicted and conservation strategy of threatened species could be adopted in advance to save them from any illeffect (Qin et al., 2017; Setyawan et al., 2018).

WORKING PLAN AND GIS

Forest management activities are guided by WP in any forest division of Indian states and union territories. However, WP preparation is a stupendous task based on huge data collection and its analysis guiding the future course of management actions based on scientific prescription. The task becomes easy by applying techniques and methods based on geomatics in order to get precision and objectivity leading to significant saving of cost and time. This provides a spatial database on forest in GIS which is a set of multi-layered geo-referenced data of particular areas. Each layer is devoted to a particular theme or an entity on the ground e.g., administrative boundaries, road network, forest cover, forest types, altitude, operational guidelines, etc. Such layers in combination or independently, could be converted into maps. As per map policy of the Government of India; Open Series Maps (OSM), unlike Defence Series Maps, are used to support planning of developmental activities in the country. These maps are prepared by using the coordinate system based on International Terrestrial Reference Frame (ITRF), a geocentric reference frame. They have Universal Transverse Mercator (UTM) projection on WGS-84 Datum.

Mapping

Traditionally WP maps were being prepared manually using Survey of India toposheets which were being copied from the previous maps and feasible modifications were being added. This possibly added some errors which were being amplified in the long run resulting in aberrations. A case study given in box 1 may explain the reality in some but not in all the states of India. This is not to criticize but bring forth the necessity of use of unbiased technology which is available in the form of GIS. However, basic maps of the WP which can be prepared efficiently using modern technology are as below:

i. Stock map (1:12,500): This is a specialized map, contains the details of blank areas, crop composition, density, quality, age classes and regeneration status. Different colours and symbols may be used in the preparation of stock maps. This is normally a forest division level map.

ii. Management map (1:50,000): This is also a specialized map having divisional, range, block, compartment, and sub-compartment boundaries including boundary pillars with their numbers. Other features like roads, transmission lines, railway lines, water bodies, etc. make this map complete. This map also contains in colours the working circles, felling series, periodic blocks and coupe numbers.

iii. Working plan map (1:12,500): This is another specialized map prepared for forest range level use. This is similar to management maps and contain administrative and physiographic features in addition to silvicultural details like, working circle, felling series and periodic blocks.

iv. Reference map (1:125,000): This is a generalized map, like ready reference, containing detailed information of working plan and management maps to facilitate WP understanding easy and convenient while going through it. It may contain the ranges, roads, canals, forest rest houses, neighbouring towns and villages along with main boundaries of the forests and other relevant features which can be read with ease of clarity.

Above mentioned maps are basic requirement of WP with some listed features like, administrative, physiographic and silvicultural features or layers in GIS terms like, ranges, roads, canals, forest rest houses, neighboring towns and villages along with main boundaries of the forests, working circles, felling series, periodic blocks and coupe numbers. However, with the orientation of WP and SFM and mandatory use of GIS as per NWPC 2014 possibility of generating n number of map layers in consonance with SFM criteria and indicators with precision could be explored and make use of. Although all the indicators cannot be monitored through mapping, an attempt has been made to summarize this in Table 1.

Evaluation

One of the merits of WP is that the prescription of future management is based on the performance of preceding plan prescription. The same management prescription either continues as proposed in the previous plan or is changed from it, if required. However, assessment of such performance is generally qualitative and lacks quantitative evaluation. This gives rise to criticism by ecologists/scientists whose perspective is different from foresters/managers. They argue that the performance of the previous prescription is not evaluated in quantitative terms, hence, the prescription remains a hypothesis for want of testing. Therefore, the Working Plan, a much-acclaimed scientific document, stands as a Working Hypothesis not a Working Plan. Irrespective of the counter arguments of foresters who are the implementers of working plan prescription, that the forester's role or perspective is of a manager not the scientist, it appears imperative that this aspect of performance evaluation should be given due weightage so that procedural credibility of WP be further enhanced based on scientific outlook.

Monitoring and GIS

Application of GIS through Remote Sensing, GPS and IT has proved to be the most efficient, effective and transparent monitoring tool in forest management. Multi temporal satellite imagery plays a vital role in quantifying spatial and temporal phenomena which is otherwise not possible to attempt through conventional mapping (Rawat & Kumar, 2015). This can be used in monitoring changes in landcover-landuse spatiotemporally. Land use/cover change is a widespread and accelerating process, mainly driven by natural phenomena and anthropogenic activities, which in turn drive change that would impact natural ecosystem (Ruiz-Luna & Berlanga-Robles, 2003; Turner & Ruscher, 2004). The example given in Figure 6 amply indicates that over 20 years the vegetation of a given parcel changed in composition. Some of the vegetation from initial status changed to non-vegetation cover and at the same time some of the non-vegetation area changed to vegetation cover. Overall change in this case was positive.

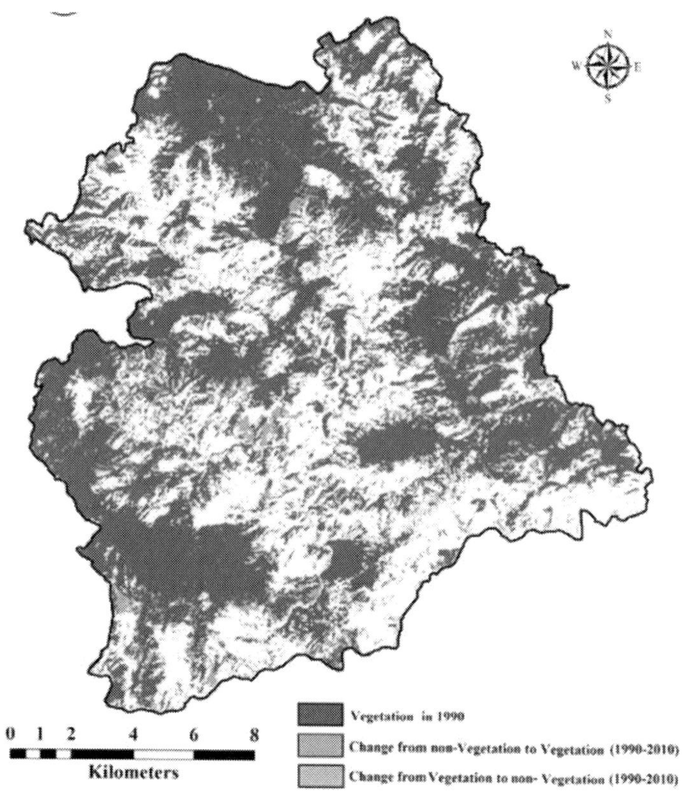

Source: Rawat and Kumar (2015).

Figure 6. Vegetation Land use/cover) change in two decades (1990 - 2010) in Hawalbagh, Almora Forest Division, Uttarakhand. This map is based on Landsat Thematic Mapper Satellite Imagery. Light green and light blue area may be noted for vegetation related changes.

Similar detection is possible in embankment erosion, fire damage, land encroachment, crop density change, logging impact etc. which directly affects forest management. However, this spatiotemporal change detection may be used in management prescription in a WP of particular forest divisions for the next plan period.

One of the most challenging management issues of current times is the encroachment over forest land for agriculture or habitation which means deforestation and continued loss of opportunity of carbon sequestration and other ecosystem services (Sharma et al., 2015). ET (2015) reported 18990 km^2 encroached forest land (>2% forest area of the country) in India. There was a spurt in land grabbing post FRA 2006 promulgation. However, the cases were dealt with, at least in Maharashtra state, very efficiently and transparently using GIS. Claims were settled using CARTOSAT images and Google Earth images by reading and comparing land use signatures of pre and post claim dates (Jha and Chaudhry, 2018).

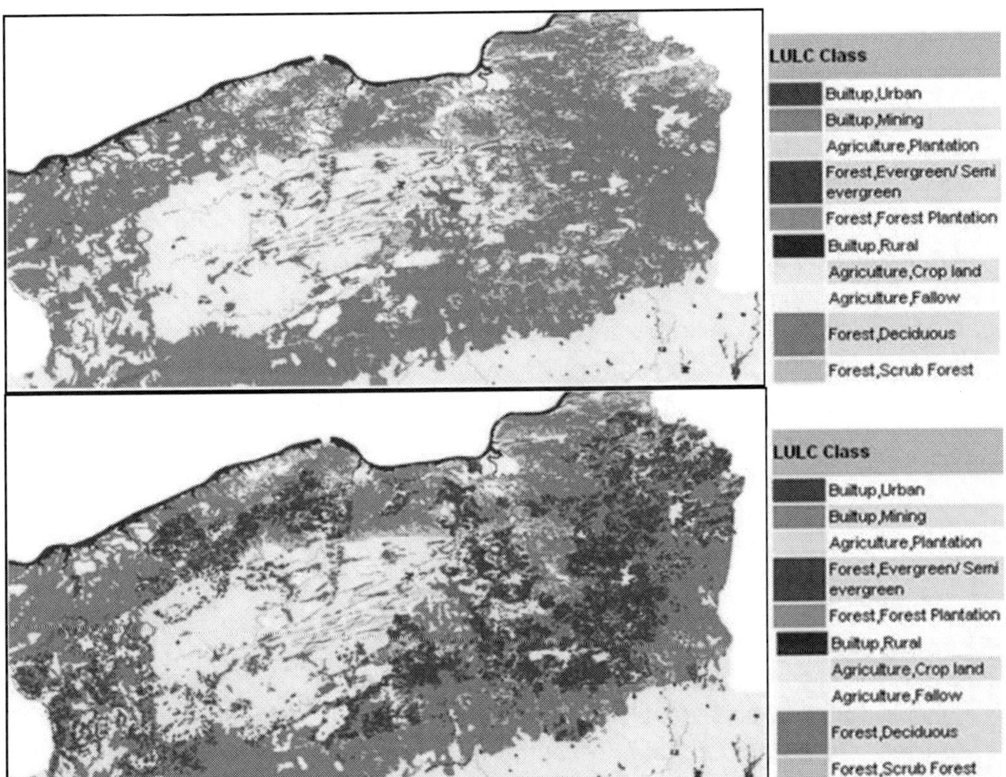

Source: Study of Changes in Vegetative Cover and Land use in forest with Claims and Forest Rights finally Recognized under the Forest Rights Act 2006. http://www.indiaenvironment portal.org.in.

Figure 7. GPS measurements overlaid on NRSC LULC map 2005-2006. GPS measurements upload between 01.01.2011 to 01.08.2012 on NRSC LULC map showing spatial distribution of cases overlaid is around 17,200. Most of the cases are falling in different categories of forest.

Figure 7 illustrates an example of 17200 claims, mostly false, settled by Maharashtra government using NRSC (National Remote Sensing Centre) LULC map showing spatial distribution of cases overlaid in Nadurbar district. On the other side, the Indian government recognised, as per the progress report of Ministry of Tribal Affairs till 31st May 2014, and a total of 14,36,290 titles had been distributed to the right holders containing 223800 km^2 forest lands (Sharma et al., 2015). This can be seen as the latest manifestation of a paradigm shift.

CONCLUSION

Forest management is returning to the Vedic period (ancient Indian) philosophy of serving the people, of course in spirit only, when resources were plenty and consumers were few with access to the resources without any restriction or injustice. It saw phases of good management for the people, then negligence, then restriction and finally lifting the control for tribal and forest dwellers. The marginalised communities have been given the opportunity to manage the resources with their involvement. Recognition of their rights over bioresources and fair and equitable sharing of the benefits arising out of utilization of genetic resources has been ensured. Due to high demands and very low supplies of goods and services, sustainable forest management has been adopted to improve the demand and supply ratio. Management processes have also been taken care of by improving WP preparation and implementation by orienting it towards SFM Criteria and Indicators. Also, the latest and efficient technology of geoinformatics has been adopted through legal compulsion. Quantitative evaluation of the management performance is advocated as further modification of WP preparation. Though the demerits of socio-cultural and socio-political conditions in the country will continue to impede the process, it is hoped that forest management will improve in the desired direction to fulfil the aspirations of current policy.

REFERENCES

Abolmaali, M. R. S., Tarkesh, M. & Bashari, H. 2018. MaxEnt modeling for predicting suitable habitats and identifying the effects of climate change on a threatened species, *Daphne mucronata*, in central Iran. *Ecological Informatics,* 43:116-123. https://doi.org/10.31195/ejejfs.428381.

Altan, M. D., Toz, F. G., Can, Z., & Tastan, H. 1991. Database-and Geographic Information Systems in Geodesy and Photogrametry. Engineering (in Turkish) *Journal of Istanbul Technical University,* 49(4): 35-44. https://doi.org/10.1016/S0924-2716(01)00025-9.

Balaji, S. 2002. *Forest Policy in India – in Retrospect and Prospect.* Paper presented at the IUFRO Science/Policy Interface Task Force regional meeting held in

Chennai, India at the M.S. Swaminathan Research Foundation, 16-19 July 2002. Retrieved from https://pdfs.semanticscholar.org/caf1/5c42ec386035f14e185c42ae0fcc2b66e4f8.pdf.

Balaji, S. & Rai, R. S. V. 1998. *Biodiversity – crisis and compulsions*. Van Vigyan, 36 (2,3&4).

Bhattacharya, P., Pradhan, L. & Yadav, G. 2010. Joint forest management in India: Experiences of two decades. *Resources, Conservation and Recycling,* 54: 469–480. https://doi.org/10.1016/j.resconrec.2009.10.003.

Crain, I. K. & MacDonald, C. L. 1983. From land inventory to land management: The evolution of an operational GIS. In Weller, B. S. (Ed.), Automated Cartography: International Perspectives on Achievements and Challenges, Vol. 1, Ottawa: *Steering Committee for the Sixth International Symposium on Automated Cartography*, 41-50.

ET. 2015. *Economic Times*, December 22, 2015. http://Economictimes.indiatimes.com. Accessed on August 15, 2019.

Ghildiyal, M. C. 2000. Participatory Forest Management: Implications for Policy and Human Resources' Development in Uttarakhand Himalaya, Uttar Pradesh. In Rana, G. M., Murray, A. B. & Joshi, S. M. (Eds.), *Participatory Forest Management: Implications for Policy and Human Resources' Development in the Hindu Kush-Himalayas:* Volume IV - India. International Centre for Integrated Development, Kathmandu Nepal. http://lib.icimod.org/record/23435/files/Participatory%20forest%20mgt%20volu% 20 IV(2000)Low.pdf. Accessed August 09, 2019.

Hernandez-Baz, F., Romo, H., Gonzalez, J. M., Hernandez, M. J., & Pastrana, R. G. 2016. Maximum Entropy niche based modelling (MaxEnt) of potential geographical distribution of *Coreura albicosta* (Lepidoptera: Erebidae: Ctenuchina) in Mexico. *Florida Entomologist,* 376-380. https://journals.flvc.org/flaent/article/view/87733/84979.

Holder, G., Van Wyngaarden, R., Pala, S. & Taylor, D. 1990. *Flexible analysis through the integration of a fire growth model using an analytical GIS*, Proceedings, GIS '90 Symposium, Vancouver, BC: Forestry Canada, 153-157.

Jha, K. K. & Chaudhry, P. 2018. Unravelling the Complexity of Protected Area Management in Two Developing Countries: Issues of Human Displacement and Wildlife Conflict. In Campbell. M. (Ed.), *Geomatics and Conservation Biology.* New York, NY: Nova Science Publishers, Inc.

Johnson, E. A. 1992. *Fire and vegetation dynamics: Studies from the North American Boreal forest.* Cambridge, UK: Cambridge University Press.

Jordan, G. A. & Erdle, T. A. 1989. Forest management and GIS: What have we learned in New Brunswick? *GISM Journal* (Canadian Institute of Surveying and Mapping), 43(3): 287-295. https://doi.org/10.1139/geomat-1989-0031.

Jordan, G. A. & Baskent, E. Z. 1991. GISFORMAN – A next generation wood supply model. In: *Applications in a Changing World*, GIS '91 Symposium Proceedings, Vancouver, Canada, February 12-15, 1991, 149-156. Vancouver, BC: Forestry Canada.

Keya, Z. Y., Faryadi, S., Yavari, A., Kamali, Y., & Shabani, A. A. 2016. Habitat Suitability & Connectivity of Alborz Wild Sheep in the East of Tehran, Iran. *Open Journal of Ecology:* 325-342. https://doi.org/10.4236/oje.2016.66032.

Kothari, A., Pathak, N., Anuradha & Yarreja, B. 1999. *Communities and conservation - Natural resources management in south and central Asia*, New Delhi, Sage Publishers.

Kumar, B. M. 2008. Forestry in ancient India: Some literary evidences on productive and protective aspects. *Asian Agri History*, 12(4): 299-206. Retrieved from http://www.indiaenvironmentportal.org.in/files/Forestry%20in%20ancient%20India.pdf.

Kumar, S., & Stohlgren, T. J. 2009. Maxent modeling for predicting suitable habitat for threatened and endangered tree Canacomyrica monticola in New Caledonia. *Journal of Ecology and Natural Environment,* 1(4): 94-98. http://citeseerx.ist.psu.edu/viewdoc/ download?doi=10.1.1.521.4231&rep=rep1&type=pdf.

Lal, J. B. 1989. *India's Forests: Myth & Reality.* Dehradun: Natraj Publishers.

Linser, S., Wolfslehner, B., Asmar, F., Bridge, S. R. J., Gritten, D., Guadalupe, V., Jafari, M., Johnson, S., Laclau, P. & Robertson, G. 2018. 25 Years of Criteria and Indicators for Sustainable Forest Management: Why Some Intergovernmental C&I Processes Flourished While Others Faded. *Forests* 2018, 9, 515. https://doi.org/10.3390/f9090515.

Loukas, M., Lanteri, A., Ferrauiola, J., Tubbs, R. S., Maharaja, G., Shoja, M. M., Yadav, A. & Rao, V. C. 2010. Anatomy in ancient India: a focus on the Susruta Samhita. *Journal of Anatomy,* 217: 646–650. https://doi.org/10.1111/j.1469-7580.2010.01294.x.

Mallick, C. R. 2018. *Working Plan Baratang and Mayabandar Foret Divisisons, 2017-18 to 2026-27.* Andaman and Nicobar Administration, Department of Environment and Forests. Port Blair.

McKendry, J. E. & Eastman, R. 2000. *Applications of GIS in Forestry: A Review.* Retrieved from https://pdfs.semanticscholar.org/7fcc/f3c6957b9377eb999845 5406670d3f79ed 6d.pdf?_ga=2.34359970.262720609.1592945489-1738787424.1592945489.

Moore, T. G. E. & Lockwood, C. G. 1990. *The HSG Wood Supply Model: Description and User's Manual*, Chalk River, Ontario: Petawawa National Forestry Institute, Forestry Canada, Information Report PI-X-98.

Negi, S. S. 1986. Forest Policy and Five-Year Plan. In *A Handbook of Forestry.* 102-120. Dehradun, IBH.

NWPC. 2014. *National Working Plan Code 2014 (For sustainable management of forests and biodiversity in India.* Ministry of Environment and Forests, Government of India, New Delhi.

Olivelle, P. 2013. *King, Governance, and Law in Ancient India: Kauṭilya's Arthaśāstra.* ISBN 978-0199891825. Oxford UK: Oxford University Press.

Olsson, L. 1985. *An Integrated Study of Desertification: Applications of Remote Sensing, GIS and Spatial Models in Semi-Arid Sudan,* Monography No. 13, Series C, General and Mathematical Geography, Lund University, Sweden.

Prasad, R. 2001. Criteria & indicators for sustainable management of dry forests in Asia. *The International Forestry Review,* 3(4):307-316. http://www.jstor.org/stable/42609407.

Prasad, R. & Kotwal, P. C. 2001. *Progress of implementation of Sustainable Management of Dry Forests in Asia with special emphasis on Indian initiative.* Paper contributed for the International expert meeting on Monitoring, Assessment and Reporting (MAR) on the progress toward Sustainable Forest Management, 5-8 November 2001, Yokohama, Japan.

Prasad, R., Raghavan, S., Phukan, B. R. & Joshi, B. 1999. *Proceedings of the national technical workshop on evolving C&I for SFM in India.* Indian Institute of Forest Management, Bhopal, January 21-23, 1999.

Qin, A., Liu, B., Guo, Q., Bussmann, R. W., Ma, F., Jian, Z., Xu, G. & Pei, S. 2017. Maxent modeling for predicting impacts of climate change on the potential distribution of *Thuja sutchuenensis* Franch., an extremely endangered conifer from southwestern China. *Global Ecology and Conservation,* 10: 139-146. https://doi.org/10.1017/S0021859619000686.

Rawat, T. S., Menaria, B. L., Dugaya, D. & Kotwal, P. C. 2008. Sustainable forest management in India. *Current Science,* 94(8): 996-1002. Retrieved from https://pdfs.semanticscholar.org/9f34/5d2947f652a27355b34615a906767bebcc3b.pdf.

Rawat, J. S. & Kumar, M. 2015. Monitoring land use/cover change using remote sensing and GIS techniques: A case study of Hawalbagh block, district Almora, Uttarakhand, India. *The Egyptian Journal of Remote Sensing and Space Sciences,* 18: 77–84. https://doi.org/10.1016/j.ejrs.2015.02.002.

Ruiz-Luna, A. & Berlanga-Robles, C. A. 2003. Land use, land cover changes and costal lagoon surface reduction associated with urban growth in northwest Mexico. *Landscape Ecology,* 18: 159–171. https://doi.org/10.1023/A:1024461215456.

Saravanan, V. 2018. Impact of Forest Rights Act, 2006. In: *Environmental History and Tribals in Modern India.* Singapore, Palgrave Macmillan.

Schreier, H., Brown, S., Shah, P. B. & Kennedy, G. 1991. Nepal 2000, a GIS evaluation of the major land resources, In: *Applications in a Changing world*, GIS "91 Symposium Applications of GIS in Forestry: A Review. Proceedings, Vancouver, Canada, February 12-15, 1991. Vancouver, BC: Forestry Canada, 221-225.

Setyawan, A. D., Supriatana, J., Nisyawati, Sutarno, Sugiyarto, & Nursamsi, I. 2018. Predicting impacts of future climate change on the distribution of the widespread selaginellas (*Selaginella ciliaris* and *S. plana*) in Southeast Asia. *Biodiversitas,* 1960-1977.

Sharma, A. O. (2018). *Working plan for Narmada Forest Division, 2017-18 to 2026-27.* Government of Gujarat. Rajkot, Government Press.

Sharma, J. V., Gokhle, Y., Chauhan, S., Tyagi, A. 2015. Forest Right Act and Climate Change Vulnerability: Impact on Forests and Forest Dwelling

Communities in Maharashtra. *Indian Forester,* 141(12): 1230-1236. Retreived from http://www.indianforester.co.in/index.php/ indianforester/article/view/85541.

Singh, M. P., Reddy, S. R. & Ashraf, J. 2014. Revised National Working Plan Code in India, News and Notes. *Indian Forester,* 140(12): 1267-1270. http://www.indian forester.co.in/index.php/indianforester/article/view/55300.

Singh, T. 2005. 'Lies about Tribal Rights', Seminar, 552, p. 38.

Thakur, K. 1997. *Environmental Protection Law and Policy in India*. 441. New Delhi: Deep & Deep Publications, Internet access: 31/07/2019. https://journals. sagepub.com/ doi/abs/10.1177/097172189900400116.

Turner, M. G. & Ruscher, C. L. 2004. Change in landscape patterns in Georgia, USA. *Landscape Ecology,* 1(4): 251–421. Retrieved from http://citeseerx.ist.psu. edu/viewdoc/download?doi=10.1.1.4.4133&rep=rep1&type=pdf.

UN (United Nations). 1992. Non-Legally Binding Authoritative Statement of Principles for a Global Consensus on the Management, Conservation and Sustainable Development of All Types of Forests; *Report of the United Nations Conference on Environment and Development*; United Nations: Rio de Janeiro, Brazil, 1992.

Upadhyay, M. 2009. *Making GIS work in forest management*. Institute of Forestry, Pokhara, Nepal.

Vansickle, G. A. 1989. GIS – A tool in forest pest management, in *GIS Applications in Natural Resources.* Heit, M. & Shortreid, A. (Eds.), Colorado, GIS World, Inc.

Wulder, M. A. & Franklin, S. E. 2007. *Understanding forest disturbance and spatial pattern: Remote sensing and GIS approaches.* Boca Raton: CRC Press.

In: Critical Research Techniques in Animal …
Editors: Kaushalendra Kumar Jha et al.
ISBN: 978-1-53619-720-4

Chapter 2

BIOGEOGRAPHY AS AN ECOLOGICAL SCIENCE AND METHODOLOGY

Michael O'Neal Campbell*

Simon Fraser University, British Columbia, Canada

ABSTRACT

This brief chapter examines biogeography in its relationship with ecology as a research methodology, and its importance as ecological techniques evolve in the explanation of biophysical relations and ever greater spatial interconnections and networks. Biogeography is a spatial science, looking at the distribution and related issues of biological phenomena, and hence is closely related to ecology, especially spatial issues and the factors from which they emerge. All the topics in this book consider issues that are potentially variable in spatial relations. From another angle, biogeography as a science has been criticized as duplicating some aspects of spatial ecology, while some specialists emphasize the unique contribution of the former science and the justification for its status as an equal science. Part of the tension between the two disciplines is their institutional distinction: ecology courses and research are usually based in biology departments in universities, while equivalent biogeography work is often based in geography departments. However, current evidence indicates a general spatialization of ecological topics and priorities, while the development of spatially oriented technologies like geomatics has increasingly welded biogeographical and ecological topics. It is concluded that ecology and biogeography have a joint future, the better to analyze the increasingly complex environmental issues of the 21st century.

---

* Corresponding Author's E-mail: ecol55@hotmail.com.

INTRODUCTION

Biogeography and ecology have historically been seen as similar disciplines, but sometimes with some recognizable differences, these mostly based on the parameters of scale, time and human action (Campbell, 2017, 2018, 2020). For example, Jenkins and Ricklefs (2011, 2331) remark that "both biogeography and ecology seek to understand the processes that determine patterns in nature but do so at different spatial and temporal scales. The two disciplines were not always so different and are recently converging again at regional spatial scales and broad temporal scales." These authors further note, in distinguishing between the two disciplines (or subdisciplines, depending on one's viewpoint) that ecology "primarily focuses on adaptive evolution, population processes, and abiotic and species interactions, to explain coexistence of species in local areas. Modern ecology gains its inferential power inherently by its reliance on experimentation and statistical models (plus advances in computation) but does so within restricted spatial scale and temporal extent" (ibid.). Biogeography, by contrast has more focus on spatial relations. Particularly, "biogeography addresses evolutionary, climatic and geological processes (e.g., plate tectonics) to explain the distribution of diversity over the surface of the Earth and across its oceans. Biogeography remains primarily descriptive, but modern approaches incorporating phylogenetic (historical) information endeavour to test hypotheses about underlying processes (e.g., cladistic biogeography)" (ibid.). The authors conclude that "As such, the two disciplines represent recently overlapping ends of a spatio-temporal gradient in our views of the natural world and how it works" (Jenkins & Ricklefs, 2011, 2331). Both sciences relate to several basic and studies, as shown in the model diagrams below (Figure 1), derived from Campbell (2020).

By these definitions, ecology appears more focussed on narrower spatial contexts than biogeography. Definitions of ecology have varied through time. For example, Cary Institute of Ecosystem Studies defines ecology as "the scientific study of the processes influencing the distribution and abundance of organisms, the interactions among organisms, and the interactions between organisms and the transformation and flux of energy and matter." They also point out that "ecology was originally defined in the mid-19th century, when biology was a vastly different discipline than it is today" the start being that of Ernst Haeckel (1834 to 1919), who defined ecology as "the study of the relationship of organisms with their environment" (Cary Institute of Ecosystem Studies, 2020). Later definitions have embraced broader principles, such as that of Odum (1971), which focussed to a greater extent on a structured environmental system, termed the ecosystem. Cowling (1992) argues the emphases of ecology are largely microscale issues such as plant reproduction (le Maitre & Midgely, 1992), plant/animal relationships (Johnson 1992), competition between organisms (Bond et al., 1992), plant structure and function (Stock et al., 1992) and ecosystem functional interactions (Stock & Allsopp, 1992). Biogeographical studies in Cowling's (1992) work emphasise the broader issues of history (Linder et al. 1992) vegetation form and associations (Cowling & Holmes 1992), plant diversity and endemism (Cowling et al. 1992) and time regimes (Deacon et al., 1992). Equivalents may be found in zoological studies (Campbell, 2020).

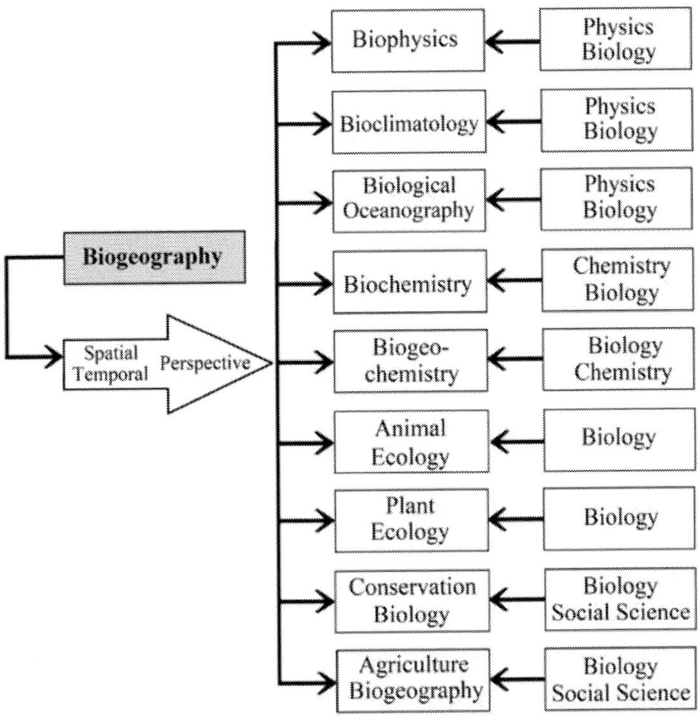

Figure 1. Supporting Sciences of Biogeography (see also Campbell. 2020).

Currently, ecology has expanded into broader, more flexible definitions, one of which sometimes moves in opposition, or in contrast to the ecosystem concept, and concerns the less structured paradigm of the "New Ecology" (Campbell, 1998). Some of these new paradigms include those that emphasise non-linear disequilibrium models (Zimmerer, 1994, 108) as counterpoints to the classical ecosystem position of linear, mechanistic change exemplified by 'natural' vegetation, climatic climaxes, succession, ecological equilibrium and human-nature separation and associated links with animal ecology" (Pickett et al. 1992; Christiansen 1993; Solbrig, 1993; Scoones, 1997, 1999). The ecological equilibrium refers to a generalized steady state, based on relations between biological and physical interactions. This ecological equilibrium position favoured ecosystem theory, mechanistic succession and vegetation climaxes, because there was an inherent assumption in the power of generalisation and scientific universalism (Burrows, 1990; Pimm, 1991; Tallis, 1991; Tudge, 1991; Beeby, 1993; Simmons, 1993; Allaby, 1994; Edwards et al., 1994; Begon et al., 1996; Campbell, 2017).

According to Solbrig (1993) the contrasted "non-equilibrium hypothesis" has developed incrementally over the years, examples being the writings of Holling (1973), Noy-Meyer (1975), Weins (1984) and Westoby et al. (1989). The positions taken by some researchers, such as Solbrig (1991), Stott (1991; 1994), Pickett et al. (1992), Scholes and Walker (1993) and Khasa and Dancik (1997) may be termed disequilibrium analyses, because they emphasise the dynamics of interrelationships within complex and unpredictable socioenvironmental contexts. Chaotic fluctuations in socioenvironmental contexts are frequently emphasised (Zimmerer 1994, Campbell, 1998), challenging the equilibrium position of classical ecology (as argued for example by Tansley 1935, Whittaker 1953; Odum 1971; Begon et al. 1996).

Biogeography has been influenced by some of these developments in ecology, but generally uses a more spatial approach to the study of biological phenomena and may include considerations of human actions. Partly, this is because biogeography is both a biological science and a geographical science, the latter involving concepts such as location, spatial distribution, ecological variation, spatial statistics and geographical information systems, and even nature-society relations and more human centered ecologies (Campbell 2018). Definitions of biogeography tend to include such concepts, such as that of Encyclopædia Britannica (2020a); "Biogeography, the study of the geographic distribution of plants, animals, and other forms of life. It is concerned not only with habitation patterns but also with the factors responsible for variations in distribution."

Several current strands emphasize the breadth and universality of biogeography as a science and research methodology, where it is not merely a branch of biology or geography, but an approach that can encompass the entire field of biology. The rationale is that as biology studies all forms of life, and these forms are invariably located in a spatial context, spatial distribution is universally relevant. Therefore, biogeography is related to any biological science that studies parameters that may contribute to (and be influenced by) spatial analyses. This universality, breadth and flexibility of biogeography and its possible role as a method across biological phenomena is underscored by several definitions. For example: "biogeography is a scientific approach to understanding the distribution and abundance of living things, the biota, on our planet" (Mueller-Dombois, 2001, 565); "Biogeography is a biological as well as a geographical discipline" (Muller-Hohenstein, 1201); biogeography is "the study of spatial and temporal distributions of organisms and their habitats, and the historical and biological factors that produced them" (Monaco et al., 2003); and most broadly, biogeography is relevant to "GIS-based land-use studies, to GPS-based studies of animal tracking, to molecular biology-based studies of plants, animals and insects" and the distribution of species including our own in space and time and the geological, physical, biological and even cultural forces that explain them" (Smithsonian Tropical Research Institute, 2020).

The spatial parameters may relate to the interfaces between biology and other sciences (see Figure 1). For example, in the field of biophysics, the main techniques are those of physics such as electric currents, temperature, stress and entropy, which are used to study biological phenomena, and may have regional significance. These may produce variations in the studied phenomena, which may vary spatially and temporarily. Other hybrid disciplines, such as bioclimatology (the study of interactions between climate and biological regions, spatially and temporarily), biological oceanography (a more environmentally focussed science than marine biology), biochemistry (the application of the chemistry techniques to the study of biological phenomena and processes) and biogeochemistry (the study of the chemical, physical, geological, and biological processes that link the physical and biological spheres of the earth) may uncover spatial and temporal differentiation and hence be amenable to biogeographical study (Campbell, 2020).

Biogeography's relationship with ecology is similarly complex, not least because both biologists and geographers may study biogeography. Muller-Hohenstein (2001) notes that "biogeography is a biological as well as a geographical discipline. Biologists prefer a more historical view and underline the importance of the evolution of species and communities. Geographers accentuate more the ecological background of the distribution of plants and animals, including the manifold influences of man, and they underline the importance of scales in time and space. Both try to understand the distribution of patterns of organisms and biocenosis." The inclusion of geographical research tools, such as geomatics may yield new approaches for ecological studies and even new findings. For example, the related subdisciplines of animal and plant ecology when defined within a spatial model and studied using spatial technology, may transpose into zoogeography (mostly spatial animal ecology) and phytography (mostly spatial plant ecology), whether studied by biologists or geographers (Katinas & Crisci, 2018; Campbell, 2017, 2018, 2020).

One consequence of such integration of research models is the possibility of the duplication of research efforts (Campbell, 2017, 2018). For example, Vitt and Caldwell (2014, 353) point out that "ecology, biogeography, and conservation biology are difficult topics to separate because of the integrated nature of each." Jenkins and Ricklesf (2011, 2331) add that part of the similarity emanates as "both biogeography and ecology seek to understand the processes that determine patterns in nature but do so at different spatial and temporal scales" and that "the two disciplines were not always so different and are recently converging again at regional spatial scales and broad temporal scales" (ibid.). The similarities, differences and integrations may be evaluated using certain common parameters, such as the four common components of niche (an environmental context favored by a species), comparative ecology (an arguably vague subdiscipline comparing defined ecologies), macroecology (ecological relationships at large spatial scales), community assembly (processes shaping the identity and abundance of species in ecological communities), and diversity (Peters et al., 1991; Campbell, 2018, 2020).

A comparison between biogeography and ecology is shown in the tables below, using organismal, population and community indices, the better to avoid duplication (Jenkins & Ricklefs, 2011, see also Campbell, 2020). The first table examines the spatial issues, while the second table examines the related temporal issues. The third table looks at units of study of the related biological sciences. The fourth table examines some related biological processes. Both biogeography and ecology are currently evolving and therefore their relation is changing. For example, Jenkins and Ricklefs (2011, 2332-2333) describe developments in both biogeography and ecology, and the current distinctions between these disciplines. For biogeography, current trends indicate the rise of evolutionary, climatic and geological processes as factors for biodiversity. There are also developments in phylogenetic (historical) information for hypothesis testing of phenomena such as processes in cladistic biogeography (here, cladistic biogeography refers to the study of historical and evolutionary relationships between species, using their particular distribution patterns), and the use of GIS and remote sensing techniques to study spatial patterns in biological diversity or macroecology (Campbell, 2018).

Table 1. Biogeography and Ecology; Key Approaches to Top Issues

Attribute	Biogeography	Ecology
Spatial Scales	*Global* e.g., range of all the *Accipitriformes*, global except for Antarctica, forest, savanna, grassland, desert and coastal ecosystems. *Regional* e.g., the range of the cougar through regions such as include the Amazonian, Central American and Central Andean regions.	*Regional* e.g., a study of an ecological region in western Brazil or northern Mexico, or a region of Turkey vultures in Northern Trinidad. *Local* e.g., town, patch or field sized studies, such as studies of parks in a Caribbean island, or deforestation in fields in El Salvador.
	Examples from biogeography journals	**Examples from ecology journals**
	Guo et al. 2020. Activity niches outperform thermal physiological limits in predicting global ant distributions. Weigelt et al., 2020. GIFT - A Global Inventory of Floras and Traits for macroecology and biogeography. Biber et al., 2020. A comparison of macroecological and stacked species distribution models for ecological prediction	Zhang et al. 2020. Climate change-induced migration patterns and extinction risks of Theaceae species in China. Krupa et al., 2020. Interactions of carnivorous plants, sheet-web spiders, and ground-running spiders as predators in a wet meadow Sharma et al., 2020. Estimating occupancy of Chinese pangolin (Manis pentadactyla) in a protected and non-protected area of Nepal.

Adapted from Jenkins and Ricklefs (2011), examples from Campbell (2015, 2017, 2018, 2020), Journal of Biogeography; Annual Review of Ecology, Evolution, and Systematics; Proceedings of the Royal Society B Biological Sciences, Nature Ecology and Evolution, Frontiers in Ecology and the Environment, Global Environmental Change and Global Ecology and Biogeography.

From the ecological perspective, the focus of some changes is primarily on "adaptive evolution, population processes, and abiotic and species interactions, to explain coexistence of species in local areas" (Jenkins & Ricklefs, 2011: 2332-2333). Ecology, like some aspects of biogeography may include experimentation, statistical models and computation, but ecological studies may still focus on smaller case studies, and shorter time periods (Campbell, 1998, 2017, 2018, 2020). The authors conclude that "the two disciplines represent recently overlapping ends of a spatio-temporal gradient in our views of the natural world and how it works" (Jenkins & Ricklefs (2011: 2332-2333).

Table 2. Biogeography and Ecology; Key Approaches to Top Issues

Attribute	**Biogeography**	**Ecology**
Temporal Scales	*Million years* + e.g., the extinction the sabre-toothed cats of North America, the change in their habitats and the evaluation of their ecological settings through time *Thousand years* + e.g., the decline in wolf populations in Western Europe, habitat destruction and the role of human civilization.	Generation cycles: population cycles with a period of roughly one generation. E.g., the impact of chemicals or ecological change on a species, comparing one generation with another. Population cycles: populations change over a predictable period. For example, population cycles longer for larger than smaller mammals
	Examples from biogeography journals	**Examples from ecology journals**
	Leone et al., 2020. Pliocene colonization of the Mediterranean by Great White Shark inferred from fossil records. Voris, 2001. Maps of Pleistocene sea levels in Southeast Asia: shorelines, river systems and time durations. Marin & Hedges, 2016. Time best explains global variation in species richness of amphibians, birds and mammals.	Spagopoulou, 2020. Transgenerational maternal age effects in nature: Lessons from Asian elephants. Koch et al., 2018. Maternal age effects on fecundity and offspring egg-to-adult viability are not affected by mitochondrial haplotype. Lahdenperä et al. 2019. Capture from the wild has long-term costs on reproductive success in Asian elephants.

Source: Adapted from Begon et al. (2006); Hartl (2007); Jenkins and Ricklefs (2011); Campbell (2015); Engler et al. (2016).

There are also subdisciplines that integrate biogeography and ecology. For example, ecological biogeography involves "the relationship between the ecology of clades and their distribution" (Heads, 2015, 456). This subdiscipline includes geographical factors such as climate, as instruments for biological clade distributions. This approach is justified on the grounds that "in many cases, species distribution models - ecological niche models - based on the current environment of a species (its environmental envelope) fail to predict the actual distribution of the species" and they may even "over-predict distributions" (Heads, 2015, 456). The limitations of niche theories have been observed by some ecologists, possibly due to the study areas being too small for reliable generalization. Arguably, such small case studies may not yield enough evidence for the understanding of biogeographical and evolutionary processes. Local patterns may be undetected, and those local patterns that are detected may not be representative of the wider systems understudy.

Ecological biogeography, by using wider patterns and case studies, may be used as an integrating discipline that bridges this gap of knowledge (Heads, 2015, 456).

Table 3. Biogeography and Ecology; Key Approaches to Top Issues

Attribute	Biogeography	Ecology
Units of Study	Clade, or a monophyletic group of organisms with a common ancestor. Examples a genus or only two species have a common ancestor.	Individuals, one member of a species, with variable or generalizable behavior. Useful for intra-species behavioral studies.
	Species, the largest group of organisms where opposite sexes can produce fertile offspring, genetic interspecies differences	Populations, all organisms of one species, in a geographical area, inter-breeding is possible between any pair
	Range, the geographical area in which a species can be found, which may have requirements for the species existence/ health	Communities, an assemblage of populations of different species, with intra- and interspecies competition, predation and mutualism.
	Distribution, how the members of a species is arranged, within the range, usually dependent habitat features	Note: individuals, populations of each species are influenced by individuals, populations of other species.

Source: Adapted from Begon et al. (2006); Hartl (2007); Jenkins and Ricklefs (2011); Campbell (2015); Engler et al. (2016).

Table 4. Biogeography and Ecology; Key Approaches to Top Issues

Processes of Interest	Speciation, the evolutionary process where populations evolve into 2+ species, by geographical separation, isolation, partial separation or non-separation	Abiotic and biotic interactions, density, i.e., non-living variables (temperature, wind, water, soil) and living organisms affecting species or individual density
	Extinction, local (extirpation), wildland (still in zoos or domesticated) or functional (too few for breeding) total global extinction of a species	Abiotic and biotic interactions, distribution, i.e., non-living variables (temperature, wind, water, soil) and living organisms
	Range expansion or retraction, increase or reduction of species habitats, due to biological, physical or human factors.	Note: density refers to the number of individuals or species per unit area.

Source: Adapted from Begon et al. (2006); Hartl (2007); Jenkins and Ricklefs (2011); Campbell (2015); Engler et al. (2016).

CONCLUSION

Biogeography is clearly a highly meritorious subject and is also an effective approach to the evaluation of ecological information and assessments. In an increasingly networked and interconnected world, regionalism is emerging as an embedded concept for accurate ecological study. The key paradigms it espouses, spatial processes and variation, spatial statistics and supportive analytic computer software, and extremely complementary to the recent developments in ecology. Biogeography's strong connections and supports from human geography as part of its home in geography (unlike ecology it largely has two homes, with advantages from both) also give a methodological advantage, especially as with the rise of the human fringes of ecology (the New Ecology, Environmental Science and Environmental Studies), ecology is increasingly seen as straddling the natural and social sciences. This is prudent, as human action is increasingly overtaking natural processes and unaltered natural environments are increasingly rare. Ultimately, the importance of biogeography as a methodological tool in ecology boils down to whether spatial analysis is relevant to current ecological issues. This is obviously a truism.

REFERENCES

Allaby, M. 1994. The Concise Oxford Dictionary of Ecology. Oxford: Oxford University Press.

Beeby, A. 1993. *Applying Ecology*. London: Chapman and Hall.

Begon, M., Harper, J. L. & Townsend, C. R. 1996. *Ecology: Individuals, Populations and Communities* (3rd ed.), Oxford: Blackwell Science Ltd.

Begon, M., Townsend, C. R. & Harper, J. L. 2006. *Ecology: From Individuals to Ecosystems.* Malden: Blackwell Publishing.

Biber, M., Voskamp, A., Niamir, A., Hickler, T., & Hof, C. (2020). A comparison of macroecological and stacked species distribution models to predict future global terrestrial vertebrate richness. *Journal of Biogeography*, 47: 114-129. https://doi.org/10.1111/jbi.13696.

Bond, W. J., Cowling, R. M. & Richards, M. B. 1992. Competition and Coexistence. In: Cowling, R. M. (Ed.) *The Ecology of Fynbos: Nutrients, Fire and Diversity*, 206-225, Cape Town: Oxford University Press.

Burrows, C. J. 1990. *Processes of Vegetation Change*. London: Unwin Hyman.

Campbell, M. 1998. *Interactions between biogeography and rural livelihoods in the coastal savanna of Ghana*. (Unpublished doctoral dissertation). University of London, Imperial College.

Campbell, M. 2015. *Vultures: Their Evolution, Ecology and Conservation*. Boca Raton, FL: CRC Press.

Campbell, M. 2017. *Biological Conservation for the 21st Century: A Conservation Biology of Large Wildlife.* New York, NY: Nova Science Publishers.

Campbell, M. 2018. *Geomatics and Conservation Biology*. New York, NY: Nova Science Publishers.

Campbell, M. 2020. *The Physical Geography of El Salvador: A Geophysical and Ecological Approach*. New York, NY: Nova Science Publishers.

Cary Institute of Ecosystem Studies 2020. *Definition of Ecology*. Cary Institute of Ecosystem Studies I Millbrook, New York.

Christiansen, N. L. 1993. Fire Regimes and Ecosystem Dynamics. In: Crutzen, P.J. & Goldhammer, J. G. (Ed.), *Fire in the Environment: The Ecological, Atmosphere, and Climatic Importance of Vegetation Fires*. Chichester: John Wiley and Sons.

Cowling, R. M. & Holmes, P. M. 1992. Flora and Vegetation. In: Cowling, R. M. (Ed.), *The Ecology of Fynbos: Nutrients, Fire and Diversity*, 23-61. Cape Town: Oxford University Press.

Cowling, R. M. 1992. *The Ecology of Fynbos: Nutrients, Fire and Diversity*. Cape Town: Oxford University Press.

Cowling, R. M., Holmes, P. M. & Rebelo, A. G. 1992. Plant Diversity and Endemism. In: Cowling, R. M., (Ed.) *The Ecology of Fynbos: Nutrients, Fire and Diversity*, 62-112. Cape Town: Oxford University Press.

Deacon, H. J., Jury, M. R. & Ellis, F. 1992. Selective Regime and Time. In: Cowling, R. M. (Ed.) *The Ecology of Fynbos: Nutrients, Fire and Diversity*, 6-22. Cape Town: Oxford University Press.

DeAngelis, D. L. & Waterhouse, J. C. 1987. Equilibrium and Nonequilibrium Concepts in Ecological Models. *Ecological Monographs*. 57:1-21.

Edwards P. J., May, R. M. & Webb, N. B. 1994. *Large Scale Ecology and Conservation Biology*. Oxford: Blackwell Scientific Publications.

Engler, J. O., Secondi, J., Dawson, D. A., Elle, O. & Hochkirch, A. 2016. Range expansion and retraction along a moving contact zone has no effect on the genetic diversity of two passerine birds. *Ecography*, 39: 884-893. https://doi.org/10.1111/ecog.01520.

Encyclopædia Britannica 2020. *Biogeography*. Retrieved from https://www.britannica.com/science/biogeography.

Guo, F., Guénard, B., Economo, E., Deutsch, C., & Bonebrake, T. C. (2020). Activity niches outperform thermal physiological limits in predicting global ant distributions. *Journal of Biogeography*, 47: 829-842. https://doi.org/10.1111/jbi.13799.

Hartl, D. L. & Clark, A. G. 2007. *Principles of Population Genetics*. Sunderland, MA: Sinauer Associates.

Heads, M. 2015. The relationship between biogeography and ecology: envelopes, models, predictions: Biogeography and ecology. *Biological Journal of the Linnean Society*, 115(2): 456-468. https://doi.org/10.1111/bij.12486.

Holling, C. S. 1973. Resilience and Stability of Ecological Systems. *Annual Review of Ecological System*, 4: 1-24.

Jenkins, D. G., & Ricklefs, R. E. 2011. Biogeography and ecology: two views of one world. Philosophical transactions of the Royal Society of London. *Series B, Biological Sciences,* 366(1576): 2331–2335. https://doi.org/10.1098/rstb.2011.0064.

Johnson, H. 1992. Rural Livelihoods: Action From Below. In: Berstein, H., Crow, B. & Johnson H., *Rural Livelihoods: Crises and Responses*, 274-306. Oxford: Oxford University Press.

Katinas, L. & Crisci, J. V. 2018. Agriculture Biogeography: An emerging discipline in search of a conceptual framework. *Progress in Physical Geography: Earth and Environment,* 42(4): 513-529. https://doi.org/10.1177/0309133318776493.

Khasa, P. D. & Dancik, B. P. 1997. Sustaining Tropical Forest Biodiversity. *Journal of Sustainable Forestry,* 5 (1-2): 217-234.

Koch, R. E., Phillips, J. M., Camus, M., & Dowling, D. 2018. Maternal age effects on fecundity and offspring egg-to-adult viability are not affected by mitochondrial haplotype. *Ecology and Evolution*, 8, 10722-10732. https://doi.org/10.1002/ece3.4516.

Krupa, J. J., Hopper, K. R., Gruber, S. B., Schmidt, J. M. & Harwood, J. D. 2020. Plant–animal interactions between carnivorous plants, sheet-web spiders, and ground-running spiders as guild predators in a wet meadow community. *Ecology and Evolution* 10:1-11. https://doi.org/10.1002/ece3.6230.

Lahdenperä, M., Jackson, J., Htut, W., & Lummaa, V. 2019. Capture from the wild has long-term costs on reproductive success in Asian elephants. *Proceedings of the Royal Society B,* 286. https://doi.org/10.1098/rspb.2019.1584.

le Maitre, D. C. & Midgley, J. J. 1992. Plant Reproductive Ecology. In: Cowling, R. M. (Eds.), *The Ecology of Fynbos: Nutrients, Fire and Diversity*, 135-174. Cape Town: Oxford University Press.

Leone, A., Puncher, G. N., Ferretti, F., Sperone, E., Tripepi, S., Micarelli, P., Gambarelli, A., Sarà, M., Arculeo, M., Doria, G., Garibaldi, F., Bressi, N., Dall'Asta, A., Minelli, D., Cilli, E., Vanni, S., Serena, F., Díaz-Jaimes, P., Baele, G., Cariani, A., & Tinti, F. (2020). Pliocene colonization of the Mediterranean by Great White Shark inferred from fossil records, historical jaws, phylogeographic and divergence time analyses. *Journal of Biogeography*, 47, 1119-1129. https://doi.org/10.1111/jbi.13794.

Linder, H. P., Meadows, M. E. & Cowling, R. M. 1992. History of the Cape Flora. In: Cowling, R. M. (Ed.), *The Ecology of Fynbos: Nutrients, Fire and Diversity*, 113-134, Cape Town: Oxford University Press.

Marin, J. & Hedges, S. B. 2016. *Time best explains global variation in species richness of amphibians, birds and mammals*, 43(6): 1069-1079. https://doi.org/10.1111/jbi.12709.

Monaco, M. E., Kendall, M. S., Higgins, J. L. Alexander, C. E. & Tartt, M. S. 2003. *Biogeographic assessments of NOAA National Marine Sanctuaries: The integration of ecology and GIS technology*. EEE Xplore. https://www.doi:10.1109/OCEANS.2003.178086.

Muller-Hohenstein, K. 2001. Biogeography. *International Encyclopedia of the Social & Behavioral Sciences,* 1201-1206.

Mueller-Dombois, D. 2001. Island Biogeography. *Encyclopedia of Biodiversity*, 565-580. https://doi.org/10.1016/B0-12-226865-2/00166-8.

Noy-Meyer, I. 1975. Stability of grazing systems: An application of predator prey graphs *Journal of Ecology* 63: 459-81.

Odum, E. P. 1971. *Fundamental of Ecology*. 3rd Edition. Philadelphia: W. B. Saunders, Philadelphia.

Peters, R. H., Armesto, J. J., Boeken, B., Cole, J. J., Driscoll, C. T., Duarte, C. M., Frost, T. M., Grime, J. P., Kolasa, J., Prepas, E., Sprules, W. G. 1991. On the Relevance of Comparative Ecology to the Larger Field of Ecology. In: Cole J., Lovett G. & Findlay S. (Eds.) *Comparative Analyses of Ecosystems*, 46-63. Springer, New York, NY. https://doi.org/10.1007/978-1-4612-3122-6_4.

Pickett, S. T. A., Parker, V. T. & Fiedler, P. L. 1992. The New Paradigm in Ecology: Implications for Conservation Biology above the Species Level. In: Fiedler, P. L. & Jain, K., (Eds.), *Conservation Biology: The Theory and Practice of Nature Conservation, Preservation and Management,* 66-87 New York: Chapman and Hall.

Pimm, S. L. 1991. *Balance of Nature: Ecological Issues in the Conservation of Species and Communities*. Chicago: University of Chicago Press.

Scholes, R. J. & Walker, B. H. 1993. An African Savanna: Synthesis of the Nylsvley Study. Cambridge: Cambridge University Press.

Scoones, I. 1997. The Dynamics of Soil Fertility Change: Historical Perspectives on Environmental Transformation from Zimbabwe. *The Geographical Journal* 63(2): 161-169.

Scoones, I. 1999. New ecology and social sciences: What prospects for a fruitful engagement? *Annual Review of Anthropology* 28: 479-507.

Sharma, S., Sharma, H. P., Chaulagain, C., Katuwal, H. B. & Belant, J. 2020. Estimating occupancy of Chinese pangolin (Manis pentadactyla) in a protected and non-protected area of Nepal. *Ecology and Evolution*, 10, 4303-4313. https://doi.org/10.1002/ece3.6198.

Simmons, I. G. 1993. *Environmental History: A Concise Introduction*. Oxford: Basil Blackwell Publishers.

Smithsonian Tropical Research Institute 2020. *Calycophyllum candidissimum (Vahl) DC.* Retrieved from https://biogeodb.stri.si.edu/biodiversity/bci/species/29473/.

Solbrig, O. T. 1991. Savanna Modelling for Global Change. *Biology International* (Special Issues), 24:1-47.

Solbrig, O. T. 1993. Ecological Constraints to Savanna Land Use. In: Young, M. D. & Solbrig, O. T., (Eds.) *The World's Savannas: Economic Driving Forces, Ecological Constraints and Policy Options for Sustainable Land Use*, 21-47. Paris: The Pantheon Publishing Group.

Spagopoulou, F. 2020. Transgenerational maternal age effects in nature: Lessons learnt from Asian elephants. *The Journal of Animal Ecology*, 89(4): 936-939. https://doi.org/10.1111/1365-2656.13218.

Stock, W. D., Van der Heyden, F. & Lewis, O. A. M. 1992. Plant Structure and Function. In: Cowling, R. M. (Ed.) *The Ecology of Fynbos: Nutrients, Fire and Diversity,* 226-240. Cape Town: Oxford University Press.

Stock, W. D. & Allsopp, N. 1992. Functional Perspective of Ecosystems. In: Cowling, R. M. (Eds.) *The Ecology of Fynbos: Nutrients, Fire and Diversity*, 241-259. Cape Town: Oxford University Press.

Stott, P. A. 1994. Savanna Landscapes and Global Environmental Change. In: Roberts, N. (Ed.), *The Changing Global Environment*, 287-303. Cambridge: Basil Blackwell Limited.

Stott, P. A. 1991. Recent Trends in the Ecology and Management of the World's Savanna Formations. In: *Progress in Physical Geography* 15: 18-28.

Tallis, J. H. 1991. *Plant Community History: Long Term Changes in Plant Distribution and Diversity*. London: Chapman and Hall.

Tansley, A. G. 1935. The Use and Abuse of Vegetational Concepts and Terms. *Ecology* 16: 284-307.

Tudge, C. 1991. Global Ecology. London: Natural History Museum Publications.

Vitt, L. & Caldwell, J. 2013. *Herpetology: An Introductory Biology of Amphibians and Reptiles* (4th ed.) Cambridge, MA: Academic Press.

Voris, H. 2000. Maps of Pleistocene sea levels in Southeast Asia: shorelines, river systems and time durations. *Journal of Biogeography*, 27: 1153-1167. https://doi.org/10.1046/j.1365-2699.2000.00489.x.

Weigelt, P., Koenig, C., & Kreft, H. (2020). GIFT - A Global Inventory of Floras and Traits for macroecology and biogeography. *Journal of Biogeography*, 47, 16-43. https://doi.org/10.1111/jbi.13623.

Weins, J. 1984. On Understanding a Non-Equilibrium World: Myth and Reality in Community Patterns and Processes. In: Strong, D. R. (Ed.), *Ecological Communities: Conceptual Issues and the Evidence,* 439-457. Princeton: Princeton University Press.

Westoby, M., Walker, B. & Noy-Meir, I. 1989. Opportunistic Management for Rangelands not at Equilibrium. *Journal of Range Management*, 42: 264-274.

Whittaker, R. H. 1953. A Consideration of Climax Theory: the Climax as a Population and Pattern. *Ecological Monographs*, 23: 41-78.

Zhang, Y., Meng, Q., Wang, Y., Zhang, X. & Wang, W. 2020. Climate change-induced migration patterns and extinction risks of Theaceae species in China. *Ecology and Evolution*. 10(10): 4352-4361. https://doi.org/10.1002/ece3.6202.

Zimmerer, K. S. 1994. Human Geography and the 'New Ecology': The Prospect of Integration. In: *Annals of the Association of American Geographers* 84 (1): 108-125.

In: Critical Research Techniques in Animal ... ISBN: 978-1-53619-720-4
Editors: Kaushalendra Kumar Jha et al.

Chapter 3

LANDUSE-LANDCOVER DYNAMICS ANALYSIS USING REMOTE SENSING AND GIS: A CASE OF CENTRAL HIMALAYAN TERAI, INDIA

***Piara Lal Loi*[1] *and Kaushalendra Kumar Jha*[2,3*]**
[1]State Forest Research Institute, Uttar Pradesh, Kanpur, India
[2]Indian Institute of Forest Management, Bhopal, India
[3]Department of Forest, Environment and Climate Change, Uttar Pradesh, Lucknow, India

ABSTRACT

Natural resources are under high anthropogenic pressure, which results in serious environmental problems with impacts on socio-economic development. Remote Sensing (RS) and Geographical Information Systems (GIS) are widely accepted tools for the estimation and analysis of change dynamics in natural resources or landuse-landcover (LULC). One of the underdeveloped Terai districts within the Central Himalayan region of India was selected for studying LULC dynamics and their driving factors. The LULC was classified into five classes (forest land, agricultural land, waterbodies, wasteland and built-up areas) and 15 subclasses using LISS I and LISS III images of 1989 and 2002, respectively. In the span of 13 years forest land, wasteland and waterbody got reduced while agriculture land expanded. Some of the drivers identified were urbanization, infrastructure, low/high monsoon, encroachment, livelihood security, plantation/regeneration/orchard, overexploitation and wood-based industry. This study generated vital information which could be useful in district level planning elsewhere. It was recommended that LULC change-detection based planning of development should be prepared at the district level. Since the district has plenty of forest this kind of planning must be incorporated in Forest Working Plan in order to achieve sustainable management of forest, water and wildlife resources.

* Corresponding Author's E-mail: jhakk1959@gmail.com.

INTRODUCTION

Natural resources like forest, waterbodies and soil comprise the landcover on the earth's surface, while landuse refers to changes in surface structure for human uses such as agriculture, mining, irrigation canals, etc. However, these two terms are often used interchangeably (Dimyati et al., 1996 in Kumar et al., 2016) perhaps due to their variable form and the wide use of natural resources for development. Rawat & Kumar (2015) argue that land use affects land cover and changes in land cover affect land use. However, Lillesand & Kiefer (2000) have precisely defined land use as the human activity or economy-related function associated with a specific piece of land, while land cover relates to the type of feature present on the surface of the earth. As they form the basis of the life support system, natural resources are extremely important in every sphere of life. Ever increasing human populations on earth have contributed to an increase in the demands for natural resources, which has resulted in changes in landuse-landcover (LULC) over varying periods. LULC keeps changing, at least partially, into variable forms, depending on the development activities (Mallupattu & Reddy, 2013). Therefore, it is crucial to closely monitor resources which are being rapidly depleted. This necessitates an integrated and unbiased study of the resources so that these are conserved through their sustainable use with proper planning, utilization and management (Asselman & Middelkoop, 1995).

Change detection is an increasingly common application of Remote Sensing (RS), this involving a process of identifying differences in the state of an object or phenomenon by observing it at different times (Singh, 1989). This provides a holistic picture of the earth's surface which can be indispensable means for the integrated development of the region. Therefore, satellite monitoring can be used to compare the changes that have taken place over a period and for the detection and evaluation of damage to natural resources over short periods of time as well. Remote sensing is a highly potent technology for accurate, cost effective and holistic information on LULC. RS techniques coupled with Geographic Information Systems (GIS) provide effective tools for analyzing the land use dynamics of the region as well as for monitoring, mapping and management of natural resources (Kachhwaha, 1985; Poongothai et al., 2014). In particular, some notable Indian studies focus on the identification of LULC (Rai & Kumar, 2013), watershed and hydrology (Tiwari et al., 2017; Koneti et al., 2018), urban areas (Rawat et al., 2013; Rawat et al., 2014; Khan & Jhariya, 2016), coal mining impacts (Chitade et al., 2010; Rai & Kumar, 2013; Baruah et al., 2016) and, in general, on landscape change (Roy et al., 2002; Singh et al., 2010; Mageswaran et al., 2015; Rawat et al., 2015; Roy et al., 2015; Roy & Inamdar, 2018).

The *Terai* is a moist region that lies in a long parallel strip along the foothills of the Himalayan mountains. Most of the *Terai* districts are dominated by agriculture within a moist deciduous forest landscape. One such district selected for the present study is Lakhimpur Khiri with 21.3% natural forest resource (reserve forests, a national park, and a wildlife sanctuary). Another important resource of this region is the fertile alluvial soil with a river system which regularly floods the landscape. This is a resource use concentrated region, with a high human population and undeveloped amenities, that requires urgent aid to improve living standards and amenities, and to alleviate the pressure on resources. Therefore, the main objective of this study was to utilize the application of modern technology (RS & GIS) for three purposes: (i) map and quantify the landuse-landcover classes; (ii) evaluate resource change dynamics in a selected time period (1989 to 2002); and (iii) understand the drivers of change of natural resources (LULC) and their role in developmental activities.

MATERIALS AND METHODS

The Study Area

Lakhimpur Khiri district (7,680 km^2) is within the larger North Central part of Uttar Pradesh Province (Figure 1). Geographically (27.6°N 80.34°E – 28.6°N 81.3°E) it is within the central Himalayan *Terai* (moist) region. This district consists of a vast alluvial plain with a gentle south-eastern slope, within which perennial and seasonal rivers carry large amounts of silt and other debris from the hills of adjoining Nepal. The average altitude of the district ranges from 150 m to 182 m above sea level (asl). The major forest types are Northern Tropical Semi-Evergreen Forests and North Indian Moist-deciduous Forests (Champion & Seth, 1968). Local wildlife includes tigers (*Panthera tigris*, Linnaeus, 1758), rhinoceroses (*Rhinoceros unicornis,* Linnaeus, 1758), elephants (*Elephas maximus*, Linnaeus, 1758) and smaller animals such as the civet (*Viverridae*, Gray, 1821), mongoose (*Herpestes*, Illiger, 1811), hyena (*Hyaena hyaena*, Linnaeus, 1758), bear (*Ursus thibetanus*, G. Cuvier, 1823), deer (*Cervinae*, Goldfuss, 1820), antelopes (*Bovidae*, Gray, 1821) and reptiles (*Reptilia*, Laurenti, 1768) etc. A considerable part of the study area is the cultivated land.

Data Source

Satellite Data LISS-III (Linear Imaging Self-Scanning Sensor) of IRS-1D (Indian Remote Sensing Satellite) with a ground resolution of 23.5 m, and LISS-I with a ground resolution of 72 m of IRS -1A were acquired, respectively, on 3rd March 2002 and 6th February 1989. These were used for the purpose of comparison and change detection in land use. Survey of India Topo-sheets (Maps on 1:50,000 scales) were used for base line field surveys and for delineating other benchmarks such as roads, settlement locations and water bodies etc.

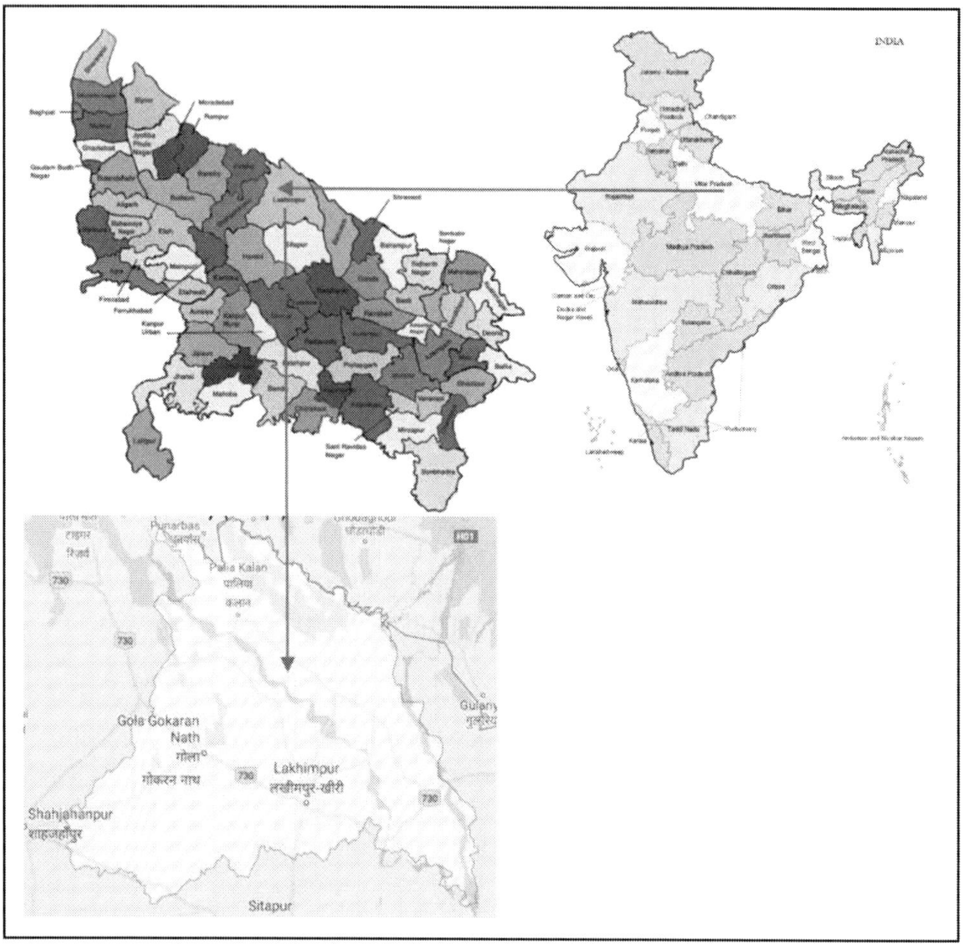

Source: Internet image.

Figure 1. Clock wise from bottom: Lakhimpur Khiri district located in Uttar Pradesh, a north Indian state, situated at the foothills of Nepal. The agriculture dominant district has forests in northern part and a river system running north-south.

Image Analysis

Satellite data 1989 was rectified using Ground Control Points (GCPs) identified on the image and ground with the help of GPS (Global Positioning System). Satellite data (2002) was rectified by image-to-image registration with (1989) data. Image Enhancement was carried out to provide informative image classification (Howard, 1991). This involved the use of software programs to ensure the photographic products best suited for the field. Image restoration (corrections) was done to the raw satellite data to eliminate radiometric and geometric distortions. The geometric accuracy of the scene was further improved for thematic mapping to be used multi-temporarily in monitoring changes. The next step was interactive image classification using enhancement techniques. Earth Resources Data Analysis System (ERDAS) image processing software was used for standard supervised classification. Change detection in the forest cover and other land use categories was achieved by comparison of classified outputs of 1989 and 2002 data using ArcInfo GIS software.

Collection of Secondary Data

In order to utilize the remote sensing data for the management of the natural resources of the study area, it was necessary to obtain other related data from various sources for the purpose of comparison and linking. In order to find out the seasonal as well as temporal variations in the ground water level, the data was obtained from the Ground Water Department of the State Government. Similarly, meteorological data was obtained from the working plans of North and South Khiri Forest Divisions and from the management plan of Dudhwa National Park (2001). Demographic data for 1991 and 2001 census was obtained from the Collector/District Magistrate of Lakhimpur Khiri District. In order to see the impact of industrialization in the district, data relating to industries was obtained from the office of Director of Industries, Kanpur. Wildlife census data was obtained from the Director, Dudhwa National Park. Forest fire data was obtained from North and South Forest Divisions, Lakhimpur Khiri and Directorate, Dudhwa National Park. Data relating to rivers, swamps and wetlands was obtained from the book titled Ecology of Dudhwa by Dr. R. L. Singh (2002).

Results

The LULC maps depicting five LULC classes and fifteen subclasses were prepared for the two periods viz. 1989 and 2003 (Figure 2). Although LULC categories have been shown differently in various studies (Singh et al., 2010; Mageswaran et al., 2015; Baruah et al., 2016; Tiwari et al., 2017; Bhatti et al., 2019; Roy & Inamdar, 2019 etc.), the five major LULC classes of the present study were forest land, agriculture land, wasteland, water bodies, and built-up area. The present study endeavored to classify the forest into dense forest, open forest, scrub forest, forest blank, grassland and one more category of forest land encroachment. The agriculture included cropland and plantation/orchard. Wasteland covered land with scrub, land without scrub, and waterlogged/wetland area. Water bodies included river/stream/surface water body, sandy area/river sand canal. Built-up area was a mono subclass. For better visual retrieval of data, close up of a section of Figure 2 (District LULC map) is given in Figure 3. The area covered under each of the classes and subclasses and the changes over the study period is presented in Table 1. Area extent assessment of these classes and subclasses and detected change are described in the following paragraphs.

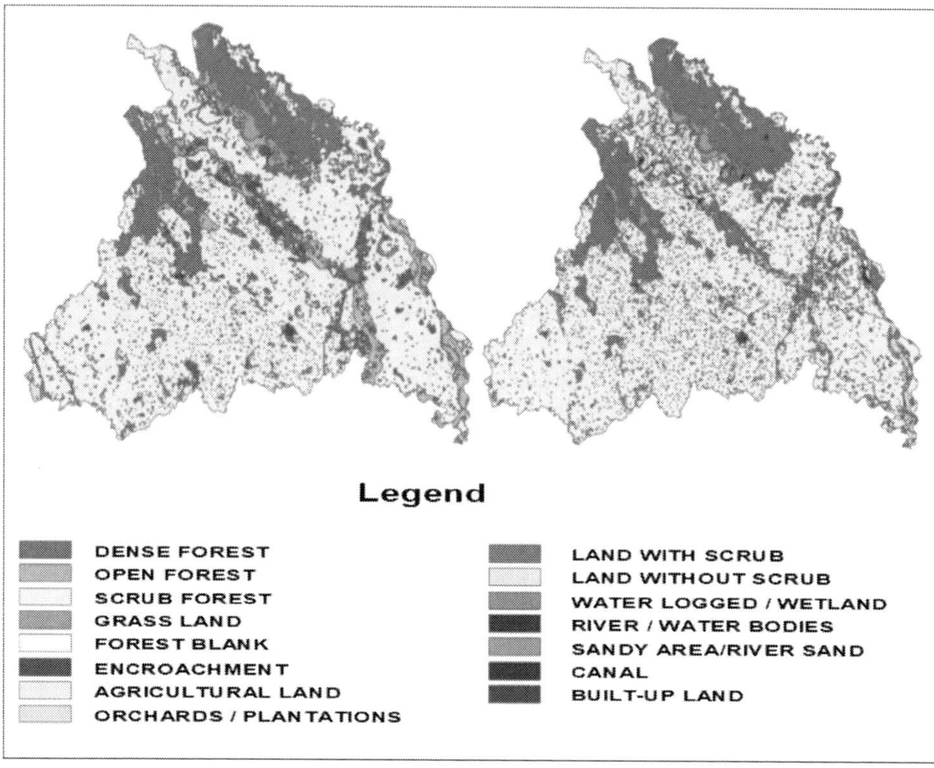

Figure 2. Temporo-spatial Land use-Landcover maps of Central *Terai* District, Lakhimpur Khiri. An agragrian district having moist deciduous forest cover and rainfed perennial river system. Land use-Landcover map of 1989 (left) and 2002 (right).

Forest Area

As evident from Table 1 area under forest land (1308.10 km^2) in 1989, was reduced (1278.29 km^2) in 2002. There was a net reduction by 29.81 km^2 over a period of 13 years. Thus, the change in the forest area was 0.4% of the total geographical area of the district and 2.3% of the total forest land. Sub-category wise changes in forest area were as follows. Dense forest area was increased (from 943.44 km^2 in 1989 to 1009.03 km^2 in 2002) by 65.59 km^2. Open forest area had an appreciable negative trend, reducing from 152.45 km^2 in 1989 to 94.00 km^2 in 2002 resulting a loss of 58.45 km^2 of area. This was 4.5% of the total forest land and 38% of open forest as it existed in 1989. Scrub forest area had increased by 13.80 km^2 (from 75.90 km^2). Forest blank area decreased by 8.44 km^2 (from 10.84 km^2 in 1989). An additional area of about 1.21 km^2 of forest land had been found to be encroached. A reverse trend was observed in open grassland areas. This category of land cover had declined by 43.52 km^2 (from 124.87 km^2) over a period of about 13 years, which was 3.3% of total forest land and 35% of open grassland.

Table 1. Area (km^2) of various land use-land cover categories in Lakhimpur Khiri district, percentage of geographical area and change in land use pattern from the year 1989 to 2002

SN	Classes and sub-classes	1989		2002		Area Change
		Area	% Area	Area	% Area	
1	Forest Land	1308.1	17.0	1278.29	16.7	-29.81
1.1	Dense Forest	943.44	12.3	1009.03	13.2	65.59
1.2	Open Forest	152.45	2.0	94.00	1.2	-58.45
1.3	Scrub Forest	75.90	1.0	89.70	1.2	13.80
1.4	Forest Blank	10.84	0.1	2.40	-	-8.44
1.5	Encroachment	0.60	0.0	1.81	-	1.21
1.6	Open Grass Land	124.87	1.6	81.35	1.1	-43.52
2	Agricultural Land	5291.9	69.0	5628.21	73.2	336.29
2.1	Crop Land	5218.8	68.0	5519.52	71.8	300.75
2.2	Plantations/ Orchards	73.15	1.0	108.69	1.43	35.54
3	Waste Land	521.22	6.8	296.60	3.9	-224.62
3.1	Land with scrub	348.39	4.6	119.78	1.6	-228.61
3.2	land Without Scrub	70.97	0.9	40.88	0.5	-30.09
3.3	Water Logged/Wetland	101.86	1.3	135.94	1.8	34.08
4	Water Bodies	408.84	5.3	327.00	4.3	-81.84
4.1	R/S/SW Body	52.86	0.7	127.02	1.7	74.16
4.2	Sandy Area/River sand	335.12	4.4	179.12	2.3	-156.00
4.3	Canal	20.86	0.3	20.86	0.3	0
5	Built-Up-Land	142.98	1.9	142.98	1.9	0
	Total	7673.1	100	7673.1	100	0

Note: R/S/SW Body refers to Rivers/Steams/Surface Water Body.

Agriculture Land

The results showed that the area under agriculture land increased from 5291.92 km^2 in 1989 to 5628.21 km^2 in 2002. Thus, a total increase of 336.29 km^2 was recorded under agriculture land, which was an appreciable increase by about 4% of the total geographical area of the district and 6.4% of agriculture land in 1989. Sub-category wise situation was as follows. Cropland increased from 5218.77 km^2 to 5519.52 ha km^2. A net gain of 300.75 km^2 area under this category was 5.8% of the crop land in 1989. Plantation/orchards area increased by 35.54 km^2. This increase in orchards was estimated as 48.6% of that in 1989.

Wasteland

It was observed that the area under wasteland in Lakhimpur district was reduced from 521.22 km^2 of 1989, indicating a net loss of 224.62 km^2 in a period of 13 years. This was 2.9% of the total geographical area of the district and 43% of the wasteland in 1989. Sub-category wise position was as follows. Land with scrub was reduced considerably from 348.39 km^2. This decline by 228.61 km^2 was 65.6% change in this sub-class. Area of land without scrub was also reduced by 30.09 km^2. This amounted to 42.4% reduction from the year 1989 to 2002. However, the area under waterlogged/wetland condition increased by 34.08 km^2 over the study period.

Water Bodies

Considering both dry areas and water bearing areas of stagnant water bodies, there was a declining trend in this category. The area under this class was 408.84 km^2 in the year 1989 which reduced by 81.84 km^2 in 13 years. This reduction was 1% of the total geographical area of the district and 20% of the water bodies as in 1989. Sub-category wise, river/stream/surface water bodies was found to increase by 74.16 km^2 from 1989 to 2002. However, sandy area/river beds were considerably reduced. There was a total loss of 156.00 km^2 in the dry part of water bodies over a period of 13 years (46.6% of sandy area/river sand in 1989). The river embankment forest was found prone to erosion as a lot of driftwood was recovered during the period.

Built-Up Area

Based on the resolution of satellite data used in the present study, changes in the area under canal and built-up land could not be detected/depicted.

Forest Fire

In the present study, it was found that satellite detected 19 cases of forest fires in Lakhimpur Khiri district while only four cases of forest fire were reported by the department. All the 19 cases reported were verified on the ground. This showed a tendency of under reporting of forest fires. However, grasslands and nearby forest area were reported to be more sensitive to fire.

Soil and Hydrology

Remote sensing data coupled with ground verification yielded very useful information regarding soil structure and status of soil erosion. The low alluvial soils consisted mainly of pure sand to recent loamy sand. This, being less fertile, supported very little vegetation and remained prone to erosion. Middle alluvial soil, which is very fertile land, had a fair amount of humus, good quality of vegetation and was less prone to soil erosion. High alluvial soils with stiff clay and large amount of decayed or decaying organic matter caused waterlogging due to low permeability and were devoid of good vegetation.

Groundwater was an important source of water supply in the study area. This was extracted from dug wells and from boreholes for domestic and agriculture use. With Lakhimpur Khiri being a *Terai* area, the water table was not very deep, but it varied from place to place ranging between 1.33 m to 6.9 m. The average ground water level (GWL) for pre-monsoon and post-monsoon was 4.74 m and 2.57 m, respectively in 1989. Similarly, the average GWL pre- and post-monsoon in 2002 was 4.63 m and 3.68 m, respectively. Thus, the pre-monsoon water table was not affected but post-monsoon showed a fall in level by 1.11 m over 13 years.

Forest-Based Industries

One hundred and eighteen wood based industrial units were established in the district during the study period of 15 years (1988 and 2002). This included 99 wooden furniture units, 16 paper units and 3 plywood units. On an average, eight industrial units were established each year resulting in additional demand on forest resources.

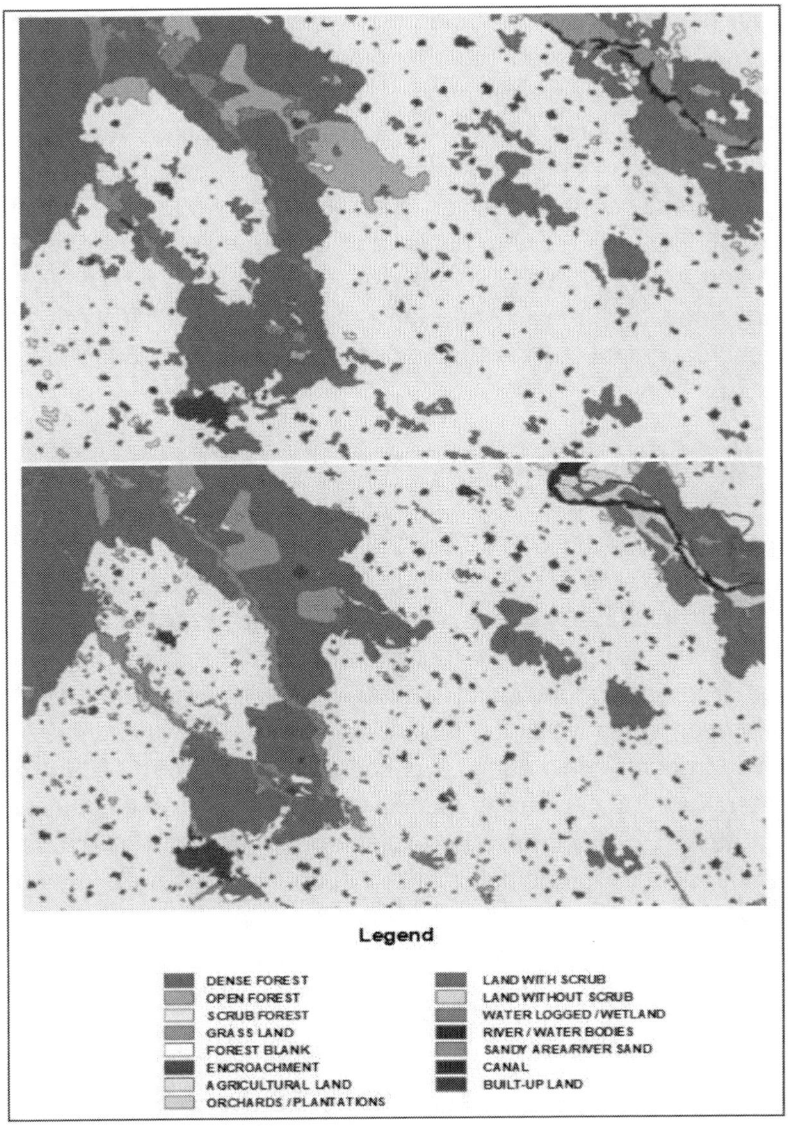

Figure 3. Magnified part of the district (from Figure 2). Top: 1989, Bottom: 2002. Please note the distinct changes in Land use-Landcover (Open forest, dense forest, waterlogged & sandy area, etc.) Top map from 1989, bottom map from 2002.

Discussion

The total forest area was estimated to be 1278.29 km^2 in 2002 for the current study, while Forest Survey of India published this as 1314.00 km^2 during same time (SFR, 2003). This difference of 35.71 km^2 may be due to the difference in the tools and technologies used in the two studies. The present study was more detailed, applied to a single administrative unit whereas the SFR was for the entire country. Therefore, our result may be considered more accurate and reliable. In addition, our categorization of the forest area could be more useful from a management point of view as a result of the finer classification. SFR (2003) classified forest land into three categories: very dense forest (366.00 km^2), moderately dense forest (502.00 km^2), and open forest (446.00 km^2). The current study demarcated five clear classes of forest land. This showed the area of dense forest as 1009.03 km^2 whereas SFR 2003 reported 868.00 km^2 of very dense and moderately dense forests. The open forest of 446.00 km^2 described by SFR 2003 was further sub-categorized in the present study into open forest (94.00 km^2), scrub forest (89.70 km^2), forest blank (2.40 km^2), encroachment (1.81 km^2) and open grass land (81.35 km^2).

Landuse-Landcover Pattern

A comparative study of the 1989 and 2002 data, indicated an overall reduction in the area under forest land. However, there had been an increase in the area under dense forest. This increase was mainly due to an improvement in the density of open forest, that means a part of open forest had changed to dense forest. This change may be due to the regeneration of natural vegetation and gap filling by way of plantations in open areas within the open forest. As a result, a decreasing trend had been observed in open forest during a period of 13 years from 1989 to 2002. Apart from this there was a decreasing trend in forest blank areas, but scrub forest increased in area. This could also be explained by the fact that scrubby growth in some of the forest blank areas had resulted in an increase in the area under scrub forest. But open grassland declined, mainly in a National Park area. Overall reduction in the forestland was mainly due to water logging and it was the grassland area which was most affected. This also explained the increase in the area classified under the waterlogged/wetland category.

Our change detection result, showing that the area of forest, wasteland and waterbodies decreased while agriculture area increased, contrasts with other similar studies. For example, Tiwari et al., (2017) found that the forest, water body, and wasteland increased, and agriculture/other vegetation decreased. Rawat & Kumar (2015) found that vegetation increased while agriculture decreased. This variation may be due to different driving forces prevalent in the respective areas. In other studies, the driving factors of decrease in forest area were mining and associated activities (Areendran et al., 2013), urbanization (Anchan et al., 2018), agriculture expansion (Prakasam, 2010), etc. Anchan et al., (2018) also reported a reduction in waterbody area due to urbanization. Nevertheless, increases in dense forest area like in the present study was also reported elsewhere due to the favourable climate in the examined period, the protection system of the local area, the reforestation programs and the regeneration activities (Barakat et al., 2018). The relationship between LULC and their drivers of change detected in the present study is depicted in Figure 4, based on field observation and discussions with stakeholders.

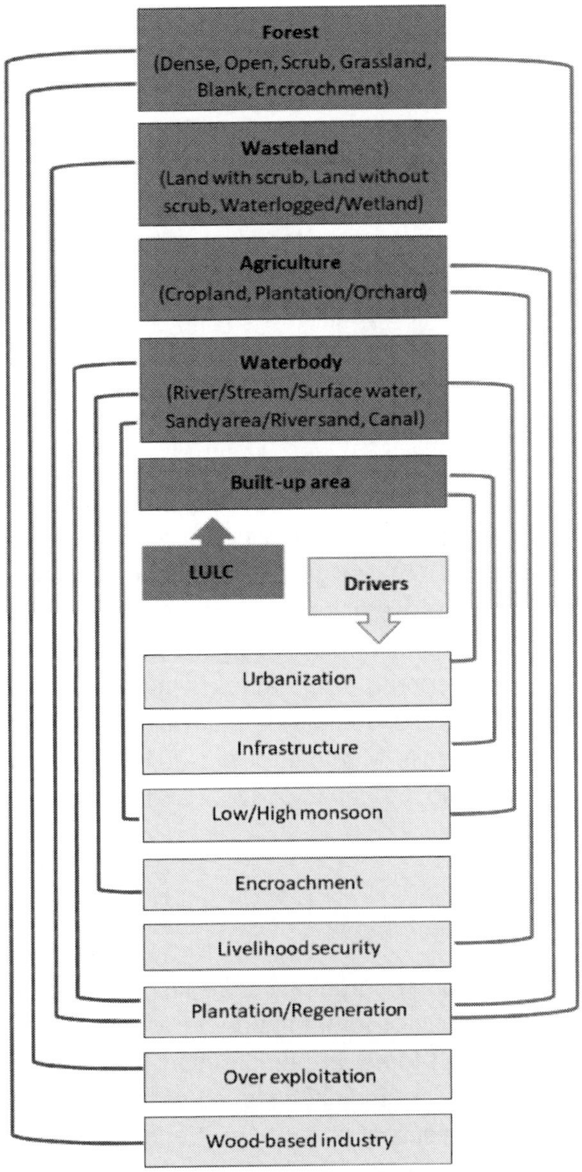

Figure 4. Relations between LULC/Resources and Change Drivers. Green connectors show positive effect/increase and red connecters show negative/decrease in LULC.

This study showed an increase in the agriculture area mainly under crops and also in the area under Plantation/Orchards. Selected field interaction revealed that large-scale horticulture and plantation activity was undertaken by the farmers in this district during the study period. It appeared that some area of the cultivable wasteland was converted to agricultural crops by way of land improvement. This also explained reduction in the area under land with or without scrub. As explained above, increase in the wasteland class of waterlogged /wetland was mainly in open grassland area of a National Park. The governing factors for increase in agriculture area was livelihood securing due to population pressure having no opening for employment. This was possible due to favourable soil and good precipitation in the area. Elsewhere, there is a contrasting report of decreasing agriculture area due to erratic monsoon in semi-arid basin (Roy & Inamdar, 2019) with low ground water table and water stressed agricultural activity (Ratna, 2012).

Over a period of 13 years (1989 to 2002), while the area of surface water in the water bodies had increased, the area under sandy area/river sand (dry part of water bodies) had reduced considerably. As a result, there was a net reduction in the area of water bodies. However, the increase in the area under the surface water bodies could be due to increase in the rainfall in 2002 as compared to the year 1989. It appeared that some of the peripheral area of water bodies (dry part) had been put to use for agricultural purpose, which could explain decline in the area under sandy area/river sand and increase in the area under cropland.

Similar to our study, decrease in wetland and waterbody area was reported in several other studies (Poongothai et al., 2014; Mageswaran et al., 2015; Baruah et al., 2016 etc.). Butt et al., (2015) assigned the reason of decrease in water class landuse to easy accessibility of water resulting in depletion of water and leading to the drying up of tributaries and their replacement by either compact surfaces or barren land. However, water spread area decrease reported by Mallupattu & Reddy (2013) was due to the gradual conversion of water spread area into builtup area or human developmental area while Munthali et al. (2019) attributed this decrease to the practice of agriculture along the banks of rivers and streams.

Vegetation Dynamics

Though overall area of forest shrunk @ 2.29 $km^2\ a^{-1}$, the quality of forest was found to improve as the density of forest improved and open forest reduced @ 4.5 $km^2\ a^{-1}$. Additionally, the scrub forest area increased @ 1.06 $km^2\ a^{-1}$ and the forest blanks decreased @ 0.65 $km^2\ a^{-1}$ mainly because of plantation activities carried out by the forest department, particularly outside the National Park area. It takes four to five years for the plantations to be captured through remote sensing techniques. Much of the increase in scrub forest and subsequent decrease in forest blanks and open grass land was mainly due to afforestation activities. This vegetation dynamics might have played a positive role in wildlife conservation as the secondary data on wild animal census showed that the number of many wild animals (Cheetal, Swamp

Deer, Hog Deer, Sambhar, Barking Deer, Blue Bull, Rhesus Monkey, Langur, Jackal, Crocodile and Bengal Florican) increased during the study period. This corroborates with habitat improvement strategy which is a well-known tool for endangered animal conservation (Bro et al., 2004).

Forest Fire and Fire Prone Areas

Under reporting of forest fire cases in the forest division could possibly be for two reasons. One reason could be that the surface fire which burns only the litter and minor shrubs (Roy, 2003) causes minimal or no commercial damage, therefore being of very little consequence, and is therefore ignored. Another reason could be that since the administrative fallout of damage or property loss on the lowest field staff is the heaviest, there could be a tendency of hiding the incident (Pers. Com. Field staff: Confession on anonymity). Other types of fire reported by Roy (2003) like ground fire is non occurrent and crown fire is rare in *Terai*. Satellite provided data even for the cases of controlled burning of fire lines inside the forest and for the fires occurring outside forest areas, particularly in agricultural fields. This unbiased and early detection data could be used for the management of forest fire. It is, therefore, suggested that remote sensing technique could be an integral part of any district fire management plan.

Human Factors

Human population in the district has increased @ 2.5% a^{-1} during the study period. Apart from this the natural resource had pressure from the northern side as well, especially across the porous international border. Moreover, the negative role of the large population of the district cannot be ignored. They had converted encroached forest land to agriculture cropland and removed unrecorded amount of semidry fuelwood and fodder from the forest. Because of increased prices of timber and wildlife products, the unemployed youth were attracted towards illicit felling and illegal trade of wildlife products. Study of five-year sample data showed that on an average 101 cases of illicit felling occurred in Dudhwa National Park, in which 233.4 numbers of trees were felled annually. Bose & Leitmann (1996) and Greenpeace (2011) in Areendran et al., (2013) confirmed elsewhere that people unable to find employment in agriculture or industry turned to tree cutting because of high wood fuel prices and strong demand. However, Singh et al., (1991) and Areendran et al., (2013) recorded in the same area that encroachment on forest land might have resulted into cropland increase.

Soil and Hydrology

Soil quality and GWL regulate vegetation cover in the region. Low and high alluvial soil have the disadvantage of erosion and waterlogging which is a form of soil degradation (Peng et al., 2008). Therefore, it is necessary to implement suitable soil conservation practices in such areas to prevent numerous environmental, economic, social and ecological consequences (AbdelRehman et al., 2016). India is the biggest user of groundwater in the world. Groundwater caters to more than 60% for agricultural activities and more than 85% for drinking water (TWB, 2010 in Senthilkumar et al., 2019). Along similar patterns the local populace also depended totally on groundwater for domestic use and partially for irrigation. Groundwater withdrawal over the years has resulted in the deepening of water levels and drying of shallow groundwater abstraction structures (CGWB, 2015 & 2016). However, the lowering of GWL at a lower rate (@ 8.5 cm a^{-1}) in the district was still a concern. It was probably caused by increased demand and exploitation for agriculture coupled with lower recharge on account of higher run off. This was an indication of the requirement of long-term planning for groundwater recharge effort.

Impact of Wood-Based Industry

The 36.99% annual rate of increase in wood-based industries was very high. Though there was no established system to estimate and know the exact demand of the raw material, it was evident that these industries could not thrive solely on legal supplies from the reserve forest alone. It was observed that there was large scale production of raw material from the social forestry sector (tree outside forest). These wood-based industries were increasingly supplemented by outside resources also.

Conclusion

The present study revealed that remotely sensed data analyzed with the help of GIS technique are extremely useful from the point of view of quantifying LULC, identifying the dynamics of change on a spatial and temporal scale. The data obtained from the study for this aspect served as a benchmark for comparison with another year. These LULC transformation data, like depletion of forest land, loss of water bodies, increase in waterlogging etc. posed a serious threat to the district resources. Drivers of change, like land encroachment in forests and riverine areas, overexploitation of forest products in the form of legal and illegal timber, fuelwood etc. needed to be curbed and controlled. This study also provided reliable information on potential areas of degradation (erosion prone and fire sensitive area) indicating timely remedial action to be taken to minimize the hazards. Forest based industries needed to be regulated and stopped beyond the carrying capacity of the region. All these indicate that there should be proper management of LULC to optimize resource utilization for more effective socioeconomic development. Afforestation, forest regeneration, social forestry plantation, orchard establishment and other methods need to be prioritized in the district. It is suggested that LULC change detection-based planning of development should be prepared at the district level or this must be incorporated, at least in the Forest Working Plan, since the district has plenty of forest, water and wildlife resources to be managed sustainably.

Acknowledgments

One of the authors (PLL) is thankful to Dr Ashwani Kumar, IFS (Retired), Former DG, ICFRE, Dehradun and PCCF, UP Lucknow for conceiving the idea and guiding through the research. Dr TS Kachhwaha, Dr Phool Kunwar and Dr V Rajamani from RSAC Lucknow deserve due appreciation for providing help in procuring RS images and helping analyze the data. Dr Sonal Jha, Assistant Professor, IIT Bhilai is also thanked for language improvement.

References

AbdelRahman, M. A. E., Natarajan, A., Srinivasamurthy, C. A. & Hegde, R. 2016. Estimating soil fertility status in physically degraded land using GIS and remote sensing techniques in Chamarajanagar district, Karnataka, India. *The Egyptian Journal of Remote Sensing and Space Sciences*, 19: 95–108. https://doi.org/10.1016/j.ejrs.2015.12.002.

Anchan, S. S., Shetty, A., Bhat, H. G. & Chadaga, M. 2018. Land use and land cover change detection through spatial approach: A case study of Mangaluru Taluk, Karnataka. *Journal of Geomatics*, 12(2): 167-173. Retrieved from https://isgindia.org/wp-content/uploads/2018/11/Pap_10_Volume_2_Oct_2018.pdf.

Areendran, G., Rao, P., Raj, K., Mazumdar, S. & Puri, K. 2013. Land use/land cover change dynamics analysis in mining areas of Singrauli district in Madhya Pradesh, India. *Tropical Ecology,* 54(2): 239-250. https://doi.org/10.31195/ejejfs.428381.

Asselman, N. E. M. & Middelkoop, H. 1995. Floodplain sedimentation: quantities, patterns and processes. *Earth Surface Processes & Landforms,* 20(6): 481–499. https://doi.org/10.1002/ esp.3290200602.

Barakat, A., Khellouk, R., Jazouli, A.E., Touhami, F. & Nadem, S. 2018. Monitoring of forest cover dynamics in eastern area of Béni-Mellal Province using ASTER and Sentinel-2A multispectral data. *Geology, Ecology and Landscapes*, 2(3): 203–215. https://doi.org/10.1080/24749508.2018.1452478.

Baruah, J., Baruah, B. K., Kalita, S. & Choudhury, S. K. 2016. Impact Analysis of Open Cast Coal Mining on Land Use/Land Cover using Remote Sensing and GIS Technique in Ledo-Margherita Region of Assam, India. *Imperial Journal of Interdisciplinary Research*, 2(8): 671-676. http://www.onlinejournal.in/IJIRV2I8/110.pdf.

Bhatti, N. B., Siyal, A. A., Qureshi, A. L. & Bhatti, I. A. 2019. Land Covers Change Assessment After Small Dam's Construction Based on the Satellite Data. *Civil Engineering Journal,* 5(4): 810-818. http://dx.doi.org/10.28991/cej-2019-03091290.

Bose, R. K. & Leitmann, J. 1996. Environmental profile of the Singrauli region, India. *Cities*, 13: 71-77. https://doi.org/10.1016/0264-2751(95)00127-1.

Butt, A., Shabbir, R., Ahmad, S.S. & Aziz, N. 2015. Land use change mapping and analysis using Remote Sensing and GIS: A case study of Simly watershed, Islamabad, Pakistan. *The Egyptian Journal of Remote Sensing and Space Sciences,* 18: 251–259. https://doi.org/10.1016/j.ejrs.2015.07.003.

Bro, E., Mayot, P., Corda, E. & Reitz, F. 2004. Impact of habitat management on grey partridge populations: assessing wildlife cover using a multisite BACI experiment. *Journal of Applied Ecology*, 41: 846–857. https://doi.org/10.1111/j.0021-8901.2004.00939.x.

Champion, H. G. & Seth, S. K. (1968). *A revised survey of the forest types of India.* Manager Publication, Delhi.

Chitade, A. Z. & Katyar, S.K. 2010. Impact analysis of open cast coal mines on land use/land cover using remote sensing and GIS technique: A case study. *International Journal of Engineering Science & Technology*, 2: 7171–7176. Retrieved from http://www.idc-online.com/technical_references/pdfs/civil_engineering/IMPACT%20ANALYSIS.pdf.

CGWB. 2015. *Aquifer Mapping and Ground Water Management*. Chennai: Central Ground Water Board.

CGWB. 2016. *Ground Water Scenario in India.* Faridabad: Central Ground Water Board.

Dimyati, M., Mizuno, K. & Kitamura, T. 1996. An analysis of land use/cover change using the combination of MSS Landsat and land use map: a case study in Yogyakarta, Indonesia. *International Journal of Remote Sensing*, 17: 931–944. https://doi.org/10.1080/0143116 9608949056.

Greenpeace. 2011. *Singrauli: The Coal Curse - A Fact Finding Report on the Impact of Coal Mining on the People and Environment of Singrauli.* Greenpeace India Society. [Online] Retrieved from http://www.greenpeace.org/india/Global/india/report/Fact-finding-report-Singrauli-Report.pdf.

Kachhwala, T. S. 1985. Temporal monitoring of forest land for change detection and forest cover mapping through satellite remote sensing. In: *Proceedings of the 6th Asian Conference on Remote Sensing*. Hyderabad.

Khan, R. & Jhariya, D. C. 2016. Land Use Land Cover Change Detection Using Remote Sensing and Geographic Information System in Raipur Municipal Corporation Area, Chhattisgarh. *International Journal of Geo Science and Geo*

Informatics 3(1): 1-4. Retrieved from https://www.researchgate.net/profile/Dalchand_Jhariya/publication/306324178_Land_Use_Land_Cover_Change_Detection_Using_Remote_Sensing_and_Geographic_Information_System_in_Raipur_Municipal_Corporation_Area_Chhattisgarh/links/57b85d8e08ae6f173766079c.pdf.

Koneti, S., Sunkara, S.L. & Roy, P.S. 2018. Hydrological Modeling with Respect to Impact of Land-Use and Land-Cover Change on the Runoff Dynamics in Godavari River Basin Using the HEC-HMS Model. *International Journal of Geo-Information*, 7: 206. https://doi.org/10.3390/ijgi7060206.

Kumar, K., Kumar, V. & Kumar, D. 2016. Land Use and Land Cover Change Detection in Gagas River Valley Watershed Using Remote Sensing and GIS. *International Journal of Research in Engineering and Applied Sciences,* 6 (5): 31-37.

Lillesand, T. M. & Kiefer, R. W. (2000). *Remote Sensing and Image Interpretation*, IV Edition. USA, John Wiley & Sons Inc.

Mageswaran, T., Sachithanandam, V., Sridhar, R., Thirunavukarasu, E. & Ramesh, R. 2015. Mapping and monitoring of landuse/landcover changes in Neil Island (South Andaman) using geospatial approaches. *Indian Journal of Geo-marine Sciences,* 44(11): 1762-1768. Retrieved from http://nopr.niscair.res.in/bitstream/123456789/35018/1/IJMS% 2044% 2811%29%201762-1768.pdf.

Mallupattu, P. K. & Reddy J. R. S. 2013. Analysis of Land Use/Land Cover Changes Using Remote Sensing Data and GIS at an Urban Area, Tirupati, India. *The Scientific World Journal*, 2013: 6. ID 268623. http://dx.doi.org/10.1155/2013/268623.

Munthali, M. G., Botai, J. O., Davis, N., Adeola, A. M. 2019. Multi-temporal Analysis of Land Use and Land Cover Change Detection for Dedza District of Malawi using Geospatial Techniques. *International Journal of Applied Engineering Research*, 14(5): 1151-1162. http://hdl.handle.net/2263/71103.

Peng, W., Zhoub, J., He, Z., & Yang, C. 2008. Integrated Use of Remote Sensing and GIS for Predicting Soil Erosion Process. *The International Archives of the Photogrammetry, Remote Sensing and Spatial Information Sciences*. Vol. XXXVII. Part B4. Beijing 2008.

Poongothai, S., Sridhar, N. & Shourie, R. A. 2014. Change Detection of Land use/Land Cover of a Watershed using Remote Sensing and GIS. *International Journal of Engineering and Advanced Technology*, 3(6): 226-230. Retrieved from https://www.ijeat.org/wp-content/ uploads/papers/ v3i6/F3401083614.pdf.

Rai, B. & Kumar, K.E.M. 2013. Mapping of Mining Areas in Aravalli Hills in Gurgaon, Faridabad & Mewat Districts of Haryana Using Geo-Informatics Technology. *International Journal of Remote Sensing & Geoscience*, 2(1): 1-8.

Ratna, S. B. 2012. Summer monsoon rainfall variability over Maharashtra, India. *Pure & Applied Geophysics,* 169: 259 - 273. https://doi.org/10.1007/s00024-011-0276-4.

Rawat, J. S. & Kumar, M. 2015. Monitoring land use/cover change using remote sensing and GIS techniques: A case study of Hawalbagh block, district Almora,

Uttarakhand, India. *Egyptian Journal of Remote Sensing Space Science,* 18: 77–84. https://doi.org/10.1016/ j.ejrs.2015.02.002.

Rawat, J. S., Biswas, V. & Kumar, M. 2013. Changes in land use/cover using geospatial techniques: A case study of Ramnagar town area, Nainital district, Uttarakhand, India. *Egyptian Journal of Remote Sensing & Space Science*, 16: 111–117. https://doi.org/10.1016/ j.ejrs.2013.04.002.

Rawat, J. S., Kumar, M. & Biswas, V. 2014. Land use/cover dynamics using multi-temporal satellite imagery: a case study of Haldwani Town area, district Nainital, Uttarakhand, India. *International Journal of Geomatics & Geoscience*, 4 (3): 536–543.

Roy, A. & Inamdar, A B. 2019. Multitemporal Land Use Land Cover (LULC) change analysis of a dry semi-arid river basin in western India following a robust multisensory satellite image calibration strategy. *Heliyon,* 5: (2019) e01478. http://dx.doi.org/ 10.1016/j.heliyon.2019. e01478.

Roy, P. S. 2003. Forest Fire and Degradation Assessment Using Satellite Remote Sensing and Geographic Information System. In Sivakumar, M. V. K., Roy, P. S., Harmsen, K. & Saha, S. K. (Eds.), *Satellite Remote Sensing and GIS Applications in Agricultural Meteorology Proceedings of the Training Workshop* 7-11 July, 2003, Dehra Dun, India Indian Institute of Remote Sensing (NRSA), Dehra Dun. *Internet document*, accessed on August 25, 2019. http://www.wamis.org/agm/pubs/agm8/Paper-18.pdf.

Roy, P. S., Dutt, C. B. S. & Joshi, P. K. 2002. Tropical forest resource assessment and monitoring. *Tropical Ecology*, 43(1): 21-37. Retrieved from http://citeseerx.ist.psu.edu/viewdoc/download?doi=10.1.1.530.4090&rep=rep1&type=pdf.

Roy, P. S., Roy, A., Joshi, P. K., Kale, M. P., Srivastava, V. K., Srivastava, S. K., Dwevidi, R. S., Joshi, C., Behera, M. D., Meiyappan, P., Sharma, Y., Jain A. K., Singh J. S., Palchowdhuri, Y., Ramachandran R. M., Pinjarla B., Chakravarthi V., Babu N., Gowsalya, M. S.,Thiruvengadam P., Kotteeswaran, M.,Priya, V., Yelishetty, K. M. V. N., Maithani, S., Talukdar, G, Mondal I., Rajan, K. S., Narendra, P. S., Biswal, S., Chakraborty, A., Padalia, H., Chavan, M., Pardeshi, S. N., Chaudhari, S. A., Anand, A., Vyas, A., Reddy, M. K., Ramalingam, M., Manonmani, R., Behera, P., Das, P., Tripathi, P., Matin, S., Khan, M. L., Tripathi, O. P., Deka J., Kumar, P. & Kushwaha, D. 2015. Development of Decadal (1985–1995–2005) Land Use and Land Cover Database for India. *Remote Sensing*, 7(3): 2401–2430.

Senthilkumar, M., Gnanasundar, D. & Arumugam, R. 2019. Identifying groundwater recharge zones using remote sensing & GIS techniques in Amaravathi aquifer system, Tamil Nadu, South India. *Sustainable Environment Research*, 29: 15. https://doi.org/10.1186/s42834-019-0014-7.

SFR. 2003. State Forest Report. *Forest Survey of India*, Dehradun 2003.

Singh, A. 1989. Digital change detection techniques using remotely sensed data. *International Journal of Remote Sensing*, 10: 989-1003. https://doi.org/ 10.1080/01431168908903939.

Singh, C. K., Shashtri, S., Avtar, R., Mukherjee, S. & Singh, S. K. 2010. Monitoring change in land use and land cover in Rupnagar district of Punjab, India using Landsat and IRS LISS III satellite data. *Ecological Questions*, 13: 73 – 79. https://doi.org/10.2478/v10090–010–0018–8.

Singh, J. S., Singh, K. P. & Agrawal. M. 1991. Environmental degradation of the Obra - Renukoot -Singrauli area, India and its impact on natural and derived ecosystems. *The Environmentalist*, 11: 171-180. https://doi.org/10.1007/BF01263230.

Tiwari, J., Sharma, S. K. & Patil, R. J. 2017. An Integrated Approach of Remote Sensing and GIS for Land Use and Land Cover Change Detection: A Case Study of Banjar River Watershed of Madhya Pradesh, India. *Current World Environment,* 12(1): 157-164. http://dx.doi.org/10.12944/CWE.12.1.18.

TWB. 2010. *Deep Wells and Prudence: Towards Pragmatic Action for Addressing Groundwater Overexploitation in India*. Washington, DC: The World Bank.

In: Critical Research Techniques in Animal …
Editors: Kaushalendra Kumar Jha et al.
ISBN: 978-1-53619-720-4

Chapter 4

GEOINFORMATICS, CLIMATE CHANGE, HABITAT DYNAMICS AND A CASE OF VULTURES IN CENTRAL INDIA

Kaushalendra Kumar Jha*
Indian Institute of Forest Management, Bhopal, India
Department of Forest, Environment and Climate Change,
Uttar Pradesh, Lucknow, India

ABSTRACT

Climate crisis and its impact on biodiversity has been the real concern of recent times as habitats are modified and species individuals decline in numbers. This necessitates the understanding of habitat dynamics and possible drivers of change with greater amount of accuracy in the knowledge. Therefore, this chapter discusses habitat (niche) modelling or species distribution with the help of various types of models (heuristic models, statistical model, predictive law optimization and machine learning) and how accuracy could be maximised and measured. Apart from current habitat projection future habitat status modelling is also discussed with the help of climate models along with options of selection in the case of different scenarios of emission. Constraints aside, it is advisable to go for ensemble modelling for Species Distribution Models (SDMs) and Global Circulation Models (GCMs). The role of geographical information systems has also been discussed in brief in model development and data analysis. For better understanding of the theoretical concept of geoinformatics, climate change and habitat dynamics, the case of the vulture, a highly imperilled species of the Indian subcontinent, in a vulture stronghold of India has been presented along these lines to assess the habitat and climate change impact on it. The global vulture habitat which has seen changes with continued deterioration has been discussed. The encouraging positive result of habitat change of vulture needs to

* Corresponding Author's E-mail: jhakk1959@gmail.com.

be taken with caution. The habitat has been categorised according to suitability for vulture which has shown some increase. Based on these initial results vulture conservation planning could be taken up. Further research for enhanced accuracy will result in better protection of the species.

INTRODUCTION

The ecological niche of a species is the interaction of space and conditions where it can survive and continue its reproductive ability to remain as a viable population (Choudhury et al., 2016). Ecological niches play a central role in the explanation of species origin, persistence, distribution and capacity of competition (Silvertown, 2004; Shabani et al., 2018b). Such niches or habitat of the species are known to be affected by any environmental change in climatic conditions (temperature and precipitation), topography, soil types (in the case of plants), landcover change (in the case of animals) and water proximity (Botella et al., 2017). Geographical Information Systems/Science (GIS), has provided efficient assessment of species habitat or niche and projection of future shape due to climate change impact. The following paragraphs detail the relationship between GIS, Climate change and Habitat dynamics sub-heads.

Geographic Information System

Geoinformatics or geomatics is the technology dealing with the structure and character of spatial information. This field uses remote sensing data and GIS to portray and record features of the Earth's surface on paper or a computer screen (Jha & Chaudhry, 2018). The evolution of GIS, the Global Positioning System (GPS), and Remote Sensing (RS) technologies has enabled the collection and analysis of field data in ways that were not possible before the arrival of computers (Sonti, 2015). GIS can provide crucial information about resources and can make planning and management of resources easier, for example, ecosystem management, and landscape and habitat planning (Upadhyay, 2009). In the field of forest ecology this technology has been applied to study the habitats of several plant and animal species through modelling. After some early studies (Bustamante, 1997; Suarez-Seoane et al., 2002; Gibson et al., 2004; Franco & Sutherland, 2004), modelling has increased in recent times with further development of GIS software and new powerful statistical techniques. This has facilitated the use of ecological niche modelling (ENM)/ species distribution modelling (SDM) to generate potential geographic distributions of species in ecology, conservation and evolutionary biology (Terribile et al., 2009 in Sourabh et al., 2018). This is also linked with locating populations of rare, endangered and threatened species as an important consideration in biodiversity conservation and for developing effective strategies for in-situ and ex-situ conservation (Menon et al., 2010). Additionally, GIS tools have the capability of using large digital cartographic databases and extracting potential predictor variables (de Frutos et al., 2004).

Climate Change

The world is changing rapidly due to the cascading and intertwined effects of human population growth, widespread poverty, economic globalization, land-use change, and, looming over it all, climate change (Wiens et al., 2009). These are a result of industrialization, intensive farming and anthropogenic activities leading to higher greenhouse gas emission (IPCC 2007). The direct consequences of increasing greenhouse effect are changes in temperature, precipitation patterns and other climate variables. Climate change models predict that while certain areas will be subject to increasing temperatures, others will experience decreasing temperatures, and changes in precipitation will cause a much greater impact than changes in temperature in other regions (Moller et al., 2010). These changes have profound effects on biological diversity which will only become more dire as the pace of environmental change quickens (Wiens et al., 2009). Cordellier et al. (2012) reiterates that the 21st century is predicted to witness rapid changes in climate, therefore, impact on biodiversity at multiple levels is inevitable.

Humans are dependent on global biodiversity and ecosystems which are under major threat due to climate change (Rosenzweig et al., 2008). As reviewed by Shabani et al. (2018b), there is strong evidence that climate change has already caused changes to distribution of species, and will continue to do so (Shabani et al., 2012; VanDerWal et al., 2013; Shabani et al., 2013 & 2015; Paterson et al., 2017; Shabani et al., 2018b), through alterations in phenology (Thackeray et al., 2010), physiology (Rosenzweig et al., 2008), morphology (Walsh et al., 2016), demography and community composition (Bellard et al., 2012) and the nature of ecological interactions (Parmesan, 2006).

In response to these changes, species distribution could be shifting to follow changing environments (with species adapting to changing conditions in a place). If a species is unable to do so, it may remain in isolated pockets of an unchanged environment called 'refugia'. If none of the two happens then the species is more likely to become extinct (Holt, 1990). Sony et al. (2019) have argued that at the very least a severe reduction in species range size and abundance are the major predicted consequences of this climate change (Thomas et al., 2004; Malcolm et al., 2006; Warren et al., 2013). Nevertheless, attention has been paid to these worrisome conditions and efforts are being made towards habitat dynamics study through modelling of niche or species distribution (Pearson & Dawson, 2003; Thuiller et al., 2005; Huntley et al., 2008; Morin & Lechowicz, 2008; Lawler et al., 2009). Possible impacts of climate change on habitat are also being studied for different species individually and species ensembles in different regions in search of strategies for land managers to mitigate the potential threats. This has made use of SDMs or ENMs very popular (Abdelaal et al., 2019; Ahmad et al., 2019; Alcala-Canto et al.,

2019; Ashrafzadeh et al., 2019; Bahadur et al., 2019; Barragan-Barrera et al., 2019; D'Addario et al., 2019; Dew et al., 2019; Gorel et al., 2019; Habibzadeh & Ludwig, 2019; Islam et al., 2019; Lepcha, 2019; McDonald et al., 2019; Mohammadi et al., 2019; Zhang et al., 2019a,b etc. Table 1).

Habitat Dynamics

Habitat dynamics may be defined as habitat turnover due to natural perturbations or human activities. They are commonplace, particularly in intensively used landscapes. (Teeffelen et al., 2012). Human activities like industrialization, fossil fuel combustion, fertilizer production and use, and land use change (intensive farming etc.) are driving up concentrations of CO_2 and other greenhouse gases in the atmosphere. The emissions of these gases, which are accelerating, are trapping heat and altering the Earth's climate (IPCC 2007; Rustard et al., 2016). The resulting climate change affects the distribution and abundance of many wildlife species in the region through changes in habitat, food availability, thermal tolerance, species interactions such as competition, and susceptibility to parasites and disease (Rustard et al., 2009). Climate change is expected to place an additional burden on many species of conservation concern which are already declining within their current ranges due to a range of human activities, like habitat destruction, habitat fragmentation, pollution, invasive species, disease, and others (Miller-Rushing et al., 2010). For the sake of brevity, the term habitat dynamics is confined in this text to habitat change of avian fauna/raptors due to climate change. Global climate change is significantly altering the structure and functioning of many ecosystems and, consequently, temporal and spatial patterns of population and species abundance (Stenseth et al., 2005; Rosenzweig et al., 2008). Several species are expected to increase/decrease and/or to shift their geographic distribution in response to climate change (Elith & Leathwick, 2009; Lawler, 2009; Mohammadi et al., 2019).

The rampant human growth on the planet following the industrial revolution has led to a new epoch, the "Anthropocene," which involves rapid disappearance of species, the so-called "sixth extinction" (Barnosky et al., 2011; Dirzo et al., 2014; Donazar et al. 2016). Most tropical bird species vulnerable to climate change are not currently considered threatened with extinction, often due to lack of knowledge. Systematically and regularly gathering information on the ecology and current and future distributions of these species is an urgent priority (Sekercioglu et al., 2012). Long-distance migratory bird species that have a higher range of mobility are expected to suffer fewer extinctions from climate change than sedentary species (Sekercioglu et al., 2008). Higher mobility provides migratory birds with a higher capacity to deal with global change than sedentary (non-migratory) birds. Such sedentary species are already 2.6 times more threatened with extinction than are long distance migratory bird species (Sekercioglu, 2007). Most raptor species, including vultures, fall in these two categories. This makes the Indian subcontinent an important area as it harbours resident as well as migratory vultures in different types of habitats (Jha, 2018) which have already faced decimating population crisis due to several environmental factors (Prakash et al., 2007).

GLOBAL VULTURE HABITATS

Trees, cliffs and grasslands are the main habitat components of resident and migratory Indian vulture species. As reviewed by Botha et al. (2017), the habitat of different vultures in India and elsewhere are as follows. The Bearded Vulture occupies remote mountainous areas, with precipitous terrain, usually above 1,000 m in Europe and Asia in particular areas where large predators such as wolves, snow leopards and golden eagles are present. There are herds of mammals around, such as mountain goats, ibex, and sheep (Ferguson-Lees & Christie, 2001). The Egyptian Vulture in most parts of its breeding range, inhabits arid woodlands and semiarid bush country, especially canyons and rocky areas, often near villages and along roads. It usually occurs singly or in pairs. It is found less commonly in small groups, and rarely in large groups of more than 100. It roosts on cliff faces or in dead trees and is rarely found far from nesting cliffs. It is more tolerant of humans than other vultures. The wintering habitat includes mainly sub-deserts and savanna in the Sahel zone (Meyburg et al., 2004; Oppel et al., 2015) where birds often roost on pylons (Arkumarev et al., 2014).

Red-headed Vultures occur in a wide variety of habitats, including open countryside, cultivated areas, savanna woodland and foothills usually below 2,500 m (del Hoyo et al., 1994, BirdLife International 2017). The Himalayan Griffon inhabits mountainous areas, mostly at 1,200–4,500 m, but has been recorded at heights up to 6,000 m (Ferguson-Lees & Christie, 2001). In winter it moves lower down, with juveniles wandering into open plains and grasslands. It has also been observed foraging on rubbish dumps (BirdLife International 2016). The White-rumped Vulture occurs in a wide range of open country habitats, as well as near villages, towns and cities and the recent remaining breeding populations are mainly in more tree-covered habitats, but also include the city centre of Ahmedabad in Gujarat. In the Himalayan foothills, it occurs up to about 1,500 m where it utilises sparsely wooded areas, open areas and human settlements (del Hoyo et al., 1994).

Indian Vultures were previously found in many cities, towns and villages across its range, as well as in a wide range of agricultural habitats and wooded areas. This species nests primarily on cliffs and suitable ruined buildings; the belief that it will also nest in trees (del Hoyo et al., 1994) may be mistaken, referring to the similar Slender-billed Vulture (which certainly nests in trees) before the taxonomy was clarified distinguishing the two species. Slender-billed Vultures are found in dry open country and forested areas, although often rely on human habitation for nesting sites and carrion. In South-east Asia it is primarily a lowland species. The Griffon Vulture roosts and nests on large cliffs and soars over the surrounding open countryside in search for food and avoids woodlands. The landscape should support the formation of thermals (Mebs & Schmidt, 2006) as large vultures prefer to save energy by gliding and soaring overactive flight. Generally, it occurs from sea level up to an elevation of 1,500 m and occasionally as high as 2,500 m (Slotta-Bachmayr et al., 2006 in Botha et al., 2017). The Cinereous Vulture prefers arid hilly and mountainous habitat, including wooded areas and semi-desert, areas above the treeline, and agricultural habitats with patches of forest. Vultures spend much time soaring overhead in search of food. They perch more often on trees than on cliff faces or on the ground. Although not numerous, in places of abundant food they may congregate in large flocks (Flint, 1984).

Vultures in India could be categorised as highland (Bearded Vulture and Himalayan Griffon) and lowland (Cinereous Vulture, Egyptian vulture, Eurasian Griffon, Long-billed vulture, Red-headed Vulture, Slender-billed Vulture and White-rumped Vulture) species. They use waterbody interspersed mountainous region, forested land, grassland and agriculture area. Unfortunately, the main components of vulture habitat landscape, such as water bodies and vegetated land are shrinking, wood density is decreasing and above all climatic conditions (precipitation and temperature) are also getting modified with expected adverse impact on biodiversity. Sloan & Sayer (2015) analysed the Global Forest Resources Assessment (FRA) of the Food and Agricultural Organization of the United Nations (FAO) and concluded that despite positive global trends forest loss and degradation persist in poor tropical countries. Botkin & Keller (2000) observed that perhaps as much as 90% of freshwater wetlands have disappeared (Hardlife et al., 2014). Wetlands, amongst the world's most economically valuable ecosystems and essential regulators of the global climate, are disappearing three times faster than forests (UNCC 2018). India has lost over 1.6 million hectare of tree cover between 2001 and 2018, according to a new study released by the World Resources Institute (WRI), a nonprofit global research organisation (Nandi, 2019).

Habitat Suitability Projection

Suitable habitat may be considered as an envelope surrounding a well-maintained population of an organism. The envelope is a complex of environmental factors interacting among themselves with positive or negative impact. Therefore, a suitable habitat may be considered as the one which favours the population to grow, survive and develop well. Any organism lives in small localities spread over a large space like hypothetical nodes in a large network. The nodes are suitable locations but all the area in the mess may or may not be suitable. This suitability depends on a combination of positive factors which broadly may be climate (temperature and precipitation) and other biophysical features (topography, soil, vegetation, water availability etc.). The value of different variables/factors could be calculated in an area in the form of index which may determine the suitability of the area. Such an index could be projected over a large space of geographical choice and could be categorised differently, like not suitable, suitable, highly suitable etc.

Thus, habitat suitability projections could be achieved conceptually. Practically, models (Habitat Suitability Model) have been developed which aim at defining the 'envelope' for any chosen species that best describes its spatial range limits by identifying those environmental variables that limit its distribution (Guisan & Thuiller, 2005; Soberon & Peterson, 2005). They are built by relating current species' distributions to current environments. Future species' biogeographical ranges are modelled by projecting these relationships to selected environmental change scenarios (Peterson, 2003b; Thuiller et al., 2005). Habitat suitability modelling can be subjected to spatial analysis such as area calculation (Di Marco et al., 2016), importance of variables in the geographic distribution (Marino et al., 2011) and interaction of the variables with each other (Hirzel & Le Lay, 2008; Minoli et al., 2018).

However, the environmental variables for birds like climate, landscape structure and heterogeneity, resources and biotic competitors exert direct or indirect effect on the species which is reflected in the influences on them. This can be grouped into limiting factors, disturbances and resources. The limiting factors are defined as factors controlling species' ecophysiology (minimum winter temperature) or appearance (competition and facilitation). The disturbances are defined as all types of perturbations affecting environmental systems (fire frequency). The resources are defined as all materials that can be assimilated by organisms (availability of seeds or insects). All the direct and indirect relationships between an avian organism and its surroundings making a complex could be spatially scaled but such a scale dependent relationship is not easy to transform into mathematical equations replicating the nature (Thompson & McGarigal, 2002; Graf et al., 2005, Thuiller & Munkemuller, 2010). Nevertheless, bioengineers have been able to develop many successful models of various degrees which are discussed in brief in other sections of this text. Keeping in view all these existing conditions, the suggestion of Miller-Rushing et al. (2010) about knowledge and management of habitat and species itself become very relevant, for example, (1) identifying and monitoring species at risk from climate change, (2) managing existing populations and the habitats in which they live to maintain these populations, (3) acquiring new protected areas in places where ecological communities are expected to shift in response to climate change, (4) restoring degraded habitats of species at risk, and (5) transporting endangered bird species to suitable locations where they can establish new populations.

Species Distribution or Ecological Niche Modelling

Species Distribution Models (SDMs) and Ecological Niche Models (ENMs) might differ slightly in their emphasis—whether the geographical distribution or the niche is the focus—but are regularly used as synonyms (Peterson & Soberón, 2012). Modelling ecological niches or species distribution for prediction of the geographic occurrence of a species is a growing field in ecology (Scott et al., 2002). The recent development of computational and statistical tools allowed the emergence of new techniques to model ecological niches and potential species distribution (Phillips et al., 2006; Rangel et al., 2006; Warren et al., 2008; Thuiller et al., 2009; Monahan, 2009; Muscarella et al., 2014; Hattab et al., 2017). Such modelling made it possible to evaluate the potential ecological niche of a species by associating geo-referenced occurrence data and environmental variables. It is also capable of identifying the geographical areas that are favourable to its presence (Guisan & Thuiller, 2006; Elith et al., 2006; Elith & Leathwick, 2009; Guisan et al., 2013). Species distribution models (Elith et al., 2006) and ENMs (Peterson et al., 1999), which have been variously termed have become increasingly important in the last few decades for the study of biodiversity, macro ecology, community ecology and conservation ecology. Although the interpretation of model output may vary, the methodological approach is essentially the same on following accounts: (i) The study area is modelled as a map composed of grid cells at a specified resolution; (ii) The dependent variable is the current known species' distribution; (iii) A suite of environmental variables is collated to describe the characteristics of each cell; and (iv) A function of the environmental variables is calibrated so as to classify the degree to which each cell is either suitable or unsuitable for the species (Hirzel et al., 2002; Raxworthy et al., 2007).

These models (SDMs or ENMs, depending on the source reference) represent useful approaches to predict changes in the distribution of species in relation to climatic predictors (Peterson et al., 2011; Anderson, 2013). They permit analysis of a wide variety of biodiversity phenomena, including geographic distributions (Elith & Burgman, 2002), future potential distributions under scenarios of climate change (Thomas et al., 2004), invasion of species (Peterson, 2003a), agricultural crop damage by pests (Sánchez-Cordero & Martínez-Meyer, 2000), and priorities for biodiversity conservation (Chen & Peterson, 2002). They have the potential to fill knowledge gaps regarding species distribution (Rodríguez et al., 2007; Lewison et al., 2012; Mateo et al., 2013), aid in conservation planning (Peterson 2006; Rodríguez et al., 2007), assess conservation-human conflict (Rodríguez et al., 2007), and provide insight into the impact of climate change on species distribution (Peterson, 2006).

Popular Models

The concept of niche and relating species distribution with environmental factors was in vogue in ecology before the advent of Species Distribution Models (Grinnell, 1917; Hutchinson, 1957; Voous, 1960; Stiels & Schidelko, 2018). The emergence of computational power in 1980s gave rise to the initial form of SDMs which further got developed into very strong tools for ecological studies. However, in SDM, contrary to the nomenclature's suggestion, species' distribution models are not made to predict the future distribution of a target species but to project spatially the geographic area that might be suitable to the species (Thuiller & Munkemuller, 2010). Nevertheless, populations can be calculated using density of organism times suitable area projected by the model (Zhang et al., 2019b). SDM predicts species occurrence using mathematical models based on field data and environmental variables (Phillips et al., 2006), which can indicate the suitability of habitats for developing populations of particular species or community (Ferrier, 2002; Bosch et al., 2014). The SDMs are increasingly used to forecast the impact of climate change on species geographic distribution, but the reliability of predictions is scrutinized. The main limitation of SDMs lies in their assumption that species' ranges are determined mostly by climate, which is arguable. For instance, biotic interactions, habitat and elevation may affect species ranges (Hof et al., 2015).

Before the technological advancements in GIS and data analysis (Foody, 2008; Breiman, 2001b) modelling methods were simple environmental matching techniques (BIOCLIM; Busby, 1991 and DOMAIN, Carpenter et al., 1993). Later advancements were focussed on non-linear relationships of greater complexity between species presence and its environment (Shabani et al., 2018b). However, the creation of BIOCLIM software lead the way for SDMs and subsequently DOMAIN, GARP and others followed (Booth et al., 2014; Carpenter et al., 1993; Stockwell & Peters, 1999; Pereira, 2002) (Figure 1). At present, the most frequently used ecological models in the prediction of the distribution of various species are bioclimatic modelling (BIOCLIM; Busby, 1991), domain environmental envelope (DOMAIN; Carpenter et al., 1993), ecological niche factor analysis (ENFA; Hirzel & Guisan 2002), generalized additive model (GAM; Guisan et al., 2006), genetic algorithm for rule-set production (GARP; Stockwell, 1999), and maximum entropy (Maxent; Phillips et al., 2006). Such a variety of algorithms could be grouped into heuristic models (BIOCLIM), statistical models (GLM), predictive law optimization (GARP) and machine learning (Maxent). Among those models, Maxent modelling has been widely employed, because it performs well with either incomplete data or presence-only data (Phillips et al., 2006). Maxent modelling has been widely applied to plant and animal conservation (Phillips et al., 2006; Zhang et al., 2018), endangered species management (Ashraf et al., 2016), invasive species control (Ficetola et al., 2007; Peterson et al., 2011), and in the regionalization of agriculture (Peterson et al., 2011; Hu et al., 2015; Ashraf et al., 2016; Zhang et al., 2019). A brief account of some of these popular models is given in the following paragraphs.

Generalized linear models (GLMs) are a generalization of simple linear models (McCullagh & Nelder, 1989). The common types of GLMs are linear regression, logistic regression and Poisson regression. Logistic regression is suitable for present and absent species occurrence data and Poisson regression is suitable for species count data. The explanatory variables (related linearly to dependent variables) of GLMs can contain interaction terms and polynomial terms, so they are preferable for nonlinear yet simple relationship between species and environment variables. All model parameters of GLMs can be clearly interpreted with ecological meanings (Li & Wang, 2013).

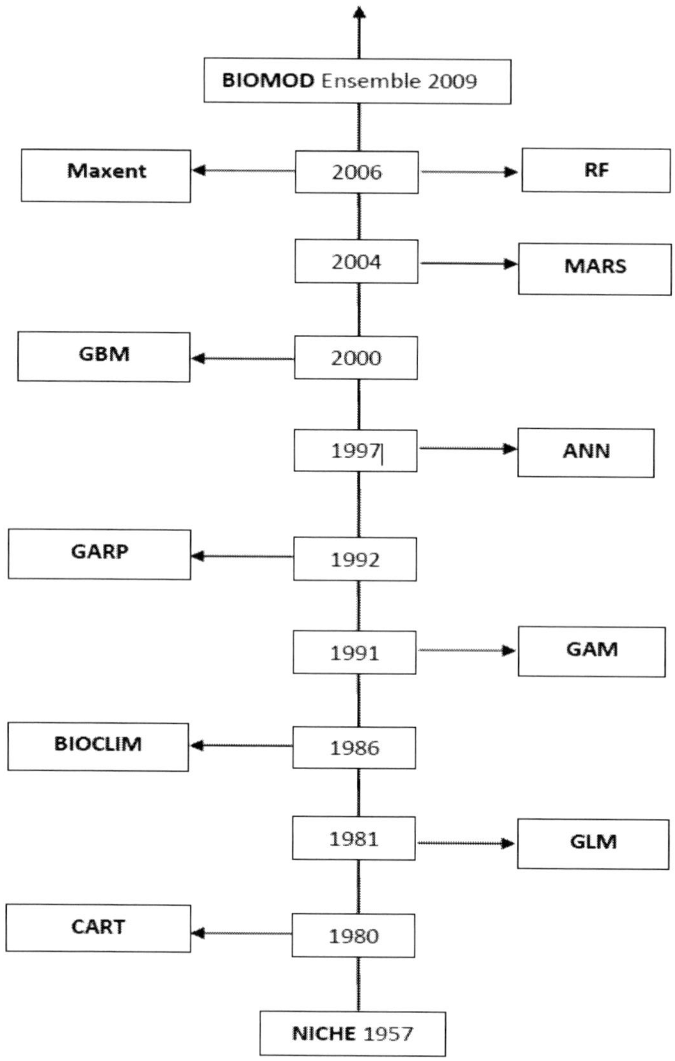

Adopted and modified from Gobeyn et al. (2019).

Figure 1. Depiction of evolutionary sequence of popular SDMs/ENMs: post niche concept given by Hutchinson in 1957. Refer text for acronyms used in this chart.

Generalized additive models (GAMs) are non-parametric extensions of GLMs, which provide the potential for better fits to data than GLMs (Hastie & Tibshirani, 1986). GAMs use data-defined smoothing functions to fit nonlinear species–environment relationships. The smooth functions are computed independently for each explanatory variable and added to build the final model. GAMs are useful when the relationship between species and environmental variables has a more complex form not easily fitted by GLMs (Yee & Mitchell, 1991; Li & Wang, 2013).

Multivariate adaptive regression splines (MARS) allow changes in coefficients (contrary to linear models where coefficients are stable) and are suitable when it is suspected that the model's coefficients have different optimal values across different levels of the explanatory variables. The breakpoints or thresholds of coefficients are termed spline knots. The regression with numerous spline knots can be thought of as a piecewise regression. In MARS, the spline knots are determined automatically and first-order interactions between variables can also be specified (Li & Wang, 2013). Models are particularly powerful when there are large numbers of explanatory variables and low-order interaction effects (Thuiller et al., 2009a).

Generalized boosting models (GBMs) have two algorithms: the boosting algorithm iteratively uses the regression-tree algorithm to construct a combination of trees. GLMs and GAMs can be used in tree building. Boosting is used to overcome the inaccuracies inherent in a single tree model. GBMs can eventually produce a good fit of the predicted values to the observed values, even if the specific nature of the relationships between the predictor variables and the dependent variable is complex (nonlinear, interacted or noisy with outliers). GBMs can be used for regression as well as classification problems, with continuous and/or categorical predictors (Li & Wang, 2013).

Artificial neural networks (ANN) are complex model systems, involving a network of simple processing elements (artificial neurons). The key feature of an ANN is that each neuron in the hidden layer receives information from each input and processes it. Models can operate like multiple regressions when the outputs are continuous variables, or like classifications when the outputs are categorical. ANNs are nonlinear models but extremely flexible enough to approximate any smooth function (Thuiller et al., 2009b; Li & Wang, 2013).

Classification and regression trees (CART) use recursive partitioning to split the data into increasingly smaller, homogenous subsets until a termination is reached (Venables & Ripley, 2002). In a CART, each group of records, represented by a 'node' in a decision tree, can only be split into two groups: regression tree or classification tree. The best tree is a trade-off between a high decrease of deviance and the smallest number of nodes. CART is able to uncover complex interactions between predictors that may be difficult or impossible to uncover using traditional multivariate techniques (Breiman et al., 1984; De'ath & Fabricius, 2000; Li & Wang, 2013).

Random forest (RF) is an ensemble classifier that consists of many decision trees. Each tree gives a classification (which is usually called the tree 'votes' for that class). The forest chooses the classification with the most votes (over all the trees in the forest). RF is not sensitive to the problem of multicollinearity; it handles correlated variables well (Breiman, 2001 a & b). Random forest is one of the most accurate learning algorithms with high performance in predicting species distributions (Iverson et al., 2008).

The genetic algorithm for rule set production (GARP) is an SDM based on a genetic algorithm for developing rule sets constraining species distribution (Stockwell & Peters, 1999). A GARP model is a random set of mathematical rules, known as genes, that can be interpreted as limiting environmental conditions (range specifications) and certain species–environment relationships (formed regression patterns). Each set of genes is combined randomly to further generate many possible models describing the potential of species occurrences. (Costa et al., 2002; Stockwell & Peterson, 2002; Li & Wang, 2013).

The Maximum Entropy model (MaxEnt) (Elith et al., 2011; Abdelaal et al., 2019), hereafter Maxent (Jafari et al., 2018; McDonald et al., 2019; etc.), is a machine learning algorithm that allows SDMs to be generated. It is one of the most reliable and statistically robust modelling softwares among well-established SDMs (Phillips et al., 2004, 2006; Elith et al., 2006; de Pando & de Giles, 2007; Wisz et al., 2008; Hernandes-Baz et al., 2016). It is also the most frequently used method for SDM/ENM and has been shown to perform better than many other models (Phillips et al., 2006; Elith et al., 2006; Fischer et al., 2011; Banag et al., 2015). This is the most popular modelling tool for use and functionality (Morales et al., 2017; Mohammadi et al., 2019).

The popularization of Maxent is evident in the review of Nursamsi et al. (2018) which records that more than a 1000 published studies used this software within few years of its release (Merow et al., 2013; Fourcade et al., 2014). Feng et al. (2019) also reported that Maxent, has been cited in tens of thousands of research papers in the past decade alone. This could be due to its higher predictive accuracy than any other method (Elith et al., 2006; Summers et al., 2012). The popularity of Maxent is on rise in the field of wildlife study as suggested by a random and quick search on wild plant and animal related studies on niche/habitat determination for the current scenario as well as the impact of climate change on future habitat/ niche (Table 1). Studies are even attempting to assess the future population using this tool (Zeng et al., 2015).

Table 1a. Details of recent studies using MaxEnt for SDM development in important animals. Additional information in Elith et al., 2010 and Baldwin, 2009

SN	Location	Study material	Study aspect	Reference
1	Mexico	*Eimeria* Species (Cattle Parasite)	Niche Prediction	Alcala-Canto et al. (2019)
2	Iran	*Neurergus kaiseri* (Endemic salamander)	Habitat/connectivity	Ashrafzadeh et al. (2019)
3	Nepal	*Neophron percnopterus* (Egyptian vulture)	Factors for presence	Bahadur et al. (2019)
4	Caribbean	*Stenella attenuata, S. frontalis, S. longirostris* (Dolphins)	Ecological Niche	Barragan-Barrera et al. (2019)
5	Mexico	*Aquila chrysaetos* (Golden Eagle)	Likely distribution	D'Addario et al. (2019)
6	Australia	*Ceratina australensis (Bee)*	Range expansion	Dew et al. (2019)
7	Iran	Caucasian grouse (*Lyrurus mlokosiewiczi*)	Potential abundance	Habibzadeh & Ludwig (2019)

Table 1b. Details of recent studies using MaxEnt for SDM development in important animals. Additional information in Elith et al., 2010 and Baldwin, 2009

SN	Location	Study material	Study aspect	Reference
8	Saudi Arabia	*Canis lupus arabs* (Pocock),	Range of wolves	Islam et al. (2019)
9	Kenya and Tanzania.	*Colobus angolensis palliatus, Colobus angolensis sharpei* (colobus monkeys)	Ecological niches	McDonald et al. (2019)
10	Iran	*Jaculus blanfordi* and *J. loftusi* (Desert jerboas)	Current and potential ranges	Mohammadi et al. (2019)
11	China	Himalayan Vulture (*Gyps himalayensis*), Saker falcon (*Falco cherrug*), Upland buzzard (*Buteo hemilasius*)	Spatial distribution	Zhang et al. (2019a)
12	Europe	*Darevskia praticola* (Meadow lizard)	Distribution	Ćorović et al. (2018)
13	Iran	*Neophron percnopterus* (Egyptian Vulture)	Potential distribution	Farashi & AlizadehNoughani (2018)
14	Java Island, Indonesia	*Nisaetus bartelsi* (Javan hawk eagle)	Suitable habitat distribution	Nursamsi et al. (2018)

Table 1c. Details of recent studies using MaxEnt for SDM development in important animals. Additional information in Elith et al., 2010 and Baldwin, 2009

SN	Location	Study material	Study aspect	Reference
15	South America	*Cyanallagma bonariense* (Endemic damselfly)	Species distribution	Pires et al. (2018)
16	Turkey	*Ips mannsfeldi*	Current/Future Distributions	Sarikaya et al. (2018)
17	Western Ghat, India	*Nilgiritragus hylocrius* (Nilgiri Tahr)	Habitat suitability	Sony et al. (2018)
18	New Mexico, USA	*Patagioenas fasciata* (band-tailed pigeon)	Habitat change	Coxen et al. (2017)
19	China, Nepal, India	*Grus nigricollis, Panthera pardus, Presbytis entellus, Pseudois nayaur, Tetraogallus tibetanus*, etc.	Suitable Habitat Changes	Liu et al. (2017)
20	Southern Africa	Cape vulture	Potential impacts on occurrence	Phipps et al. (2017)
21	USA	*Phalacrocorax auritus* (Double-crested cormorants)	Nesting habitat	Sheehan et al. (2017)
22	Brazil	Cougar	Potential land restoration	Angelieri et al. (2016)
23	Mexico	*Coreura albicosta*	Potential range	Hernandez-Baz et al., 2016
24	Iran	Wild Sheep	Geographical distribution pattern	Jafari et al. (2016)
25	Iran	Alborz Wild Sheep	Habitat Suitability & Connectivity	Keya et al. (2016)
26	Iran	Goitered gazelle	Potential distribution	Khosravi et al. (2016)

Table 1d. Details of recent studies using MaxEnt for SDM development in important animals. Additional information in Elith et al., 2010 and Baldwin, 2009

SN	Location	Study material	Study aspect	Reference
27	USA	*Gymnogyps californianus* (California condors)	Planning reintroductions	D'Elia et al. (2015)
28	India	*Ophrysia superciliosa* (Himalayan Quail)	Potential distribution	Dunn et al. (2015)
29	China	*Mergus squamatus* (Scaly-Sided Merganser)	Distribution and population	Zeng et al. (2015)
30	Spain	Wild boar	Distribution	Bosch et al. (2014)
31	California, USA	*Rana pretiosa* (Oregon Spotted Frog)	Rare Amphibian Surveys	Groff et al. (2014)
32	South America	*Melanophryniscus* and *Anura*) Red-Bellied Toads	Distribution	Zank et al. (2014)
33	Brazil	Bearded capuchin monkey	Habitat	Howard et al. (2012)
34	China	*Lobesia botrana*	Potential distributions	Lv et al. (2012)
35	Spain	*Gyps fulvus* (Griffon vulture)	Habitat Suitability	Mateo-Tomas & Olea (2010)
36	Peru and Bolivia	Andean Slope Birds	Areas of Endemism	Young et al. (2009)
37	Madagascar	*Phelsuma* (Day Geckos)	Species Delimitation	Raxworthy et al. (2007)

Table 1e. Recent studies using MaxEnt for SDM development. See also Elith et al., 2010 and Baldwin, 2009

SN	Location	Study material	Study aspect	Reference
1	India	*Parthenium hysterophorus* (Congress grass)	Invasion potential and niche dynamics	Ahmad et al. (2019)
2	Egypt.	*Rosa arabica* Crép.	Potential distribution	Abdelaal et al. (2019)
3	Southern Africa	*Erythrophleum ivorense*, *E. suaveolens*, *E. africanum*,	Bioclimatic impacts on ranges of related species	Gorel et al. (2019)
4	Sikkim, India.	*Paris polyphylla*	Species distribution	Lepcha (2019)
5	China	*Paeonia veitchii*	Potential Distribution	Zhang et al. (2019b)
6	Iran	*Daphne mucronata*	Predicting suitable habitats	Abolmaali et al. (2018)
7	Russia	*Oxytropis triphylla*	Distribution	Baikov & Krivenko (2018)
8	Sikkim, India	*Hippophae salicifolia* D. Don. (Seabuckthorn)	Habitat Distribution	Chetri et al. (2018)
9	Nepal	*Abies spectabilis*, *Betula utilis*, and *Pinus wallichiana* (Treeline spp.)	Habitat suitability	Chhetri et al. (2018)
10	Iran	*Artemisia aucheri*, *Bromus tomentellus* and *Festuca ovina*	Potential distribution	Esfanjani et al. (2018)
11	Tibet, China	*Stipa purpurea*	Distribution prediction	Ma & Sun (2018)

Table 1f. Details of recent studies using MaxEnt for SDM development in important plants. Additional information in Elith et al., 2010 and Baldwin, 2009

SN	Location	Study material	Study aspect	Reference
12	Southeast Asia.	*Selaginella ciliaris* and *S. plana*	Distribution	Setyawan et al. (2018)
13	India	*Ephedra gerardiana, Lilium polyphyllum, Crepidium acuminatum, Pittosporum eriocarpum* and *Skimmia anquetilia* (Medicinal Plants)	Reintroduction of Endangered Plants	Sourabh et al. (2018)
14	China, Nepal, India	*Alcimandra cathcartii*, and *Trillium govanianum*, etc.	Suitable Habitat Changes	Liu et al. (2017)
15	China	*Thuja sutchuenensis*	Potential distribution	Qin et al. (2017)
16	India	*Hevea brasiliensis* (Rubber tree)	Distribution prediction	Ray et al. (2017)
17	Iran	*Artemisia sieberi* and *A. aucheri*	Habitat suitability	Sanjerehei & Rundel (2017)
18	Pakistan	*Olea ferruginea*	Potential Distribution	Ashraf et al. (2016)
19	China.	*Cornus officinalis*	Habitat distribution	Cao et al. (2016)
20	Iran	*Tamarix ramosissima, Seidlitzia rosmarinus* and *Cornulaca monocantha*	Habitat distribution	Chahouki & Sahragard (2016)
21	India	*Bergenia ciliata*	Distribution Modeling	Kaliyathan et al. (2016)
22	Philippines	Ixora species	Habitat suitability	Banag et al. (2015)
23	Pakistan	*Pentatropis spiralis, Tylophora hirsuta,* and *Vincetoxicum arnottianum* (Medicinal asclepiads)	Impacts of climate change	Khanum et al. (2013)
24	India	*Justicia adhatoda*	Potential distribution	Yang et al. (2013)
25	Sinai, Egypt.	*Hypericum sinaicum*	Suitable Habitat	Khafagi et al. (2012)
26	India	*Berberis aristata*	Predictive distribution	Ray et al. (2011)
27	New Caledonia	*Canacomyrica monticola*	Suitable habitat	Kumar & Stohlgren (2009)

In any Maxent based study the model was chosen, among several SDMs algorithms, due to its numerous advantages which are listed as follows: (i) The input species data can be presence points only, thus it is very useful when presence-absence data collection is impractical (Phillips et al., 2006); (ii) This algorithm is considered among the best, when absence data for the species is not available. It seems to outperform other modelling methods in quality and predictive power when the number of geographic records is scarce (Elith et al., 2006; Phillips & Dudík 2008; Wisz et al., 2008; Corovic et al., 2018); (iii) Both, categorical and continuous environmental layers can be applied in this software; (iv) Its prediction is stable and reliable with a great accuracy even if species record is low. Thus, it is a suitable choice for prediction of distribution for rare and threatened species, (Hernandez et al., 2006; Pearson et al., 2007; Wisz et al., 2008; Garcia et al., 2013); (v) It creates a spatially explicit map for habitat suitability with an easy interpretation; (vi) It enables replicated runs to test model robustness, nonetheless threshold rule; (vii) The importance of each environmental variable can be measured using the jackknife test, in terms of gain. Gain is a unitless statistic that assesses how well the predicted distribution fits the occurrence data compared to a uniform distribution (Elith et al., 2011; Groff et al., 2014); and (viii) The Maxent model (bioclimatic envelope model) can be used to project into the future under climate change to predict habitat losses and gains within species range and thus help in planning appropriate conservation measures (Elith et al., 2011; Fois et al., 2018; Pearson et al., 2007; Phillips et al., 2006; Abdelaal et al., 2019).

The Maxent approach estimates the most uniform distribution (maximum entropy) of sampling points compared to background locations given the constraints derived from the data (Phillips et al., 2004, 2006; Grendar & Grendar, 2001). It also offers a wide variety of setting options which is different in each case and occasionally requires species-specific settings (Merow et al., 2013; Nursamsi et al., 2018). Nonetheless, Maxent is prone to overfitting, resulting in predicted distributions that are clustered around location points. Therefore, a relaxation component, called regularization, has been added to Maxent to constrain the estimated distribution thereby allowing the average value of each sampled variable to approximate its empirical average but not equal it. This regularization component can be adjusted for each sampling area. However, recent simulations have indicated that default settings perform as good as adjusted settings (Phillips & Dudik, 2008; Baldwin, 2009).

Maxent estimates species' distributions by calculating the most uniform distribution given the constraint that the expected value of each environmental variable matches the empirical average of the locality data (Phillips et al., 2006). Importantly, Maxent generates a probability distribution for habitat suitability, based on an index, across the study area (Elith et al., 2011), allowing comparison of suitability estimates among regions (Groff et al., 2014). Maxent models can be conservatively interpreted as a relative index of environmental suitability or relative density, where higher index values depict better conditions for the species (Phillips et al., 2006; D'Elia et al., 2015).

Maxent facilitates, through ENMs/SDMs, a wide range of applied uses in ecology. Some of the examples are understanding the ecological requirements or biogeography of species, finding new species or populations, identifying and prioritizing reintroduction sites, conservation planning and reserve design, predicting species invasions, predicting the effects of climate change or habitat loss, and for integrating information on movement ecology and demography to predict metapopulation dynamics in a spatially-explicit framework (As per review of Peterson, 2006; Franklin, 2009; Peterson et al., 2011 in D'Elia et al., 2015).

Ensemble Modelling

Based on current trends, uses of single models (quite a few mentioned in the above section) have been very popular for ascertaining species distributions, including the impact of climate change on habitat. These models have their own strengths but some uncertainties as well. To overcome the weaknesses and harvest the benefit of individual regression power, ensemble modelling approach has emerged (Capinha & Anastácio, 2011; Stohlgren et al., 2010; Ng et al., 2018; Stiels & Schidelko, 2018 etc.) and is expected to continue (Hao et al., 2019). From two to many SDMs may be used on BIOMOD platform depending on the priority and objectives of the study and also because combining the predictions of individual algorithms (Buisson et al., 2010; Grenouillet et al., 2011) increases the accuracy of species distribution forecasts (Araújo & New, 2007; Marmion et al., 2009; Thuiller et al., 2009b; Ahmad et al., 2019). Some examples are given in the following select random studies.

For example, Grenouillet et al. (2011) explored the usefulness of ensemble modelling approaches and tested whether species attributes influenced the outcomes of such approaches. Eight SDMs were used to model the current distribution of 35 fish species at 1110 stream sections in France. These methods included three regression methods (GLM, GAM, MARS), three machine learning methods (ANN, RF, aggregated boosted trees, ABT) and two classification methods (factorial discriminant analysis, FDA; classification tree analysis, CTA). They concluded that this study strengthened the usefulness of ensemble modelling of species distribution for conservation and biodiversity management.

In another example, Schorr et al. (2012) used GLM, GAM, GBM, and RF on latest BIOMOD platform. Their study was on integrating SDMs and phylogeography for two species of alpine *Primula*. This was keeping in view the advice of several recent studies that the combination of results from multiple modelling techniques ("ensemble forecasting"; Thuiller, 2004; Pearson et al., 2006; Araujo & New, 2007; Pearson et al., 2007) account for between-techniques uncertainty in predictions. Forester et al. (2013) also conducted modelling using the BIOMOD package (Thuiller et al., 2009) in R and Maxent, v. 3.3.3 (Phillips et al., 2006; Phillips & Dudik, 2008), while studying climate change impact on species distribution. Nine algorithms were run: GLM, GAM, MARS, CTA, FDA (flexible discriminant analysis), ANN, GBM, RF and Maxent. They supported the hypothesis that 'when these are combined using consensus methods (such as the mean of all models), they can form a more accurate projection, outperforming single models' (Marmion et al., 2009; Grenouillet et al., 2011).

Another example is the BIOMOD2 ensemble platform for species distribution modelling used to predict suitable pest habitat (Thuiller et al., 2009). The method enabled the simultaneous processing of a number of modelling techniques to create a consensus, or "ensemble," model (Araújo & New, 2007; Thuiller et al., 2009). In this study Shabani et al. (2018b) used three SDM methods, Maxent, GLM and BRT, to form an integrated prediction of the pest preferred habitat. This research has predicted areas conducive to the Dubas bug under future climate scenarios. In yet another example, Passadore et al. (2018) used an ensemble modelling approach to investigate the spatio-temporal distribution of southern Australian bottlenose dolphins (*Tursiops* cf. *australis*) in relation to a variety of ecogeographical and anthropogenic variables. This approach combined the results from five different algorithms implemented in Biomod2 package: two regression methods, GAMs and GBMs; one classification technique, CTA; and two machine learning approaches, RF and Maxent. They evaluated the overlap between current spatial management measures and important dolphin habitat with moderate to excellent precision (AUC 0.79 - 0.93).

Model Uncertainty and Averaging

Modern machine learning methods have achieved great success in predictive accuracy for supervised learning tasks but may still fall short of giving useful estimates of their predictive uncertainty (Ovadia et al., 2019). Model updating procedures are applied in order to improve the matching between experimental data and corresponding model output. The discrepancies between the output of the updated finite element model and the results of tests are due to the uncertainties that are involved in the modelling process (Goller & Schueller, 2011). SDM uncertainty can generally be classified into two categories: model uncertainty and measurement uncertainty (Elith et al., 2002). The former arises from model simplifications, limitations or assumptions in describing processes of extreme complexity. The latter arises from data imprecision and error, occurring through incorporation of incorrect geographic coordinates of species observations, or climatic datasets created inconsistently and interpolated into the mapping process (Shabani et al., 2018a).

Although good models are available, it becomes difficult to decide about the most appropriate model in many cases. Like, process-based models may differ in the specific ways they represent ecological mechanisms, without a clear empirical or theoretical reason to prefer one option over the other. Statistical analyses rarely offer a single solution, for two reasons (i) the limited amount of data allows for several plausible combinations of predictors, and (ii) different modelling approaches are available for statistical analysis (Hastie et al., 2009; Kuhn & Johnson, 2013). Model averaging seemingly solves this dilemma. It has been claimed by the proponents that calculating a weighted average of the predictions of all candidate models will reduce prediction error through reduced variance and bias (Madigan & Raftery, 1994), as well as better represent uncertainty about model parametrisation and structure (Wintle et al., 2003).

Dormann et al. (2008) working on 'prediction uncertainty of environmental change effects on temperate European biodiversity' concluded that selection of the single-best model introduces model selection error, which can be, to some extent, overcome by the multimodel approach. Weighted model averaging (Hoeting et al., 1999; Johnson & Omland, 2004) is one option to combine the different predictions of qualitatively different models for species richness. As their results show, model averaging attaches larger error to environmental change predictions. On the other hand, using multiple models also allows factors to influence predictions that have not been singled out by the best model. Further, Dormann et al. (2018) in a review on 'model averaging in ecology' have shown that model averaging can produce predictions with a smaller error than any of the contributing models by averaging away their variance and bias. Their overall recommendations stressed the importance of non-parametric methods such as cross validation for a reliable uncertainty quantification of model-averaged predictions.

For future projection of the habitat expanse or species distribution and comparing the dynamics in relation to current one, Global Climatic Models are further needed. Such models developed at various laboratories of the world have their own advantages and projection differences in climatic conditions resulting into future projected habitat. As suggested by Garcia et al. (2012) multi-model ensemble averages have often been shown to improve the outcome of climate simulations (Phillips & Gleckler, 2006; Tebaldi & Knutti, 2007; Gleckler et al., 2008; Reichler & Kim, 2008; Pierce et al., 2009; Knutti et al., 2010). However, to avoid the limitation (loss of higher order variability reflected in extreme projections, Beaumont et al., 2008), a hybrid consensus approach (Araujo et al., 2006) that groups co-varying projections before averaging them, could be used. When there is great variation in projections – as it is often the case with precipitation – this approach also avoids averaging projections that are very different or even contradictory, by placing them in different groups.

HABITAT PREDICTORS

Habitat is an important place for the survival, reproduction, and population development of organisms. Its quality can directly affect the distribution, quantity, and survival rate of species (Block & Brennan, 1993; Hall et al., 1997; Zhang et al., 2019). The species distribution can be determined on various geographic scales by climate, topography, soil characteristics, land use land cover, and biological interactions (Walter, 1985; Woodward, 1987; Ellenberg, 1988; Abolmali et al., 2018). Such habitat variables regulating the species distribution could be broadly grouped into environmental and climatic predictors, although the latter forms a part of former, as the climate has been assumed to change and will have impending modification in the habitat composition, impacting the species diversity.

Environmental Predictors

Depending on the objective of the study, data availability, expert knowledge etc. a set of environmental variables could be finalised differently. For example, climatic components, energy availability, and habitat heterogeneity are the useful variables for macrolevel species richness assessment (Evans et al., 2005; Thuiller et al., 2006; Davies et al., 2007; Rahbek et al., 2007; Whitton et al., 2007). Habitat variability has landcover, terrain ruggedness (elevation) and patch size and dominant cover as important constituents (Mouchet et al., 2015). Vegetation cover, a limiting factor for the spread of species, influences the distribution of an animal species more than any other factor since it determines the land's ability to supply food and/or shelter for animals (Herrero et al., 2006; Bosch et al., 2014).

Hof et al. (2012) reviewed and concluded, on the basis of randomly chosen literature, that half of the papers on species distribution considered elevation with almost no trend except the mammalian range. However, the migrating fauna like Monarch butterfly has shown obvious links with elevation during overwintering (Fisher et al., 2018) while Farashi & Alizaded-Noughani (2018) reported different reaction of Egyptian Vultures during summer and winter to mean elevation. Winter visitors showed a positive trend with an increase in elevation; while summer visitors showed a preference for middle elevations (500–2800 m). High elevation (>2000 m) with good forest is considered suitable habitat for the endemic Bush-shrike (Berzaghi et al., 2018). Other topographical features like slope and aspect do play their role in species distribution (Keppel et al., 2017; Bailey et al., 2018). Water plays a very important role in plant or animal physiology; therefore, distance of water body could be another variable to consider for species distribution or niche modelling (Howes, 2016; Mirzaei et al., 2017). Soil could be yet another parameter, but this is generally important for plant species only. Rural and urban built-up area, roads, mines etc., though not the part of habitat but associated with the environment, are known to affect distribution of some of the organisms. Food availability in the surrounding is another factor which regulates niche suitability.

In a case of plant species' spatial representation, Botella et al. (2018) used as many as 46 environmental descriptors which could be categorised as bioclimatic, topological, pedologic hydrographic and land cover descriptors. Excluding 19 specific variables of climate, they chose 27 other environmental variables like, potential evapotranspiration, elevation, topsoil available water capacity, base saturation of the topsoil, topsoil cation exchange capacity, soil crusting class, depth to a gleyed horizon, depth to an impermeable layer, soil erodibility class, topsoil organic carbon content, topsoil packing density, dominant surface textural class, <50 meters to fresh water, artificial area, semi-artificial area, arable land, pasture land, broad-leaved forest, coniferous forest, mixed forest, natural grasslands, moors, sclerophyllous vegetation, transitional woodland-shrub, no or little vegetation, coastal area, and ocean surface.

Dettki et al. (2003) used another set of variables for animal (Moose) habitat suitability modelling like, altitude, distance to railroads, distance to roads 7-9 m wide, distance to settlements, slope, distance to streams, distance to roads 5-7 m wide, distance to brooks, stem volume of Norway spruce, distance to roads ≤ 5 m wide, degree of stocking (%), distance to lakes and rivers, stem volume of Scots pine, mean stem diameter, stem volume of other deciduous species, stem volume of Lodgepole pine, stem volume of birch, biomass of needles and branches, trunk biomass, stand age, aspect, mean stand height, bark biomass. Not always a long list, Gaurdreau et al. (2018) selected only four (elevation in meter, percentage of forested area, percentage of wet area and linear disturbances) variables in a recent study on future distribution trend of boreal birds.

However, requirement and relationship of each organism varies with the environmental components. Thus, out of several environmental variables that are chosen based on an assumed relationship with the organism are more useful for modelling than simply throwing every available variable into the model to see what sticks. Therefore, environmental variables should generally be selected on the basis of an individual's knowledge and scientific literature review of the habitat requirements of the species of interest, along with discussions with taxonomic experts for the species (Helliwell & Chapman, 2013).

Climatic/Bioclimatic Predictors

Changing temperature and precipitation are the main regulators of ecosystem functioning. While temperature increase is the main driver of climate change caused habitat loss at higher elevations, precipitation changes are the main cause in lowlands (Enquist, 2002; Li et al., 2009 in Sekercioglu et al., 2012). Furthermore, temperature, through its effects on evapotranspiration and NPP, also has a crucial effect on vegetation change in the tropics (Delire et al., 2008). Temperature increases are likely to have a greater impact in the loss of tropical endemics than are changes in precipitation. However, past changes in bird communities suggest that major changes in precipitation patterns, such as changes in the monsoon regime, may be more important than temperature changes, particularly if warming is limited to 2°C (Tyrberg, 2010 in Sekercioglu et al., 2012).

Monthly temperature and rainfall driven variables in order to make them biologically more meaningful are considered as bioclimatic variables which are used in species distribution and other ecological modelling techniques. They represent annual trends, seasonality, and extreme limiting environmental factors. These variables along with standard code names are: (i) BIO1 = Annual Mean Temperature, (ii) BIO2 = Mean Diurnal Range (Mean of monthly (max temp - min temp)), (iii) BIO3 = Isothermality (BIO2/BIO7) (* 100), (iv) BIO4 = Temperature Seasonality (standard deviation *100), (v) BIO5 = Max Temperature of Warmest Month, (vi) BIO6 = Min Temperature of Coldest Month, (vii) BIO7 = Temperature Annual Range (BIO5-BIO6), (viii) BIO8 = Mean Temperature of Wettest Quarter, (ix) BIO9 = Mean Temperature of Driest Quarter, (x) BIO10 = Mean Temperature of Warmest Quarter, (xi) BIO11 = Mean Temperature of Coldest Quarter, (xii) BIO12 = Annual Precipitation, (xiii) BIO13 = Precipitation of Wettest Month, (xiv) BIO14 = Precipitation of Driest Month, (xv) BIO15 = Precipitation Seasonality (Coefficient of Variation), (xvi) BIO16 = Precipitation of Wettest Quarter, (xvii) BIO17 = Precipitation of Driest Quarter, (xviii) BIO18 = Precipitation of Warmest Quarter, and (xix) BIO19 = Precipitation of Coldest Quarter.

Global Climate Models

Climate data are available for use in three forms. Past conditions are downscaled global climate model output. Current conditions are the interpolation of observed data, representative of 1960-1990. Future conditions are downscaled global climate model (GCM) data from CMIP5 (IPPC Fifth Assessment). These future projection models' name along with code and institutions are listed below:

- ACCESS1-0 (AC) Australian Community, Australia,
- BCC-CSM1-1 (BC) Beijing Normal University, China,
- CCSM4 (CC) National Center for Atmospheric Research, USA,
- CNRM-CM5 (CN) Centre National de Recherches Meteorologiques, France,
- GFDL-CM3 (GF) NOAA/Geophysical Fluid Dynamics Laboratory, USA,
- GISS-E2-R (GS) NASA/Goddard Institute for Space Studies, USA,
- HadGEM2-AO (HD) Met Office Hadley Centre, UK,
- HadGEM2-CC (HG) Met Office Hadley Centre, UK,
- HadGEM2-ES (HE) Met Office Hadley Centre, (UK),
- INMCM4 (IN) Institute for Numerical Mathematics, Russia,
- IPSL-CM5A-LR (IP) Institut Pierre Simon LaPlace, France,
- MIROC5 (MC) Japan Agency for Marine Earth Science, Japan,
- MIROCESM (MR) Japan Agency for Marine Earth Science, Japan,
- MIROC-ESM-CHEM (MI) Japan Agency for Marine Earth Science, Japan,
- MRI-CGCMM3 (MG) Japan Agency for Marine Earth Science, Japan,
- MPI-ESM-LR (MP) Max Planck Institute, Germany,
- NorESM1-M (NO) Norwegian Climate Centre, Norway (https://www.worldclim.org/cmip5_30s, accessed on 29/09/2019).

These GCMs vary among themselves due to their varied data collection centres, their use and extrapolation. Since different GCMs rely on different parameters and incorporate different functions to portray the dynamics of atmospheric circulation, ocean effects, or feedback between the land surface and the atmosphere, they may project different consequences for the same level of greenhouse gas emissions (Wiens et al., 2009). For example, projections for California derived by using the National Center for Atmospheric Research (NCAR) PCM model with the A1fi emissions scenario (IPCC 2007), for example, suggest a temperature rise of 3.8°C by the end of the century from the 1990–1999 baseline, whereas the Hadley CM3 model projects a temperature rise of 5.8°C for the same emissions scenario (Heyhoe et al., 2004).

Representative Concentration Pathways

Global Climate Models projections based on Representative Concentration Pathways (RCPs), provide credible quantitative estimates of future climate. Climate modelling often uses these pathways to describe four possible climates in the near future worldwide (Vuuren et al., 2009). RCPs are the four independent pathways of greenhouse gas concentration trajectories (Moss et al., 2008; IPCC 2013). These four RCPs (RCP 2.6, RCP 4.5, RCP 6.0, and RCP 8.5) are named after the potential radiative forcing value in 2100 relative to the pre-industrial values (+2.6, +4.5, +6.0, and +8.5 W/m^2, respectively) (Weyant et al., 2009). Shabani et al. (2018b) quoting (Stocker et al., 2013) suggested that RCP 2.6 assumed the global warming increase from 0.4 to 1.6°C and 0.3 to 1.7°C between the years 2046 and 2064, and 2081 to 2100, respectively. RCP 2.6 also assumes that the mean global sea level will increase from 0.17 to 0.32 m and 0.26 to 0.55 m, over the same periods. RCP 4.5 assumed the global warming increase from 0.9 to 2.0°C and 1.1 to 2.6°C between the years 2046 and 2064 and 2081–2100, respectively. Mean global sea level will increase from 0.19 to 0.33 and 0.32 to 0.63 m over the same periods. RCP 6.0 assumed the global warming rise from 0.8 to 1.8°C and 1.4 to 3.1°C between the years 2046 and 2064 and 2081–2100, respectively. Mean global sea level will increase from 0.18 to 0.32 and 0.33 to 0.63 m over the same periods. RCP 8.5 assumed the global warming rise from 1.4 to 2.6°C and 2.6 to 4.8°C between the years 2046 and 2064 and 2081–2100, respectively. Mean global sea level will increase from 0.22 to 0.38 and 0.45 to 0.82 m over the same periods. However, these RCPs are considered optimistic (2.6), moderate/intermediate (4.5 and 6.0) and pessimistic (8.5) as they have lowest to highest GHG emission assumption (Iannella et al., 2018; Louppe et al., 2019). Manning et al. (2010) have advocated in their study that the scenario RCP 8.5 should be taken into consideration as lower emission scenarios are unlikely.

GCM Choices: Single or Multiple

Projections from climate distribution modelling depend on the chosen GCM and assumptions on future atmospheric CO_2 levels (Milanovich et al., 2010). Whereas no consensus exists as to what makes the most accurate GCM, integrating the different GCMs and RCPs to reinforce accuracy of the projections is strongly recommended (Sutton et al., 2015 in Ashrafzadeh et al., 2019). Most SDM studies use one or at the most two GCMs for future projections; there is certainly room for improvement in this component of projecting climate change impacts (Forester et al., 2013). Literature survey suggests that while doing the modelling, selection of GCM(s) and RCP(s) varied in different cases right from one to combination of many, generally, depending on, regions, species and other parameters. Some of the prominent contributions are as follows:

- Recently, Abdelal et al. (2019) and Mohammadi et al. (2019) used CCSM4 assigning the reason that this is one of the most efficient models to predict future climatic change on distribution of plant and animal species (Al-Qaddi et al., 2017; Sanjerehei & Rundel, 2017; Farashi & Erfani, 2018; Liang et al., 2018; Salas et al., 2018),
- While modelling impact of climate change on Giant Panda, Songer et al. (2012) included two general climate models; one described by the Canadian Centre for Climate Modeling and Analysis Coupled Model, version 3

(CGCM3) (Flato et al., 2000) and one from the Hadley Center for Climate Modeling Coupled Model, version 3 (HadCM3) (Gordon et al., 2000). For Predicting the effects of future climate change on the distribution of an endemic damselfly Pires et al. (2018) used another two models: CCSM4 and HadGEM2-AO in the modelling to account for inherent differences in their algorithms (Hijmans et al., 2005),

- Khanum et al. (2013) and Buisson et al. (2010) working on uncertainty in species distribution selected three models, Canadian (CCCMA/CGCM2), British (HadCM3) and Australian (CSIRO), as they were commonly used (Thuiller, 2004; Araujo et al., 2006; Tuck et al., 2006; Mika et al., 2008) and to reduce the uncertainties of single model. Louppe et al. (2019) also used three models (CNRM-CM5; GISS-E2-R; MIROC-ESM-CHEM) for climatically favourable current and future areas for the raccoon,
- Sony et al., 2018 used as many as four GCMs (HadCM3-AO, CANESM2, MIROC-ESM and CSIRO-MK) to reduce uncertainties while assessing the impact of climate change on Nilgiri Tahr population,
- For assessing the biodiversity scenario on global scale five GCMs were used [(CESM1-BGC, CMCC-CMS, CM5A-LR, MIROC5 and ESM-MR) Lindsay et al., 2014; Scoccimarro et al., 2011; Persechino et al., 2013; Watanabe et al., 2010; Giorgetta et al., 2013)] along with multiple SDMs (Thuiller et al., 2019).

Multicolinearity

Ecologists often use multiple regression to develop models that describe the regulation of particular aspects of organism, population, and community ecology (dependent or response variables) by various environmental and biological factors (independent or explanatory variables) (James & McCulloch, 1990). Statistical and inferential problems created by multicollinearity among the variables can be extremely severe under realistic ecological conditions (Graham, 2003). Multicollinearity is a statistical phenomenon in which one predictor variable in a multiple regression model can be linearly predicted from the others with a substantial degree of accuracy. This correlation is a problem because independent variables should be independent. If the degree of correlation between variables is high enough, it can cause problems when the model is used, and the results are interpreted. Therefore, colinear variables are required to be removed from the set of initially selected variables.

Multicollinearity is frequently observed in nature and more explanatory variables may be retained in the model, while the important ecological variables are eliminated (Graham, 2003). Therefore, ecological studies that used multiple regressions in presence of multicollinearity may have yielded incorrect ecological conclusions where multicollinearity was ignored (de Frutos et al., 2004). The increasing use of SDMs has raised new concerns regarding the inaccuracies, misunderstanding, and misuses of this important tool. One of those possible pitfalls – collinearity among environmental predictors – is assumed as an important source of model uncertainty, although it has not been subjected to a detailed evaluation in recent SDM studies (De Marco & Noabrega, 2018). SDMs are based on multiple regression techniques for prediction with the help of environmental variables which may be correlated. Therefore, there are several examples of species distribution studies which performed multicollinearity test and found out the correlated variables to be removed from regression (Elith et al., 2006; Elith et al., 2011; Stohlgren et al., 2011; Dormann et al., 2013; Khanum et al., 2013; Brandt et al., 2017; Iannella et al., 2018). Such a test was done using Pearson Correlation Coefficient (r), threshold value of which varied from study to study for example, <0.7 (Young et al., 2009), >±0.8 (Yang et al., 2013), >+0.8 (Abolmaali et al., 2018), <±0.9 (Heikkinen et al., 2006) etc.

Model Evaluation

Simulation models are approximate imitations of real-world systems, but they never exactly imitate the real-world due to lack of highest accuracy. Therefore, imitation requirement is sufficient accuracy of prediction which means that the model can be used as a substitute for the real system for the purposes of experimentation and analysis (Carson, 2002; Sargent, 2011). This is related to validation of the model which is used as model evaluation of performance accuracy from applied angle. As per definition evaluation is a systematic determination of a subject's merit, worth and significance, using criteria governed by a set of standards. This is generally expressed in numerical values which are different per se methods.

The most common method is to find out AUC (Area Under the Receiver Operating Characteristics Curve) values of the model which range from 0.5 for models with no predictive power to 1.0 for models with perfect predictive power. AUC values greater than 0.9 denote "very good" predictive power, values between 0.8 and 0.9 denote "good" predictive power and values between 0.7 and 0.8 indicate "useful" predictive power (Swets, 1998). Although AUC has known limitations as a measure of model performance (Franklin 2009), it still is the most used metric (Zank et al., 2014). Nevertheless, as highlighted by other studies (Jiménez-Valverde et al., 2013; Lobo et al., 2008; Peterson et al., 2008), AUC values can be misleading and unfairly denigrate the accuracy of the model (Fourcade et al., 2014). It is considered a reliable measure of discrimination ability, but it has been seen to have limitations in attaining ecological realism in modelled distribution (Jiménez-Valverde, 2014; Lobo et al., 2008).

Usage of AUC alone for evaluating model performance has been widely criticized (Austin, 2007; Lobo et al., 2008), since a high AUC score does not always reflect that the models are highly informative (Phillips et al., 2006). Other model evaluation alternative is introduction of true skill statistic (TSS) (Shabani et al., 2018b). TSS metric ranges from – 1 to + 1. In general, values of TSS below 0.40 specifies poor model performance, while values ranging from 0.40 to 0.75 indicates good model performance and the values above 0.75 are considered excellent (Allouche et al., 2006; Beaumont et al., 2016; Ahmad et al., 2019). In some of the studies both the evaluators (AUC and TSS) have been used for prediction accuracy (El-Gabas & Dormann, 2018; Iannella et al., 2018; Sony et al., 2018; Croft et al. 2019).

Modelling Limitations

Models are described as educated guesses about the future. The guesses can be narrowed down by knowing the assumptions well, evaluating their potential effects as correctly as possible, and incorporating additional detail as fine as possible (Wiens et al., 2009). Model's success depends on the steps it has climbed on the ladder of evolution in reaching the top of a biological phenomenon (nature's creation). This success is always impeded by the limitations. For example, Songer et al. (2012) reviewed that climate envelope models are widely used in climate change research prediction of future species distributions (Huntley et al., 2004; Hijmans & Graham, 2006), but this has limitations. Ibanez et al. (2006) rightly point out that this approach effectively treats the realized niche as though it were the fundamental niche. Additionally, changes in climate may also affect species dispersal, reproductive capacity, and disruption of ecosystem functioning due to inconsistent movement of a key species (Loreau et al., 2001). Another factor is the landscape mosaic and human activities which may impede or facilitate migration (Colligham & Huntly, 2000).

However, limitations are further elaborated in Banag et al. (2015) working on plant species, that many other factors beyond climate play a role in structuring species' distributions (e.g., plant to plant interactions, pollinator availability, dispersal ability, and plant plasticity) (Pearson & Dawson, 2003; Heikkinen et al., 2006; Hijmans & Graham, 2006; Oney et al., 2013). Other important environmental variables like soil type, topography, vegetation type, and land use change are suggested to be considered in future modelling studies (Trisurat et al., 2011; Ndayishimiye et al., 2012). Banag et al. (2015) also emphasized the need for improving the data situation at a sufficient spatial (and temporal) resolution in order to enhance model predictions.

Vulture Case Study: Central India

This case study is based on research by the author after 2016. There are reports of decimation of vulture population in Africa as well as South Asia especially the Indian sub-continent due to several factors, the major one of them being diclofenac poisoning. Decimation was always expressed in qualifying terms, such as 'several tens of millions of individuals', 'the commonest large bird of prey', 'remarkably high density; among others (Houston 1985 in Botha et al., 2017; Thakur & Kataria, 2012;

Thakur & Narang, 2012; Botha et al., 2017; Majgaonkar et al., 2018, etc.) but quantification was lacking. Therefore, the study of vultures with an intention for quantification frequently used total count protocol (Jha, 2018). Further, habitat projection studies were also taken up based on presence survey data from the field counts in a 2016 census. Some of the results are presented in following sub-sections.

Resident and Migratory Vultures

Several provinces of India harbour nine vulture species, but Central India or Madhya Pradesh province is considered as an Indian vulture stronghold (Jha 2018). Other than Bearded and Slender-billed vultures, seven species e.g., Indian or Long-billed, White-rumped, Red-headed, Egyptian, Cinereous, Himalayan Griffon and Eurasian Griffon vultures are found in this region (study area). The former four (Figure 2) are resident while the latter three are migratory vultures. As per the 2016 census, the total count in the state there was 7028 individuals (adults 6267 and 761 juveniles) of these species. However migratory vultures shared only 2.5% of the population. Resident vultures' contribution, in decreasing order, was 47.6% (Long-billed), 26.6% (Egyptian), 21.6% (White-rumped), and 1.3% (Red-headed). Most vultures were in forest, and Protected Area had higher density than Reserve Forest. Agriculture landscape had minimum occupancy. They were generally seen on trees. Long-billed and Egyptian favoured cliff and agriculture areas. High monuments were also used by Long-billed and Egyptian vultures. The five most frequently used trees were *Terminalia arjuna* (Arjun), *Ficus religiosa* (Peepal), *F. beghalensis* (Bargad), *Boswellia serrata* (Salai) and *Tectona grandis* (Sagaun). Migratory vultures roosted in any safe refuge (Jha 2017). Some vulture habitats are shown in Figure 3.

Source: Sanjay Shukla IFS.

Figure 2. Resident vultures in Central India. Top: left Egyptian vulture and right White-rumped vulture; Bottom: left Red-headed vultures and right Long-billed vulture with chick.

Source: Sanjay Shukla IFS.

Figure 3. Vulture roosting and habitat landscapes. Top layer: On the left dominant tree in the forest being used for roosting; On the right cliff is being used for roosting. Bottom layer: Forested area with suitable cliff on the left and a river passing by the forested hill.

Model Selection

Li & Wang (2013) reviewed that 'researchers need to select models on the basis of their research targets (data types: e.g., noisy, nonlinear and high-dimensional, lots of categorical variables) and objectives (e.g., checking variable weight, explaining species–environmental relationships, predicting distribution and estimating future range shifts). Another selection basis could be model quality. Although complex models have better prediction accuracy, simple models have their own unique advantages. GLMs are transparent; that is, all coefficients of explanatory variables (including their quadratic terms and interaction terms) can be clearly shown and well interpreted. On the contrary, other models (i.e., GAM, MARS, MDA, GBM, CART and complex models) have too many parameters, so that meaningful ecological interpretation is not possible. The complex models, such as random forest, are also called 'black boxes': they are easy to use yet hard to explain (Breiman, 2001b). However, in the case presented, Maxent model for species distribution prediction in current as well as future scenario was chosen due to enormous advantages mentioned in the model subsection.

Data Preparation

Vulture data presence record was cleaned and converted into .csv file. The shape file of study area (boundary of Madhya Pradesh) was used to extract the environmental variables by masking. The climatic variables were downloaded from www.worldclim.org. Landuse landcover (LULC) data was borrowed from MPCOST (Madhya Pradesh Council of Science and Technology), Bhopal. Elevation and NDVI (Normalised Difference Vegetation Index) were downloaded from www.earthexplorer.

usgs.gov. Pearson correlation test was applied to ascertain collinearity among the environmental variables, fixing the threshold value ± 0.85. Out of 19 climatic variables only seven were selected for modelling after eliminating the rest of the correlated ones. Along with these seven other variables like LULC, elevation and NDVI were used for the current scenario modelling. For future scenario only climatic models were used. CCSM4 GCM and 2.6 and 8.5 RCPs of 2070 were chosen for the present case study.

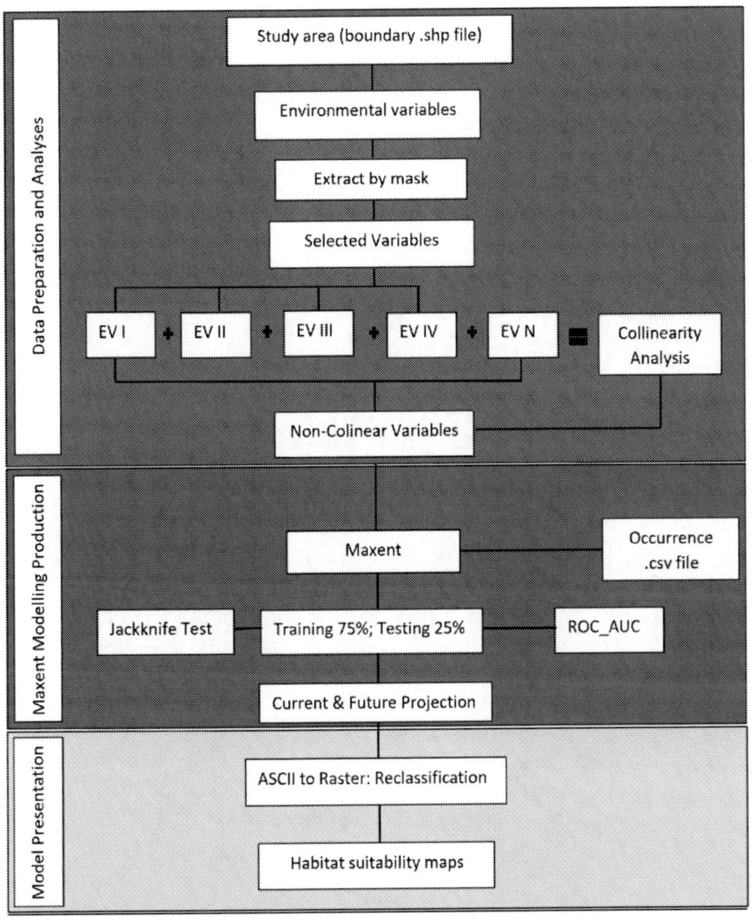

Modified from Zhang et al. (2019b) and Mwakapeje et al. (2019).

Figure 4. Flow Chart of Species Distribution of Ecological Niche Modelling (SDM/ENM) using Maxent Software (borrowed from Jha and Jha 2021). EV I to EV IV are environmental variables. N indicates any number of EV. Flow is top to bottom along connectors.

Selection of GCM and RCP was based on previous studies. For example, Abdelal et al. (2019) and Mohammadi et al. (2019) used CCSM4 assigning the reason that this is one of the most efficient models to predict future climatic change on distribution of plant and animal species (Al-Qaddi et al., 2017; Sanjerehei & Rundel, 2017; Farashi & Erfani, 2018; Liang et al., 2018; Salas et al., 2018). RCP 2.6 assumes the projected global mean temperature will rise and then stabilize due to the employment of a range of technologies and strategies for reducing greenhouse gas emissions (Collins et al., 2013; IPCC 2013). RCP 8.5 being more pessimistic assumes that these emissions continue to rise throughout the 21st century (Meinshausen et al., 2011), and will continue to rise in the future as well (IPCC 2013; Ahmad et al., 2019).

The Maxent results were analysed using AUC values, variable contribution tables and Jackknife bar charts. Habitat suitability maps were generated on the basis of maxent index using ArcGIS 10.3 software. The study area was divided into unsuitable, low suitable, moderate suitable and high suitable area for vulture as adopted by Ashraf et al. (2016); Abolmaali et al. (2018) and Zhang et al. (2019). Corresponding Maxent index value demarcating these categories was fixed at 0-0.2, 0.2-0.4, 0.4-0.6 and 0.6-1, respectively. Flow chart of the modelling is presented in Figure 4.

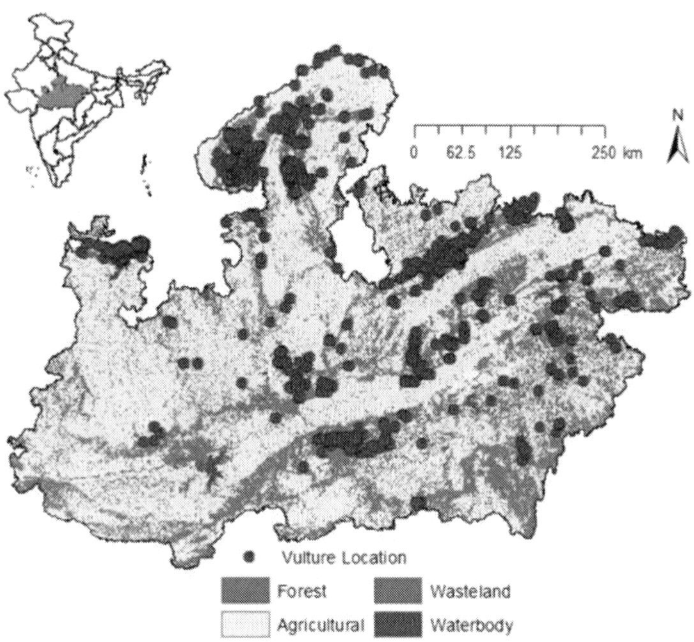

Figure 5. Vulture locations (red circles) of all seven species observed during two counts of summer and winter 2016 used as Maxent sample input after cleaning and processing. Different background colour indicates the landcover/landuse (Forest and Water body/Agriculture). Inset shows location of Madhya Pradesh as Central India.

Modelling Outcome

Vulture species were confined mostly in the forest area, nesting in cliffs or trees except a few in agriculture areas (Figure 5). Based on these sample data, projected current (2016) and future (2070) habitat is presented in Figure 6. The study area (308252 km^2) has 79.1% and 64.4% unsuitable habitat modelled by using, respectively environmental variables and climatic variables only. In the long-term prediction after impact of climate change unsuitable area remained 62.4% and

63.8%, respectively in the lowest and highest emission scenario. Habitat dynamics showed that in relation to current scenario there will be conversion of unsuitable area into suitable quality and vice versa (Figure 7). In the case of optimistic scenario (RCP 2.6) gain and loss will be 1.49% and 1.01%, respectively (net gain 0.48%). Corresponding figure for pessimistic scenario (RCP 8.5) will be 1.23% and 0.70%, respectively (net gain 0.53%).

Above results fell in model strength of higher standard as the AUC value was 0.830 to 0.862. This also provides good dependability to other observations. All variables considered, LULC was the most important variable which influenced habitat prediction by 62%. Next five important variables in order of contribution were bio-19, bio-18, bio-03, bio-09 and bio-05. The model of current scenario developed with only bioclimatic variables showed prediction role of bio-03, bio-19, bio-15, bio-05, bio-09 in decreasing order. In future scenarios also these five variables played a major role. Jackknife test (Figure 8) indicated the importance of the climatic set slightly different from the one shown in relative contributions of the Maxent table of environmental variables. The decreasing order of the importance of variables is bio-15, bio18, bio-19, bio-3, and bio-9.

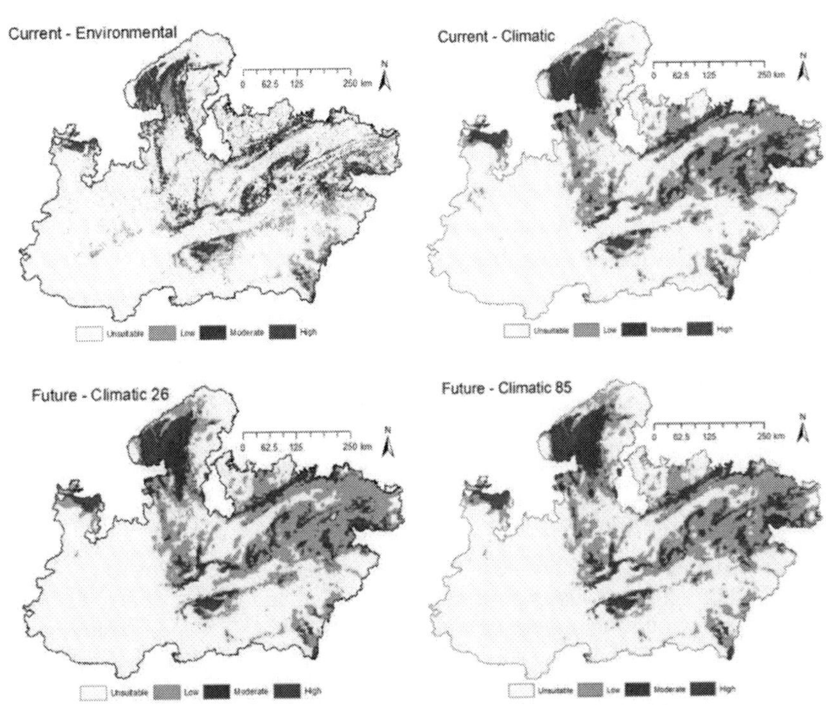

Figure 6. Projection of vulture habitat in Central India categorised into four classes of suitability. Current habitat is based on 2016 vulture data with environmental variables including climatic (top left) and only with climatic variables (top right). Lower left and right habitat prediction maps represent 2070 scenario with RCP 2.6 and RCP 8.5, respectively of CCMS4 Global Circulation Model and without environmental variables like landuse/landcover.

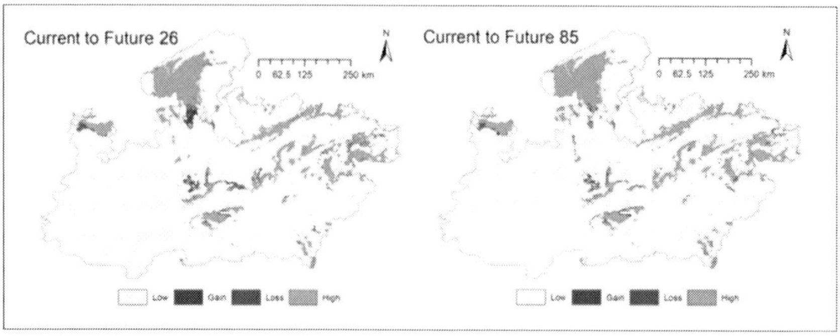

Figure 7. Impact of climate change on vulture habitat in Central India. Change scenario is of 2070 with respect to 2016 (current). Gain is conversion of Low to High suitability and Loss is conversion of High to Low suitability habitat area. Maps on left and right represent RCP 2.6 and RCP 8.5 of CCMS 4 Global circulation Model. Yellow and green area have remained unchanged.

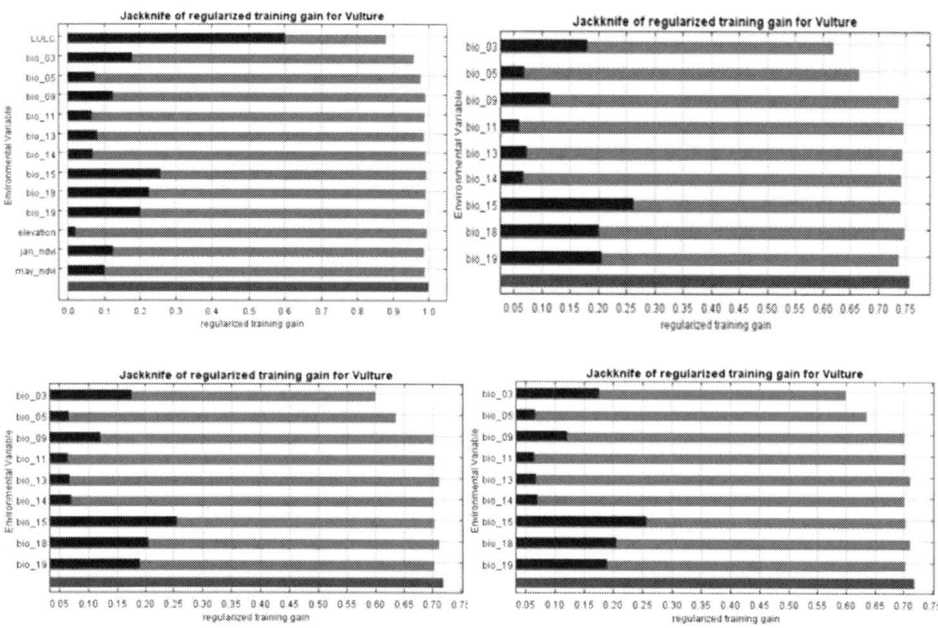

Figure 8. Current and Future Scenario. Upper row is current scenario: Environmental variables in top left bar chart includes physical attributes while top right chart has only climatic variables. Lower row is future scenario: bottom left chart is RCP 2.6 and bottom right is RCP 8.5.

Out of nine non-colinear climatic variables, five (bio-15, bio-18, bio-19, bio-3, and bio-9) were the main contributors (>75%) of habitat prediction. Precipitation (bio-15, bio-18, bio19,) and temperature (bio-3, and bio-9) related variables had almost equal contribution. Enquist (2002) and Li et al. (2009) suggested that precipitation changes are the main cause in lowlands while temperature is the main driver for habitat loss at higher elevations. Vultures are known to prefer low elevation (<1000 m; Bahadur et al. 2019). Tyrberg (2008) also reported that major changes in precipitation pattern may be more important than temperature in the case of bird communities. This correlates with the field observation that higher rainfall area had lesser vultures. Gaudreau et al. (2018) have also indicated along similar lines that the spatial distribution of species is highly correlated to climate, more particularly to precipitation and average annual temperature.

Variation in suitable habitat area under two different models (set of environmental and climate only variables) is in line with earlier findings of Preston et al. (2008) which showed that including biotic associations in climate change models reduced habitat availability for endangered California plant, butterfly, and bird species as opposed to a climate-only model. This is possibly because area with biotic associations (e.g., forests being specific part of habitat in the case of vultures) is generally lesser as compared to larger climatic envelope.

Lowering of unsuitable area, although marginally, from current to long-term low emission and further low long-term high emission scenario relates to positive impact. Lower loss in suitable habitat in the case of higher emission in future further indicates suitability of worsening climate change. Trautmann (2018) has reviewed that there are several positive aspects of climate change on a number of bird species like increasing ranges for sedentary species (Kampichler et al. 2012; Davey et al. 2012; Devictor et al. 2008), improving wintering conditions (Princé & Zuckerberg 2015; Tellería et al. 2016) etc. Nevertheless, present output should be taken up with the caution that the projection is at an early stage.

Conclusion

Broad scholarship on habitat, climate change and modelling with a case of vulture has been presented in this chapter. Predictions revealed category wise suitable habitats (Low, Moderate and High) of vulture, habitat influencing predictors (Precipitation seasonality, Precipitation of Warmest, and Coldest quarter, Iso-thermality, and Mean temperature of driest quarter) and conversion of habitat area from unsuitable to suitable and vice versa. Although most used models (Maxent and CCMS4) have been used, and model evaluation accuracy is high, the projection needs to be taken as initial indicator, since latest technology of ensemble modelling (both SDMs and GCMs) is believed to give better prediction. Also required is the incorporation of vulture habitat influencing maximum environmental factors in current as well as future prediction. Future habitat projections (forest, agriculture and water regime) may undergo change in central India due to climate crisis (EPCO 2015). However, based on some findings (basic reference point, Merow et al., 2013) some conservation planning could be taken up for vultures, such as species introductions into moderate and highly suitable areas (if needed), and habitat improvements in low suitability areas to provide the endangered species habitat against threats.

Acknowledgments

Author is thankful to Miss Radhika Jha, Lucknow University, India for her contribution in Maxent modelling. Shri Sanjay Shukla IFS, Additional Principal Chief Conservator of Forests deserves special thanks for the photographs of vultures and their habitat. Miss Sonal Jha, Assistant Professor, IIT Bhilai, deserves appreciation for going through the manuscript from language point of view.

References

Abdelaal, M., Fois, M., Fenu, G. & Bacchetta, G. 2019. Using MaxEnt modeling to predict the potential distribution of the endemic plant *Rosa arabica* Crép. in Egypt. *Ecological Informatics,* 50: 68-75. https://doi.org/10.1007/s10333-006-0047-5.

Abolmaali, S. M. R., Tarkesh, M. & Bashari, H. 2018. MaxEnt modeling for predicting suitable habitats and identifying the effects of climate change on a threatened species, *Daphne mucronata*, in central Iran. *Ecological Informatics,* 43: 116–123. https://doi.org/10.1016/j.ecoinf.2017.10.002.

Ahmad, R., Khuroo, A. A., Hamid, M., Charles, B. & Rashid I. 2019. Predicting invasion potential and niche dynamics of *Parthenium hysterophorus* (Congress grass) in India under projected climate change. *Biodiversity and Conservation,* 28: 2319–2344. https://doi.org/10.1007/s10531-019-01775-y.

Alcala-Canto, Y., Alberti-Navarro, A., Figueroa-Castillo, J. A., Ibarra-Velarde, F., Vera- Montenegro, Y. & Cervantes-Valencia, M. E. 2019. Maximum Entropy Ecological Niche Prediction of the Current Potential Geographical Distribution of Eimeria Species of Cattle, Sheep and Goats in Mexico. *Open Journal of Animal Sciences,* 9: 234-248. https://doi.org/10.4236/ojas.2019.92020.

Allouche, O., Tsoar, A. & Kadmon, R. 2006. Assessing the accuracy of species distribution models: prevalence, kappa and the true skill statistic (TSS). *Journal of Applied Ecology,* 43: 1223-1232. https://doi.org/10.1111/j.1365-2664.2006.01214.x.

Al-Qaddi, N., Vessella, F., Stephan, J., Al-Eisawi, D. & Schirone, B. 2017. Current and future suitability areas of kermes oak (*Quercus coccifera* L.) in the Levant under climate change. *Regional Environmental Change,* 17: 143–156. https://doi.org/10.1007/s10113-016-0987-2.

Anderson, R. P. 2013. A framework for using niche models to estimate impacts of climate change on species distributions. *Annals of the New York Academy of Sciences,* 1297: 8–28. https://doi.org/10.1111/nyas.12264.

Angelieri, C. C. S., Adams-Hosking, C., Ferraz, K. M. P. M. B., de Souza, M. P. & McAlpine, C. A. 2016. Using species distribution models to predict potential landscape restoration effects on Puma conservation. *PLoS ONE,* 11(1). https://doi.org/10.1371/journal.pone.0145232.

Araujo, M. B. & Guisan, A. 2006. Five (or so) challenges for species distribution modelling. *Journal of Biogeography,* 33: 1677–1688. https://doi.org/10.1111/j.1365-2699.2006.01584.x.

Araujo, M. B. & New, M. 2007. Ensemble forecasting of species distributions. *Trends in Ecology & Evolution,* 22: 42–47. https://doi.org/10.1016/j.tree.2006.09.010.

Araujo, M. B., Thuiller, W. & Pearson, R. G. 2006. Climate warming and the decline of amphibians and reptiles in Europe. *Journal of Biogeography,* 33: 1712–1728. https://doi.org/10.1111/j.1365-2699.2006.01482.x.

Arkumarev, V., Dobrev, V., Abebe, Y. D., Popgeorgiev, G. & Nikolov, S. C. 2014. Congregations of wintering Egyptian Vultures *Neophron percnopterus* in Afar, Ethiopia: present status and implications for conservation. *Ostrich,* 85: 139–145. https://doi.org/10.2989/00306525. 2014.971450.

Ashraf, U., Ali, H., Chaudry, M. N., Ashraf, I., Batool, A. & Saqib, Z. 2016. Predicting the Potential Distribution of *Olea ferruginea* in Pakistan incorporating Climate Change by Using Maxent Model. *Sustainability,* 2016, 8: 722. https://doi.org/10.3390/su8080722.

Ashrafzadeh, M. R., Naghipour, A. A., Haidarian, M., Kusza, S. & Pilliod, D. S. 2019. Effects of climate change on habitat and connectivity for populations of a vulnerable, endemic salamander in Iran. *Global Ecology and Conservation,* 19: e00637. https://doi.org/10.1016/ j.gecco.2019.e00637.

Austin, M. 2007. Species distribution models and ecological theory: a critical assessment and some possible new approaches. *Ecological Modelling,* 200: 1-19. https://doi.org/10.1016/j.ecolmodel.2006.07.005.

Bahadur, K. C. K., Koju, N. P., Bhusal, K. P., Low, M., Ghimire, S. K., Ranabhat, R. & Panthi, S. 2019. Factors influencing the presence of the endangered Egyptian vulture *Neophron percnopterus* in Rukum, Nepal. *Global Ecology and Conservation,* 20: e00727. https://dx.doi.org/10.1016/j.gecco.2019.e00727.

Baikov, K. & Krivenko, D. 2018. Climatic modeling of the distribution of *Oxytropis triphylla* (Fabaceae) by maximum entropy method. BIO Web of Conferences 11, 00002, 2018. *Prospects of Development and Challenges of Modern Botany.* https://doi.org/10.1051/bioconf/20181100002.

Bailey, J. J., Boyd, D. S. & Field, R. 2018. Models of upland species' distributions are improved by accounting for geodiversity. *Landscape Ecology,* 33: 2071–2087. https://doi.org/10.1007/s10980-018-0723-z.

Baldwin, R. A. 2009. Use of maximum entropy modeling in wildlife research. *Entropy,* 11: 854-866. https://doi.org/10.3390/e11040854.

Banag, C., Thrippleton, T., Alejandro, G. J., Reineking, B. & Liede-Schumann, S. 2015. Bioclimatic niches of selected endemic Ixora species on the Philippines: predicting habitat suitability due to climate change. *Plant Ecology,* 216: 1325–1340. https://doi.org/10.1007/s11258-015-0512-6.

Barnosky, A. D., Matzke, N., Tomiya, S., Wogan, G. O. U., Swartz, B., Ouental, T. B., Marshall, C., Mcguire, J. L., Lindsey, E. L., Maguire, K. C., Merset, B. & Ferrer, A. 2011). Has the Earth's sixth mass extinction already arrived? *Nature,* 471: 51-57. https://doi.org/10.1038/nature09678.

Barragan-Barrera, D. C., do Amaral, K. B., Chávez-Carreño, P. A., Farías-Curtidor, N., Lancheros-Neva, R., Botero-Acosta, N., Bueno, P., Moreno, I. B., Bolaños-Jiménez, J., Bouveret, L., Castelblanco-Martínez, D. N., Luksenburg, J. A., Mellinger, J., Mesa Gutiérrez, R., de Montgolfier, B., Ramos, E. A., Ridoux, V. & Palacios, D. M. 2019. Ecological Niche Modeling of Three Species of *Stenella* Dolphins in the Caribbean Basin, With Application to the Seaflower Biosphere Reserve. *Frontiers in Marine Science,* 6: 10. https://doi.org/10.3389/fmars.2019.00010.

Beaumont, L. J., Graham, E., Duursma, D. E., Wilson, P. D., Cabrelli, A., Baumgartner, J. B., Hallgren, W. & Laffan, S. W. 2016. Which species distribution models are more (or less) likely to project broad-scale, climate induced shifts in species ranges? *Ecological Modelling,* 342: 135–146. https://doi.org/10.1016/j.ecolmodel.2016.10.004.

Beaumont, L. J., Hughes, L. & Pitman, A. J. 2008. Why is the choice of future climate scenarios for species distribution modelling important? *Ecology Letters*, 11: 1135– 1146. https://doi.org/10.1111/j.1461-0248.2008.01231.x.

Bellard, C., Bertelsmeier, C., Leadley, P., Thuiller, W. & Courchamp, F. 2012. Impacts of climate change on the future of biodiversity. *Ecology Letters,* 15(4): 365–377. https://doi.org/10.1111/j.1461-0248.2011.01736.x.

Berzaghi, F., Engel, J. E., Plumptre, A. J., Mugabe, H., Kujirakwinja, D., Ayebare, S. & Bates, J. M. 2018. Comparative niche modeling of two bush-shrikes (Laniarius) and the conservation of mid-elevation Afromontane forests of the Albertine Rift. The Condor: *Ornithological Applications,* 120: 803–814. https://doi.org/10.1650/CONDOR-18-28.1.

BirdLife International 2016. *Migratory soaring birds project: energy.* Retrieved from http://migratorysoaringbirds.undp.birdlife.org/en/sectors/energy.

BirdLife International 2017. *BirdLife/IUCN Red List for birds.* Retrieved from http://www.birdlife.org.

Booth, T. H., Nix, H. A., Busby, J. R. & Hutchinson, M. F. 2014. BIOCLIM: the first species distribution modelling package, its early applications and relevance to most current MAXENT studies. *Diversity and Distributions,* 20: 1–9. https://doi.org/10.1111/ddi.12144.

Bosch, J., Mardones, F., Perez, A., Torre, A. & Munoz, M. J. 2014. A maximum entropy model for predicting wild boar distribution in Spain. *Spanish Journal of Agricultural Research,* 12(4): 984-999. http://dx.doi.org/10.5424/sjar/2014124-5717.

Botella, C., Joly, A., Bonnet, P., Monestiez, P. & Munoz, F. 2018. A Deep Learning Approach to Species Distribution Modelling. In Joly, A., Vrochidis, S., Karatzas, K., Karppinen, A. & Bonnet. P. (Eds.), *Multimedia Tools and Applications for Environmental & Biodiversity Informatics*. Multimedia Systems and Applications, 44-45. Cham: Springer. https://doi.org/10.1007/978-3-319-76445-0_10.

Botha, A. J., Andevski, J., Bowden, C. G. R., Gudka, M., Safford, R. J., Tavares, J. & Williams, N. P. 2017. *Multi-species Action Plan to Conserve African-Eurasian Vultures.* CMS Raptors MOU Technical Publication No. 3. CMS Technical Series

No. 33. Coordinating Unit of the CMS Raptors MOU, Abu Dhabi, United Arab Emirates.

Botkin, D. B. & Keller, E. A. 2000. Environmental science. *Earth: A Living Planet.* New York, NY: John Willey and Sons.

Brandt, L. A., Benscoter, A. M., Harvey, R., Speroterra, C., Bucklin, D., Romanach, S. S., Watling, J. I. & Mazzotti, F. J. 2017. Comparison of climate envelope models developed using expert-selected variables versus statistical selection. *Ecological Modelling,* 345: 10–20. https://doi.org/10.1016/j.ecolmodel.2016.11.016.

Breiman, L. 2001a. Random forests. *Machine Learning,* 45: 5–32.

Breiman, L. 2001b. Statistical modelling: the two cultures. *Statistical Science,* 16: 199–215. https://doi.org/10.1214/ss/1009213726.

Breiman, L., Friedman, J. H., Olshen, R. A. & Stone, C. J. 1984. *Classification and Regression Trees.* New York: Chapman and Hall.

Buisson, L., Thuiller, W., Casajus, N., Lek, S. & Grenouillet, G. 2010. Uncertainty in ensemble forecasting of species distribution. *Global Change Biology,* 16: 1145–1157. https://doi.org/10.1111/j.1365-2486.2009.02000.x.

Busby, J. R. 1991. BIOCLIM – a bioclimate analysis and prediction system. In Margules, C. R. & Austin, M. P. (Eds.) *Nature Conservation: Cost Effective Biological Surveys and Data Analysis*, 64–68. Canberra: CSIRO.

Bustamante, J. 1997. Predictive models for lesser kestrel *Falco naumanni* distribution, abundance and extinction in southern Spain. *Biological Conservation,* 80: 153–160. https://doi.org/10.1016/S0006-3207(96)00136-X.

Capinha, C., & Anastácio, P. 2011. Assessing the environmental requirements of invaders using ensembles of distribution models. *Diversity and Distributions,* 17: 13–24. https://doi.org/10.1111/j.1472-4642.2010.00727.x.

Carson, J. S. II. 2002. Model Verification and Validation. In Yucesan, E., Chen, C. H., Snowdon, J. L. & Charnes, J. M. (Eds.), *Proceedings of the 2002 Winter Simulation Conference, IEEE.* https://ieeexplore.ieee.org/abstract/document/1172868 Accessed 27/09/2019.

Carpenter, G., Gillison, A. N. & Winter, J. 1993. DOMAIN: A flexible modelling procedure for mapping potential distributions of plants and animals. *Biodiversity and Conservation,* 2: 667–680.

Cao, B., Bai, C., Zhang, L., Li, G. & Mao, M. 2016. Modeling habitat distribution of *Cornus officinalis* with Maxent modeling and fuzzy logics in China. *Journal of Plant Ecology,* 9(6): 742–751. https://doi.org/10.1093/jpe/rtw009.

Chahouki, M. A. Z. & Sahragard, H. P. 2016. Evaluation of MaxEnt method for habitat distribution modeling of three plant species in Garizat rangelands of Yazd province, Iran. *Range Management & Agroforestry,* 37(2): 142-147.

Choudhury, M. R., Deb, P., Singha, H., Chakdar, B. & Medhi, M. 2016. Predicting the probable distribution and threat of invasive *Mimosa diplotricha* Suavalle and *Mikania micrantha* Kunth in a protected tropical grassland. *Ecological Engineering,* 97: 23– 31.

Chen, G. & Peterson, A. T. 2002. Prioritization of areas in China for the conservation of endangered birds using modelled geographical distributions. *Bird Conservation International,* 12: 197-209.

Chettri, A., Pradhan, A., Sharma, G., Padhan, B. K. & Chhetri, D. R. 2018. Habitat Distribution Modelling of Seabuckthorn (*Hippophae salicifolia* D. Don.) in Sikkim, Eastern Himalaya, India. *Indian Journal of Ecology,* 45(2): 266-269.

Chhetri, P. K., Gaddis, K. D. & Cairns, D. M. 2018. Predicting the Suitable Habitat of Treeline Species in the Nepalese Himalayas Under Climate Change. *Mountain Research and Development,* 38(2): 153-163. URL: https://doi.org/10.1659/MRD JOURNAL-D-17-00071.1.

Collingham, Y. C. & Huntley, B. 2000. Impacts of habitat fragmentation and patch size upon migration rates. *Ecological Applications,* 10(1): 131-144.

Collins, M., Knutti, R., Arblaster, J., Dufresne, J. L., Fichefet, T., Friedlingstein, P. & Shongwe, M. 2013. Longterm climate change: projections, commitments and irreversibility. In Stocker, D., Qin, G. K., Plattner, M., Tignor, S. K., Allen, J. & Midgley, P. M. (Eds), *Climate change 2013: the physical science basis.* Working Group 1(WG1) Contribution to the Intergovernmental Panel on Climate Change (IPCC) 5th Assessment Report (AR5). 1029-1136. New York, NY Cambridge University Press.

Cordellier, M., Pfenninger, A., Streit, B. & Pfenninger, M. 2012. Assessing the effects of climate change on the distribution of pulmonate freshwater snail biodiversity. *Marine Biology,* 159: 2519–2531. http://dx.doi.org/10.1007/s00227-012-1894-9.

Corovic, J., Popovic, M., Cogalniceanu, D., Carretero, M. A. & Crnobrnja-Isailovic, J. 2018. Distribution of the meadow lizard in Europe and its realized ecological niche model. *Journal of Natural History,* 52(29-30): 1909-1925. https://doi.org/ 10.1080/00222933.2018.1502829.

Costa, J., Peterson, A. T. & Beard, C. B. 2002. Ecologic niche modelling and differentiation of populations of *Triatoma brasiliensis neiva*, 1911, the most important Chagas' disease vector in north-eastern Brazil (hemiptera, reduviidae, triatominae). *American Journal of Tropical Medicine and Hygiene,* 67: 516–20.

Coxen, C. L., Frey, J. K., Carleton, S. A. & Collins, D. P. 2017. Species distribution models for a migratory bird based on citizen science and satellite tracking data. *Global Ecology and Conservation,* 11: 298-311. https://doi.org/10.1016/j.gecco. 2017.08.001.

Croft, S., Ward, A. I., Aegerter, J. N. & Smith, G. C. 2019. Modeling current and potential distributions of mammal species using presence-only data: A case study on British deer. *Ecology and Evolution,* 9(15): 8724-8735. https://doi.org/10.1002/ece3.5424.

D'Addario, M., Monroy-Vilchis, O., Zarco-González, M. M. & Santos-Fita, D. 2019. Potential distribution of *Aquila chrysaetos* in Mexico: Implications for conservation. *Avian Biology Research,* 1–9. https://doi.org/10.1177/17581559 18823424.

Davey, C. M., Chamberlain, D. E., Newson, S. E., Noble, D. G. & Johnston, A. 2012. Rise of the generalists: evidence for climate driven homogenization in avian communities. *Global Ecology and Biogeography,* 21(5): 568–578.

Davies, R. G., Orme, C. D. L., Storch, D., Olson, V. A., Thomas, G. H., Ross, S. G., Ding, T. S., Rasmussen P. C., Bennett, P. M., Owens, I. P. F., Blackburn, T. M. & Gaston, K. J. 2007. Topography, energy and the global distribution of bird species richness. *Proceedings of the Royal society B-Biological Science,* 274(1614): 1189–1197.

De'ath, G. & Fabricius, K. E. 2000. Classification and regression trees: a powerful yet simple technique for ecological data analysis. *Ecology,* 81: 3178–92.

de Frutos, A., Oleab, P. P. & Vera, R. 2007. Analyzing and modelling spatial distribution of summering lesser kestrel: The role of spatial autocorrelation. *Ecological Modelling,* 200: 33–44.

del Hoyo, J., Elliot, A. & Sargatal, J. 1994. *Handbook of the Birds of the World.* Volume 2, New World Vultures to Guineafowl. Lynx Edicions, Barcelona.

de Pando, B. B. & de Giles, J. P. 2007. Aplicación de modelos de distribución de especies a la conservación de la biodiversidad en el sureste de la Península Ibérica. *Geofocus*, 7: 110–119. [Application of species distribution models to the conservation of biodiversity in the southeast of the Iberian Peninsula.]

D'Elia, J., Haig, S. M., Johnson, M., Marcot, B. G. & Young, R. 2015. Activity-specific ecological niche models for planning reintroductions of California condors (*Gymnogyps californianus*). *Biological Conservation,* 184: 90–99.

Delire, C., Ngomanda, A. & Jolly, D. 2008. Possible impacts of 21st century climate on vegetation in Central and West Africa. *Global and Planetary Change*, 64: 3–15.

Dettki, H., Lofstrand, R. & Edenius L. 2003. Modeling Habitat Suitability for Moose in Coastal Northern Sweden: Empirical vs Process oriented Approaches. *Ambio,* 32(8): 549-556. doi: http://dx.doi.org/10.1579/0044-7447-32.8.549.

Devictor, V., Julliard, R., Couvet, D. & Jiguet, F. 2008. Birds are tracking climate warming, but not fast enough. *Proceedings of Royal Society of London B,* 275: 2743–2748.

Dew, R. M., Silva, D. P. & Rehan, S. M. 2019. Range expansion of an already widespread bee under climate change. *Global Ecology and Conservation,* 17: (2019) e00584. https://doi.org/10.1016/j.gecco.2019.e00584.

De Marco, P. J. & Noabrega, C. C. 2018. Evaluating collinearity effects on species distribution models: An approach based on virtual species simulation. *PLoS ONE,* 13(9): e0202403. https://doi.org/10.1371/journal.pone.0202403.

Di Marco, M., Watson, J. E. M., Possingham, H. P. & Venter, O. 2016. Limitations and tradeoffs in the use of species distribution maps for protected area planning. *Journal of Applied Ecology,* 54: 402-411.

Dirzo, R., Young, H. S., Galetti, M., Ceballos, G., Isaac, N. J. B. & Collen, B. 2014. Defaunation in the Anthropocene. *Science,* 345: 401-406.

Donázar, J. A., Cortés-Avizanda, A., Fargallo, J. A., Margalida, A., Moleón, M., Morales Reyes, Z., Moreno-Opo, R., Pérez-García, J. M., Sánchez-Zapata, J. A.,

Zuberogoitia, I. & Serrano, D. 2016. Roles of raptors in a changing world: from flagships to providers of key ecosystem services. *Ardeola,* 63(1): 181-234. https://doi.org/10.13157/arla.63.1.2016.rp8.

Dormann, C. F., Calabrese, J. M., Guillera-Arroita, G., Matechou, E., Bahn, V., Barton, K., Beale, C. M., Ciuti, S., Elith, J., Gerstner, K., Guelat, J., Keil, P., Lahoz-Monfort, J. J., Pollock, L. J., Reineking, B., Roberts, D. R., Schroder, B., Thuiller, W., Warton, D. I., Wintle, B. A., Wood, S. N., Wuest, R. O. & Hartig, F. 2018. Model averaging in ecology: a review of Bayesian, information-theoretic, and tactical approaches for predictive inference. *Ecological Monographs,* 88(4): 485-504. https://doi.org/10.1002/ecm.1309.

Dormann, C. F., Elith, J., Bacher, S., Buchmann, C., Carl, G., Carre, G., Marquez, J. R. G., Gruber, B., Lafourcade, B. & Leitao, P. J. 2013. Collinearity: a review of methods to deal with it and a simulation study evaluating their performance. *Ecography,* 36(1): 27–46. https://doi.org/10.1111/j.1600-0587.2012.07348.x.

Dormann, C. F., Schweiger, O., Arens, P., Augenstein, I., Aviron, St., Bailey, D., Baudry, J., Billeter, R., Bugter, R., Bukacek, R., Burel, F., Cerny, M., De Cock, R., De Blust, G., De Filippi, R., Diekotter, T., Dirksen, J., Durka, W., Edwards, P. J., Frenzel, M., Hamersky, R., Hendrickx, F., Herzog, F., Klotz, St., Koolstra, B., Lausch, A., Le Coeur, D., Liira, J., Maelfait, J. P., Opdam, P., Roubalova, M., Schermann-Legionnet, A., Schermann, N., Schmidt, T., Smulders, M. J. M., Speelmans, M., Simova, P., Verboom, J., van Wingerden, W. & Zobel, M. 2008. Prediction uncertainty of environmental change effects on temperate European biodiversity. *Ecology Letters,* 11: 235–244. https://doi.org/ 10.1111/j.1461-0248.2007.01142.x.

Dunn, J. C., Buchanan, G. M., Cuthbert, R. J., Whittingham, M. J. & Mcgowan, P. J. K. 2015. Mapping the potential distribution of the Critically Endangered Himalayan Quail *Ophrysia superciliosa* using proxy species and species distribution modelling. *Bird Conservation International,* 25(4): 466-478. https://doi.org/10.1017/S095927091400046X.

Elith, J. & M. Burgman. 2002. Predictions and their validation: Rare plants in the Central Highlands, Victoria. In Scott, J. M., Heglund, P. J. & Morrison, M. L. (Eds.), *Predicting species occurrences: issues of scale and accuracy.* Washington, D.C. Island Press.

Elith, J., Burgman, M. & Regan, H. 2002. Mapping epistemic uncertainties and vague concepts in predictions of species distribution. *Ecological Modeling,* 157: 313-329.

Elith, J., Graham, C. H., Anderson, R. P., Dudik, M., Ferrier, S., Guisan, A., Hijmans, R. J., Huettmann, F., Leathwick, J. R., Lehmann, A., Li, J., Lohmann, L. G., Loiselle, B. A., Manion, G., Moritz, C., Nakamura, M., Nakazawa, Y., Mac, M., Overton, J., Peterson, A. T., Phillips, S. J., Richardson, K., Scachetti-Pereira, R., Schapire, R. E., Soberon, J., Williams, S., Wisz, M. S. & Zimmermann, N. E. 2006. Novel methods improve prediction of species' distributions from occurrence data. *Ecography,* 29: 129–151.

Elith, J., Phillips, S. J., Hastie, T., Dudik, M., Chee, Y. E. & Yates, C. J. 2011. A statistical explanation of MaxEnt for ecologists. *Diversity and Distribution,* 17: 43–57.

Elith, J. & Leathwick, J. R. 2009. Species distribution models: ecological explanation and prediction across space and time. *Annual Review of Ecology, Evolution and Systematics,* 40: 677–697. https://doi.org/10.1146/annurev.ecolsys.110308.120159.

Enquist, C. 2002. Predicted regional impacts of climate change on the geographical distribution and diversity of tropical forests in Costa Rica. *Journal of Biogeography,* 29: 519-534.

EPCO. 2015. Impact of Climate Change on Forests and Biodiversity of Madhya Pradesh and Climate change and vulnerability assessment for Madhya Pradesh. Environment Planning and Coordination Department, Madhya Pradesh, India.

Esfanjani, J., Ghorbani, A. & Chahouki, M. A. Z. 2018. MaxEnt modeling for predicting impacts of environmental factors on the potential distribution of *Artemisia aucheri* and *Bromus tomentellus-Festuca ovina* in Iran. *Polish Journal of Environment Studies,* 27(3):1041-1047.

Evans, K. L., Greenwood, J. J. D. & Gaston, K. J. 2005. Dissecting the species-energy relationship. *Proceedings of Royal Society B-Biological Science,* 272(1577): 2155– 2163.

Farashi, A. & Alizadeh-Noughani, M. 2018. Niche modelling of the potential distribution of the Egyptian Vulture *Neophron percnopterus* during summer and winter in Iran, to identify gaps in protected area coverage. *Bird Conservation International,* 29(3): 423-436. https://doi.org/10.1017/S0959270918000278.

Farashi, A. & Erfani, M. 2018. Modeling of habitat suitability of Asiatic black bear (*Ursus thibetanus gedrosianus*) in Iran in future. *Acta Ecologica Sinica,* 38(1): 9–14.

Feng, X., Park, D. S., Walker, C., Peterson, A. T., Merow, C. & Papes, M. 2019. A checklist for maximizing reproducibility of ecological niche models. *Nature Ecology & Evolution,* 3: 1382-1395. https://doi.org/10.1038/s41559-019-0972-5.

Ferguson-Lees, I. J. & Christie, D. A. 2001. *Raptors of the World*. Princeton/ Christopher Helm, London. Princeton University Press.

Ferrier, S. 2002. Mapping spatial pattern in biodiversity for regional conservation planning: Where to from here? *Systematic Biology,* 51: 331-363.

Ficetola, G. F., Thuiller, W. & Miaud, C. 2007. Prediction and validation of the potential global distribution of a problematic alien invasive species—the American bullfrog. *Diversity and Distribution,* 13: 476–485.

Fisher, A., Saniee, K., van der Heide, C., Griffiths, J., Meade, D. & Villablanca, F. 2018. Climatic Niche Model for Overwintering Monarch Butterflies in a Topographically Complex Region of California. *Insects,* 9, 167: 2-17. https://doi.org/10.3390/insects9040167.

Fischer, D., Thomas, S. M., Niemitz, F., Reineking, B. & Beierkuhnlein, C. 2011. Projection of climatic suitability for *Aedes albopictus* Skuse (Culicidae) in Europe under climate change conditions. *Global and Planetery Change,* 78: 54-64.

Flato, G. M., Boer, G. J., Lee, W. G., McFarlane, N. A., Ramsden, D., Reader, M. C. & Weaver, A. J. 2000. The Canadian centre for climate modelling and analysis global coupled model and its climate. *Climate Dynamics,* 16(6): 451–467.

Flint, V. E., Boehme, R. L., Kostin, Y. V. & Kuznetsov, A. A. 1984. *A field guide to birds of the USSR*. Princeton. Princeton University Press.

Fois, M., Cuena-Lombrana, A., Fenu, G. & Bacchetta, G. 2018. Using species distribution models at local scale to guide the search of poorly known species: review, methodological issues and future directions. *Ecological Modeling,* 385: 124–132.

Foody, G. 2008. GIS: biodiversity applications. *Progress in Physical Geography,* 32(2): 223-235.

Forester, B. R., De Chaine, E. G. & Bunn, A. G. 2013. Integrating ensemble species distribution modelling and statistical phylogeography to inform projections of climate change impacts on species distributions. *Diversity and Distributions,* 19: 1480–1495.

Fourcade, Y., Engler, J. O., Rodder, D. & Secondi, J. 2014. Mapping Species Distributions with MAXENT Using a Geographically Biased Sample of Presence Data: A Performance Assessment of Methods for Correcting Sampling Bias. *PLoS ONE,* 9(5): e97122. https://doi.org/10.1371/journal.pone.0097122.

Franklin, J. (2009). *Mapping Species Distributions: Spatial Inference and Prediction*. Cambridge, UK. Cambridge University Press.

Franco, A. M. A. & Sutherland, W. J. 2004. Modelling the foraging habitat selection of lesser kestrels: conservation implications of European Agricultural Policies. *Biological Conservation,* 120: 63–74.

Garcia, K., Lasco, R., Ines, A., Lyon, B. & Pulhin, F. 2013. Predicting geographic distribution and habitat suitability due to climate change of selected threatened forest tree species in the Philippines. *Applied Geography,* 44: 12–22. https://doi.org/10.1016/j.apgeog.2013.07.005.

Garcia, R. A., Burgess, N. D., Cabeza, M., Rahbek, C. & Araujo, M. B. 2012. Exploring consensus in 21st century projections of climatically suitable areas for African vertebrates. *Global Change Biology,* 18: 1253–1269, https://doi.org/ 10.1016/j.apgeog.2013.07.005.x.

Gaudreau, J., Perez, L. & Harati, S. 2018. Towards Modelling Future Trends of Quebec's Boreal Birds' Species Distribution under Climate Change. *International Journal of Geo-Information,* 7: 335. https://doi.org/10.3390/ijgi7090335.

Gibson, L. A., Wilson, B. A., Cahill, D. M. & Hill, J. 2004. Spatial prediction of rufous bristlebird habitat in a coastal heathland: a GIS-based approach. *Journal of Applied Ecology*, 41: 213–223. https://doi.org/10.1111/j.0021-8901.2004.00896.x.

Giorgetta, M. A., Jungclaus, J., Reick, C. H., Legutke, S., Bader, J., Böttinger, M., Brovkin, V., Crueger, T., Esch, M., Fieg, K., Glushak, K., Gayler, V., Haak, H., Hollweg, H. D., Ilyina, T., Kinne, S., Kornblueh, L., Matei, D., Mauritsen, T., Mikolajewicz, U., Mueller, W., Notz, D., Pithan, F., Raddatz,T., Rast, S., Redler, R., Roeckner, E., Schmidt, H., Schnur, R., Segschneider, J., Six, K. D., Stockhause, M., Timmreck, C., Wegner, J., Widmann, H., Wieners, K. H.,

Claussen, M., Marotzke, J. & Stevens, B. 2013. Climate and carbon cycle changes from 1850 to 2100 in MPI-ESM simulations for the Coupled Model Intercomparison Project phase 5. *Journal of Advances in Modeling Earth System,* 5: 572–597. https://doi.org/10.1002/jame.20038.

Gleckler, P. J., Taylor, K. E. & Doutriaux, C. 2008. Performance metrics for climate models. *Journal of Geophysical Research,* 113: D06104. https://doi.org/10.1029/2007JD008972.

Gobeyn, S., Mouton, A. M., Cord, A. F., Kaim, A., Volk, M. & Goethals, P. L. M. 2019. Evolutionary algorithms for species distribution modelling: A review in the context of machine learning. *Ecological Modelling*, 392: 179–195. https://doi.org/10.1016/j.ecolmodel.2018.11.013.

Goller, B. & Schueller, G. I. 2011. Investigation of model uncertainties in Bayesian structural model updating. *Journal of Sound and Vibration,* 330: 6122–6136. https://doi.org/10.1016/j.jsv.2011.07.036.

Gordon, C., Cooper, C., Senior, C. A., Banks, H., Gregory, J. M., Johns, T. C., Mitchell, J. F. B. & Wood, R. A. 2000. The simulation of SST, sea ice extents and ocean heat transports in a version of the Hadley Centre coupled model without flux adjustments. *Climate Dynamics,* 16(2-3): 147–168. https://doi.org/10.1007/s003820050010.

Gorel, A. P., Duminil, J., Doucet, J. L. & Fayolle, A. 2019. Ecological niche divergence associated with species and populations differentiation in *Erythrophleum* (Fabaceae, Caesalpinioideae). *Plant Ecology and Evolution,* 152 (1): 41–52. https://doi.org/10.5091/plecevo.2019.1543.

Graf, R. F., Bollmann, K., Suter, W. & Bugmann, H. 2005. The importance of spatial scale in habitat models: Capercaillie in the Swiss Alps. *Landscape Ecology,* 20: 703–717. https://doi.org/10.1007/s10980-005-0063-7.

Graham, M. H. 2003. Confronting multicollinearity in ecological multiple regression. *Ecology,* 84: 2809–2815. https://doi.org/10.1890/02-3114.

Grenouillet, G., Buisson, L., Casajus, N. & Lek, S. 2011. Ensemble modelling of species distribution: the effects of geographical and environmental ranges. *Ecography,* 34: 9– 17. https://doi.org/10.1111/j.1600-0587.2010.06152.x.

Grinnell, J. 1917. The niche-relationships of the California thrasher. *Auk,* 34: 427–433.

Groff, L. A., Marks, S. B. & Hayes, M. P. 2014. Using Ecological Niche Models to Direct Rare Amphibian Surveys: A Case Study Using the Oregon Spotted Frog (*Rana pretiosa*). *Herpetological Conservation and Biology,* 9(2): 354−368. https://doi.org/10.1371/journal.pone.0113246.

Grendár, M. Jr. & Grendár, M. 2001. Maximum entropy: Clearing up mysteries. *Entropy,* 3: 58–63. https://doi.org/10.3390/e3020058.

Guisan, A., Lehmann, A., Ferrier, S., Austin, M., Overton, J. M. C. C., Aspinall, R. & Hastie, T. 2006. Making better bio-geographical predictions of species' distributions. *Journal of Applied Ecology,* 43: 386–392. https://doi.org/10.1111/j.1365-2664.2006.01164.x.

Guisan, A. & Thuiller, W. 2005. Predicting species distribution: Offering more than simple habitat models. *Ecology Letters,* 8: 993–1009. https://doi.org/10.1111/j.1461-0248.2005.00792.x.

Guisan, A., Tingley, R., Baumgartner, J. B., Naujokaitis-Lewis, I., Sutcliffe, P. R., Tulloch, A. I. T., Regan, T. J., Brotons, L., McDonald-Madden, E., MantykaPringle, C., Martin, T. G., Rhodes, J. R., Maggini, R., Setterfield, S. A., Elith, J., Schwartz, M. W., Wintle, B. A., Broennimann, O., Austin, M., Ferrier, S., Kearney, M. R., Possingham, H. P. & Buckley, Y. M. 2013. Predicting species distributions for conservation decisions. *Ecology Letters,* 16: 1424–1435. https://doi.org/10.1111/ele.12189.

Habibzadeh, N. & Ludwig, T. 2019. Ensemble of small models for estimating potential abundance of Caucasian grouse (*Lyrurus mlokosiewiczi*) in Iran. *Ornis Fennica,* 96: 77–89.

Hall, L. S., Krausmann, P. R. & Morrison, M. L. 1997. The Habitat Concept and a Plea for Standard Terminology. *Wildlife Society Bulletin,* 25: 173-182.

Hao, T., Elith, J., Guillera-Arroita, G. & Lahoz-Monfort, J. J. 2019. A review of evidence about use and performance of species distribution modelling ensembles like BIOMOD. *Diversity and Distributions,* 25(5): 839-852. https://doi.org/10.1111/ddi.12892.

Hardlife, Z., David, C., Godfrey, M., Somandla, N. & Proud, M. 2014. The Implications for Loss and Degradation of Wetland Ecosystems on Sustainable Rural Livelihoods: Case of Chingombe Community, Zimbabwe. *Greener Journal of Environmental Management and Public Safety,* 3(2): 043-052. https://doi.org/10.15580/GJEMPS.2014.2.1212131026.

Hattab, T., Garzón-López, C. X., Ewald, M., Skowronek, S., Aerts, R., Horen, H., Brasseur, B., Gallet-Moron, E., Spicher, F., Decocq, G., Feilhauer, H., Honnay, O., Kempeneers, P., Schmidtlein, S., Somers, B., Kerchove, R. V. D., Rocchini, D. & Lenoir, J. 2017. A unified framework to model the potential and realized distributions of invasive species within the invaded range. *Diversity and Distributions*, 23: 806–819. https://doi.org/10.1111/ddi.12566.

Hastie, T. & Tibshirani, R. 1996. Discriminant analysis by Gaussian mixtures. *Journal of the Royal Statistical Society:* Series B (Methodological), 58: 155–76.

Hastie, T., Tibshirani, R. J. & Friedman, J. H. 2009. *The elements of statistical learning: data mining, inference, and prediction.* Second edition. Berlin, Germany. Springer.

Hayhoe, K., Cayan, D., Field, C. B., Frumhoff, P. C., Maurer, E. P., Miller, N. L., Moser, S. C., Schneider, S. H., Cahill, K. N., Cleland, E. E., Dale, L., Drapek, R., Hanemann, R. M., Kalkstein, L., Lenihan, J., Lunch, C. K., Neilson, R. P., Sheridan, S. C. & Verville, J. H. 2004. Emissions pathways, climate change, and impacts on California. *Proceedings of National Academy of Sciences USA,* 101: 12422-12427.

Heikkinen, R. K., Luoto, M., Araujo, M. B., Virkkala, R., Thuiller, W. & Sykess, M. T. 2006. Methods and uncertainties in bioclimatic envelope modelling under climate change. *Progress in Physical Geography,* 30: 1–27. https://doi.org/10.1177/0309133306071957.

Hernandez, P. A., Graham, C. H., Master, L. L. & Albert, D. L. 2006. The effect of sample size and species characteristics on performance of different species distribution modelling methods. *Ecography,* 29: 773–785.

Hernandez-Baz, F., Romo, H., Gonzalez, J. M., Hernandez, M. J. M. & Pastrana, R. G. 2016. Maximum entropy niche-based modeling (Maxent) of potential geographical distribution of *Coreura albicosta* (Lepidoptera: Erebidae: Ctenuchina) in Mexico. *Florida Entomologist,* 99(3): 376-380. https://doi.org/10.1653/024.099.0306.

Helliwell, R. & Chapman, J. 2013. *Use of Species Distribution Models in Conservation Biology.* Report to the Interagency Special Status and Sensitive Species Program USDA Forest Service, Region 6 and USDI Oregon/Washington Bureau of Land Management.

Hijmans, R. J., Cameron, S. E., Parra, J. L., Jones P. G. & Jarvis, A. 2005. Very high resolution interpolated climate surfaces for global land areas. *International Journal of Climatology,* 25: 1965–1978. https://doi.org/10.1002/joc.1276.

Hijmans, R. J. & Graham, C. H. 2006. The ability of climate envelope models to predict the effect of climate change on species distributions. *Global Change Biology,* 12(12): 2272–2281. https://doi.org/10.1111/j.1365-2486.2006.01256.x.

Hirzel, A. & Guisan, A. 2002. Which is the optimal sampling strategy for habitat suitability modelling. *Ecological Modelling,* 157: 331–341. https://doi.org/10.1016/S0304-3800(02)00203-X.

Hirzel, A. H., Hausser, J., Chessel, D. & Perrin, N. 2002. Ecological niche factor analysis: How to compute habitat suitability maps without absence data. *Ecology,* 83: 2027–2036. https://doi.org/10.1890/0012-9658(2002)083[2027:ENFAHT]2.0.CO;2.

Hof, A. R., Jansson, R. & Nilsson, C. 2012. The usefulness of elevation as a predictor variable in species distribution modelling. *Ecological Modelling,* 246: 86– 90. https://doi.org/10.1016/j.ecolmodel.2012.07.028.

Hirzel, A. H. & Le Lay, G. 2008. Habitat suitability modelling and niche theory. *Journal of Applied Ecology,* 45: 1372-1381. https://doi.org/10.1111/j.1365-2664.2008.01524.x.

Hoeting, J. A., Madigan, D., Raftery, A. E. & Volinsky, C. T. 1999. Bayesian model averaging: a tutorial. *Statistical Science,* 14: 382–401.

Holt, R. D. 1990. The microevolutionary consequences of climate change. *Trends in Ecology & Evolution,* 5: 311–315. https://doi.org/10.1016/0169-5347(90)90088-U.

Houston, D. C. 1974. *Breeding of the White-backed and Ruppell's Griffon vultures, Gyps africanus and G. ruppellii.* Ibis, 118: 14-40. https://doi.org/10.1111/j.1474-919X.1976.tb02008.x.

Howard, A. M., Bernardes, S., Nibbelink, N., Biondid, L., Presotto, A., Fragaszy, D. M. & Madden, M. 2012. A maximum entropy model of the bearded capuchin monkey habitat incorporating topography and spectral unmixing analysis. *ISPRS Annals of the Photogrammetry, Remote Sensing and Spatial Information*

Sciences, Volume I-2, 2012 XXII ISPRS Congress, 25 August – 01 September 2012, Melbourne, Australia.

Howes, C. G. 2016. *Power line risk to Cape (Gyps coprotheres) and White-backed (G. africanus) Vultures in southern Africa.* Master dissertation, University of Witwatersrand, Johannesburg, South Africa.

Hu, X. G., Jin, Y., Wang, X. R., Mao, J. F. & Li, Y. 2015. Predicting impacts of future climate change on the distribution of the widespread conifer *Platycladus orientalis*. *PLoS ONE,* 10: e0132326. https://doi.org/10.1371/journal.pone.0132326.

Huntley, B., Collingham, Y. C., Willis, S. G. & Green, R. E. 2008. Potential impacts of climatic change on European breeding birds. *PLoS ONE,* 3: e1439. https://doi.org/10.1371/journal.pone.0001439.

Huntley, B., Green, R. E., Collingham Y. C., Hill, J. K., Willis S. G., Bartlein, P. J., Cramer, W., Hagemeijer, W. J. M., Thomas, C. J. 2004. The performance of models relating species geographical distributions to climate is independent of trophic level. *Ecology Letters,* 7(5): 417-426. http://dx.doi.org/10.1111/j.1461-0248.2004.00598.x.

Hutchinson, G. E. 1957. Concluding remarks. *Cold Spring Harbor Symposia on Quantitative Biology*, 22: 415–427.

Ibanez, I., Clark, J. S., Dietze, M. C. Feeley, K., Hersh, M., LaDeau, S., McBride, A., Welch, N .E., Wolosin, M. S. 2006. Predicting biodiversity change: outside the climate envelope, beyond the species area curve. *Ecology,* 87(8): 1896–1906. https://doi.org/10.1890/0012-9658(2006)87[1896:PBCOTC]2.0.CO;2.

Iannella, M., Cerasoli, F., D'Alessandro, P., Console, G. & Biondi, M. 2018. Coupling GIS spatial analysis and Ensemble Niche Modelling to investigate climate change-related threats to the Sicilian pond turtle *Emys trinacris*, an endangered species from the Mediterranean. *Peer J,* 6: e4969. https://doi.org/10.7717/peerj.4969.

IPCC. 2007. Climate Change 2007: The Physical Science Basis. In Solomon, S., Qin, D., Manning, M., Chen, Z., Marquis, M., Averyt, K. B., Tignor, M. & Miller, H. L. (Eds.), *Contribution of Working Group I to the Fourth Assessment Report of the Intergovernmental Panel on Climate Change*. Cambridge, UK. Cambridge University Press.

IPCC. 2013. Climate Change 2013: The Physical Science Basis. In Stocker, T. F., D. Qin, G. K. Plattner, M. Tignor, S. K. Allen, J. Boschung, A. Nauels, Y. Xia, V. Bex & P. M. Midgley (Eds.), *Contribution of Working Group I to the Fifth Assessment Report of the Intergovernmental Panel on Climate Change*. Cambridge, United Kingdom and New York, NY, USA. Cambridge University Press.

Islam, M. Z., Boug, A., Shehri, A. & da Silva, L. G. 2019. Geographic distribution patterns of melanistic Arabian Wolves, *Canis lupus arabs* (Pocock), in Saudi Arabia (Mammalia: Carnivora). *Zoology in the Middle East,* 65(2): 95-103. http://dx.doi.org/10.1080/09397140.2019.1580931.

Iverson, L. R., Prasad, A. M., Matthews, S. N. & Peters, M. 2008. Estimating potential habitat for 134 eastern US tree species under six climate scenarios. *Forest Ecology and Management* 254: 390–406.

Jafari, A., Mirzaei, R. & Zamani-Ahmadmahmoodi, R. 2016. Species Distribution Modeling of Wild Sheep based on Improving Bias of Occurrence Records and Selecting Appropriate Environmental Predictors using Maxent. *Iranian Journal of Applied Ecology,* 5(15): 39-49. http://dx.doi.org/10.18869/acadpub.ijae.5.15.39.

James, F. C., & McCulloch, C. E. 1990. Multivariate analysis in ecology and systematics: panacea or Pandora's box? *Annual Review of Ecology and Systematics,* 21: 129–166.

Jha, K. K. 2018. Mapping and Management of Vultures in an Indian Stronghold. In Campbell, M. (Ed.), *Geomatics and Conservation Biology*. 45-75. New York, NY: Nova Science Publishers, Inc.

Jha, K. K. & Chaudhry, P. 2018. Unravelling the Complexity of Protected Area Management in Two Developing Countries: Issues of Human Displacement and wildlife Conflict. In Campbell, M. (Ed.), *Geomatics and Conservation Biology.* 257-282. New York, NY: Nova Science Publishers, Inc.

Jha, K.K. & Jha, R. 2021. Study of Vulture Habitat Suitability and Impact of Climate Change in Central India Using MaxEnt. *Journal of Resources and Ecology*. 2021 12(1): 30-42. DOI: 10.5814/j.issn.1674-764x.2021.01.004.

Jiménez-Valverde, A. 2014. Threshold-dependence as a desirable attribute for discrimination assessment: implications for the evaluation of species distribution models. *Biodiversity and Conservation,* 23(2): 369–385. http://dx.doi.org/10.1007/s10531-0130606-1.

Jiménez-Valverde, A., Acevedo, P., Barbosa, A., Lobo, J. & Real, R. 2013. Discrimination capacity in species distribution models depend on the representativeness of the environmental domain. *Global Ecology and Biogeography,* 22(4): *508–516.* http://dx.doi.org/10.1111/geb.12007.

Johnson, J. B. & Omland, K. S. 2004. Model selection in ecology and evolution. *Trends in Ecological Evolution,* 19: 101–108. https://doi.org/10.1016/j.tree.2003.10.013.

Kaliyathan, N. N., Sharma, A. K., Verma, A. K. & Singh, A. 2016. Species Distribution Modeling of *Bergenia ciliata* in Mussoorie Range of Lesser Himalayas. *Annals of Forestry,* 24(2): 149-158. https://doi.org/10.15560/11.4.1702.

Kampichler, C., van Turnhout, C. A. M., Devictor, V. & van der Jeugd, H. P. 2012. Largescale changes in community composition: determining land use and climate change signals. *PLoS One,* 7: e35272. https://doi.org/10.1371/journal.pone.0035272.

Keppel, G., Robinson, T. P., Wardell-Johnson, G. W., Yates, C. J., Niel, K. P. V., Byrne, M. & Schut, A. G. T. 2017. A low-altitude mountain range as an important refugium for two narrow endemics in the Southwest Australian Floristic Region biodiversity hotspot. *Annals of Botany,* 119: 289–300. https://doi.org/10.1093/aob/mcw182.

Keya, Y. Z., Faryadi, S., Yavari, A., Kamali, Y. & Shabani A. A. 2016. Habitat Suitability & Connectivity of Alborz Wild Sheep in the East of Tehran, Iran. *Open Journal of Ecology,* 6: 325-342. http://dx.doi.org/10.4236/oje.2016.66032.

Khafagi, O., Hatab, E. E. & Omar, K. 2012. Ecological Niche Modeling as a Tool for Conservation Planning: Suitable Habitat for *Hypericum sinaicum* in South Sinai, Egypt. *Universal Journal of Environmental Research and Technology,* 2(6): 515-524.

Khanum, R., Mumtaz, A. S. & Kumar S. 2013. Predicting impacts of climate change on medicinal asclepiads of Pakistan using Maxent modelling. *Acta Oecologica,* 49: 2331. https://doi.org/10.1371/journal.pone.0111587.

Khosravi, R., Hemami, M. R., Malekian, M., Flint, A. L. & Flint, L. E. 2016. Maxent modeling for predicting potential distribution of goitered gazelle in central Iran: the effect of extent and grain size on performance of the model. *Turkish Journal Zoology,* 40: 574-585. http://doi.org/10.3906/zoo-1505-38.

King, R. J., Batista-Navarroz, R., Nicolas, M., Hilomenx, V. & Solano, G. 2017. Ecological Niche Modelling Tool for Aquatic Life Population Distribution using Maximum Entropy Model. *Proceedings of 8th International Conference on Information, Intelligence, Systems & Applications.* 1-6. https://ieeexplore.ieee.org/document/8316390.

Knutti, R., Allen, M. R., Friedlingstein, P., Gregory, J. M., Hegerl, G. C., Meehl, G. A., Meinshausen, M., Murphy, J. M., Plattner, G. K., Raper, S. C. B., Stocker, T. F., Stott, P. A., Teng H. & Wigley, T. M. L. 2008. A review of uncertainties in global temperature projections over the twenty-first century. *Journal of Climate,* 21: 2651–2663. https://doi.org/10.1175/2007JCLI2119.1.

Kuhn, M., & Johnson, K. 2013. *Applied predictive modelling.* Berlin, Germany, Springer.

Kumar, S. & Stohlgren, T. J. 2009. Maxent modeling for predicting suitable habitat for threatened and endangered tree *Canacomyrica monticola* in New Caledonia. *Journal of Ecology and Natural Environment,* 1(4): 094-098.

Lawler, J. J. 2009. Climate change adaptation strategies for resource management and conservation planning. *Annals of New York Academy of Sciences,* 1162: 79–98. https://doi.org/10.1111/j.1749-6632.2009.04147.x.

Lawler, J. J., Shafer, S. L., White, D., Kareiva, P. M., Maurer, E. P., Blaustein, A. R. & Bartlein P. J. 2009. Projected climate-induced faunal change in the Western Hemisphere. *Ecology,* 90: 588–597.

Lepcha, D. L., Pradhan, A. & Chhetri, D. R. 2019. Population assessment and species distribution modeling of *Paris polyphylla* in Sikkim Himalaya, India. *Biodiversitas*, 20(5): 1299-1305. https://doi.org/10.13057/biodiv/d200508.

Lewison, R., Oro, D., Godley, B. J., Underhill, L., Bearhop, S., Wilson, R. P., Ainley, D., Arcos, J. M., Boersma, P. D., Borboroglu, P. G., Boulinier, T., Frederiksen, M., Genovart, M., González-Solís, J., Green, J. A., Grémillet, D., Hamer, K. C., Hilton, G. M., Hyrenbach, K. D., Martínez-Abraín, A., Montevecchi, W. A., Phillips, R. A., Ryan, P. G., Sagar, P., Sydeman, W. J., Wanless, S., Watanuki, Y., Weimerskirch, H. & Yorio, P. 2012. Research priorities for seabirds: improving conservation and management in the 21st century. *Endangered Species Research,* 17: 93–121. https://doi.org/10.3354/esr00419.

Li, J., Hilbert, D., Parker, T. & Williams, S. 2009. How do species respond to climate change along an elevation gradient? A case study of the grey-headed robin (*Heteromyias albispecularis*). *Global Change Biology,* 15: 255–26. https://doi.org/10.1111/j.1365-2486.2008.01737.x.

Li, X. & Wang, Y. 2013. Applying various algorithms for species distribution modelling. Integrative *Zoology,* 8(2): 124–135 https://doi.org/10.1111/1749-4877.12000.

Liang, Q., Xu, X., Mao, K., Wang, M., Wang, K., Xi, Z. & Liu, J. 2018. Shifts in plant distributions in response to climate warming in a biodiversity hotspot, the Hengduan Mountains. *Journal of Biogeography,* 5(6): 1334–1344. https://doi.org/10.1111/jbi.13229.

Lindsay, K., Bonan, G. B., Doney, S. C., Hoffman, F. M., Lawrence, D. M., Long, M. C., Mahowald, N. M., Moore J. K., Randerson J. T., Thornton P. E. 2014. Preindustrial-control and twentieth-century carbon cycle experiments with the Earth System Model CESM1 (BGC). *Journal of Climate,* 27: 8981–9005. http://dx.doi.org/10.1175/JCLI-D-12-00565.1.

Liu, L., Zhao, Z., Zhang, Y. & Wu, X. 2017. Using MaxEnt Model to Predict Suitable Habitat Changes for Key Protected Species in Koshi Basin, Central Himalayas. *Journal of Resource Ecology* 8(1): 77-87. http://dx.doi.org/10.5814/j.issn.1674764x.2017.01.010.

Lobo, J., Jiménez-Valverde, A. & Real, R. 2008. AUC: a misleading measure of the performance of predictive distribution models. *Global ecology and Biogeography,* 17: 145-151. https://doi.org/10.1111/j.1466-8238.2007.00358.x.

Loreau, M., Naeem, S., Inchausti, P., Bengtsson, J., Grime, J. P., Hector, A., Hooper, D. U., Huston, M. A., Raffaelli, D., Schmid, B., Tilman, D., Wardle, D. A. 2001. Ecology: biodiversity and ecosystem functioning: current knowledge and future challenges. *Science,* 294(5543): 804–808. https://doi.org/10.1126/science.1064088.

Louppe, V., Leroy, B., Herrel, A. & Veron, G. 2019. Current and future climatic regions favourable for a globally introduced wild carnivore, the raccoon *Procyon lotor. Scientific Reports*, 9: 9174. https://doi.org/10.1038/s41598-019-45713-y.

Lv, W., Li, Z., Wu, X., Ni, W. & Qv, W. 2012. Maximum entropy niche-based modeling (Maxent) of potential geographical distributions of *Lobesia botrana* (Lepidoptera: Tortricidae) in China. In V D. Li and Y. Chen (Eds.), *International Conference on Computer and Computing Technologies in Agriculture: Computer and Computing Technologies in Agriculture* CCTA 2011, Part III, IFIP AICT 370. 239–246, 2012. https://doi.org/10.1007/978-3-642-27275-2_26.

Ma, B. & Sun, J. 2018. Predicting the distribution of Stipa purpurea across the Tibetan Plateau via the MaxEnt model. *BMC Ecology,* 18: 10. https://doi.org/10.1186/s12898018-0165-0.

Madigan, D., & Raftery, A. E. 1994. Model selection and accounting for model uncertainty in graphical models using Occam's window. *Journal of the American*

Statistical Association, 89:1535–1546. https://doi.org/10.1080/01621459.1994.10476894.

Majgaonkar, I., Bowden C. G. R. & Quader, S. 2018. Nesting success and nest-site selection of White-rumped Vultures (*Gyps bengalensis*) in western Maharashtra, India. *Journal of Raptor Research,* 52(4): 431- 442. https://doi.org/10.3356/JRR-17-26.1.

Manning, M. R., Edmonds, J., Emori, S. Grubler, A., Hibbard, K., Joos, F., Kainuma, M., Keeling, R. F., Kram, T., Manning, A. C., Meinshausen, M., Moss, R., Nakicenovic, N., Riahi, K., Rose, S. K., Smith, S., Swart, R. & van Vuuren, D. P. 2010. Misrepresentation of the IPCC CO2 emission scenarios. *Nature Geoscience*, 3: 376– 377. https://doi.org/10.1038/ngeo8 80.

Malcolm, J. R., Liu, C., Neilson, R. P., Hansen, L. & Hannah, L. 2006. Global warming and extinctions of endemic species from biodiversity hotspots. *Conservation Biology,* 20: 538–548. http://dx.doi.org/10.1111/j.1523-1739.2006.00364.x.

Marino, J., Bennett, M., Cossios, D., Iriarte, A., Lucherini, M., Pliscoff, P., Sillero-Zubiri, C., Villalba, L. & Walker, S. 2011. Bioclimatic constraints to Andean cat distribution: a modelling application for rare species. *Diversity and Distributions,* 17: 311-322. https://doi.org/10.1111/j.1472-4642.2011.00744.x.

Marmion, M., Parviainen, M., Luoto, M., Heikkinen, R. K. & Thuiller, W. 2009. Evaluation of consensus methods in predictive species distribution modelling. *Diversity and Distributions,* 15: 59–69. https://doi.org/10.1111/j.1472-4642.2008.00491.x.

Mateo, R. G., De La Estrella, M., Felicísimo, Á. M., Munoz, J. & Guisan, A. 2013. A new spin on a compositionalist predictive modelling framework for conservation planning: a tropical case study in Ecuador. *Biological Conservation,* 160: 150–161. https://doi.org/10.1016/j.biocon.2013.01.014.

Mateo-Tomas, P. & Olea, P. P. 2010. Anticipating Knowledge to Inform Species Management: Predicting Spatially Explicit Habitat Suitability of a Colonial Vulture Spreading Its Range. *PLoS ONE,* 5(8): e12374. https://doi.org/10.1371/journal.pone.0012374.

McCullagh, P. & Nelder, J. A. 1989. *Generalized Linear Models.* London, Chapman and Hall.

McDonald, M. M., Johnson, S. M., Henry, E. R. & Cunneyworth, P. M. K. 2019. Differences between ecological niches in northern and southern populations of Angolan black and white colobus monkeys (*Colobus angolensis palliatus* and *Colobus angolensis sharpei*) throughout Kenya and Tanzania. *American Journal of Primatology,* 81(6): 2019; e22975. pp. 1 of 15. https://doi.org/10.1002/ajp.22975.

Mebs, T. & Schmidt, D. 2006. *Die Greifvögel Europas, Nordafrikas und Vorderasiens.* [*The birds of prey in Europe, North Africa and the Middle East*]. Stuttgart: Kosmos Verlag.

Meinshausen, M., Smith, S. J., Calvin, K., Daniel, J. S., Kainuma, M. L. T., Lamarque, J. F., Matsumoto, K., Montzka, S. A., Raper, S. C. B., Riahi, K.,

Thomson, A., Velders, G. J. M. & van Vuuren, D. P. P. 2011. The RCP greenhouse gas concentrations and their extensions from 1765 to 2300. *Climate Change,* 109: 213–241. https://doi.org/10.1007/s10584-011-0156-z.

Menon, S., Choudhury, B. I., Khan, M. L. & Peterson, A. T. 2010. Ecological niche modelling and local knowledge predict new populations of *Gymnocladus assamicus* a critically endangered tree species. *Endangered Species Research,* 11(2): 175-181. https://doi.org/10.3354/esr00275.

Merow, C., Smith, M. J. & Silander J. A. Jr. 2013. A practical guide to Maxent for modeling species' distributions: what it does, and why inputs and settings matter. *Ecography,* 36: 1058-1069. https://doi.org/10.1111/j.1600-0587.2013.07872.x.

Meyburg, B. U., Gallardo, M., Meyburg, C. & Dimitrova, E. 2004. Migrations and sejourn in Africa of Egyptian vultures (*Neophron percnopterus*) tracked by satellite. *Journal of Ornithology,* 145: 273–280. https://doi.org/10.1007/s10336-004-0037-6.

Mi, C., Falk, H. & Guo, Y. 2016. Climate envelope predictions indicate an enlarged suitable wintering distribution for Great Bustards (*Otis tarda dybowskii*) in China for the 21st century. *Peer J,* 4: e1630. https://doi.org/10.7717/peerj.1630.

Mika, A. M., Weiss, R. M., Olfert, O., Hallett, R. H. & Newman, J. A. 2008. Will climate change be beneficial or detrimental to the invasive swede midge in North America? Contrasting predictions using climate projections from different general circulation models. *Global Change Biology,* 14: 1721–1733. http://hdl.handle.net/10214/2002.

Miller-Rushing, A. J., Primack, R. B. & Sekercioglu, C. H. 2010. Conservation consequences of climate change for birds. In Møller AP, Fiedler W, Berthold P. (Eds.), *Effects of Climate Change on Birds.* New York. NY: Oxford University Press Inc.

Minoli, I., Cacciali, P., Morando, M. & Avila, L. J. 2018. Predicting spatial and temporal effects of climate change on the South American lizard genus *Teius* (Squamata: Teiidae). *Amphibia-Reptilia*, 40(3): 313-326. https://doi.org/10.1163/15685381-20181070.

Mirzaei, R., Hemami, M. R., Sari, A. E., Rezaei, H. R. & Peterson, A. T. 2017. Using ecological niche modeling to determine avian richness Hotspots. *Global Journal of Environment Science Management,* 3(2): 131-142. https://doi.org/10.22034/gjesm.2017.03.02.002.

Mohammadi, S., Ebrahimi, E., Shahriari, M. M. & Bosso, L. 2019. Modelling current and future potential distributions of two desert jerboas under climate change in Iran. *Ecological Informatics,* 52: 7–13. https://doi.org/10.1016/j.ecoinf.2019.04.003.

Moller, A. P., Fiedler, W. & Berthold, P. 2010. *Effect of Climate Change on Birds.* New York, NY: Oxford University Press.

Monahan, W. B. 2009. A mechanistic niche model for measuring species' distributional responses to seasonal temperature gradients. *PLoS One*, 4: e7921. https://doi.org/10.1371/journal.pone. 0007921.

Morales, N., Fernández, I. C. & Baca-González, V. 2017. MaxEnt's parameter configuration and small samples: are we paying attention to recommendations? A systematic review. *Peer J,* 5: e3093. https://doi.org/10.7717/peerj.3093.

Morin, X. & Lechowicz, M. J. 2008. Contemporary perspectives on the niche that can improve models of species range shifts under climate change. *Biology Letters,* 4: 573– 576. https://doi.org/10.1098/rsbl.2008.0181.

Moss, R., Babiker, W., Brinkman, S., Calvo, E., Carter, T., Edmonds, J., Elgizouli, I., Emori, S., Erda, L., Hibbard, K., Jones R., Kainuma, M., Kelleher, J., Lamarque, J. F., Manning, M., Matthews, B., Meyer, J. M. L., Mitchell, J., Nakicenovic, N., O'Neill, B., Pichs, R., Riahi, K., Rose, S., Runci, P., Stouffer, R., van Vuuren, D., Weyant, J., Wilbanks, T., van Ypersele, J. P. & Zurek, M. 2008. Towards New *Scenarios for Analysis of Emissions: Climate Change, Impacts, and Response Strategies*; IPCC Expert Meeting Report; IPCC: Geneva, Switzerland.

Mouchet, M., Levers, C., Zupan, L., Kuemmerle, T., Plutzar, C., Erb, K., Lavorel, S., Thuiller, W. & Haberl, H. 2015. Testing the Effectiveness of Environmental Variables to Explain European Terrestrial Vertebrate Species Richness across Biogeographical Scales. *PLoS ONE* 10(7): e0131924. https://doi.org/10.1371/journal.pone.0131924.

Muscarella, R., Galante, P.J., Soley-Guardia, M., Boria, R. A., Kass, J. M., Uriarte, M. & Anderson, R. P. 2014. ENMeval: An R package for conducting spatially independent evaluations and estimating optimal model complexity for Maxent ecological niche models. *Methods in Ecology and Evolution,* 5(11): 1198–1205. https://doi.org/10.1111/2041-210X.12261.

Mwakapeje, E. R., Ndimuligo, S. A., Mosomtai, G., Ayebare, S., Nyakarahuka, L., Nonga, H. E., Mdegela, R. H. & Skjerve, E. 2019. Ecological niche modeling as a tool for prediction of the potential geographic distribution of Bacillus anthracis spores in Tanzania. *International Journal of Infectious Diseases,* 79: 142–151. https://doi.org/10.1016/j.ijid.2018.11.367.

Nandi, J. 2019. India's forest cover loss in 17 years is four times the size of Goa. *Hindustan Times*, New Delhi Apr 26, 2019. Retrieved from https://www.hindu stantimes.com/india-news/indias-forest-cover-loss-in-17-years-is-four-times-the-size-of-goa/story.IY2OpSPL A7kRutBy8C XhyN.html.

Ndayishimiye, J., Greve, M., Stoffelen, P., Bigendako, M. J., De Canniere, C., Svenning, J. & Bogaert, J. 2012. Modelling the spatial distribution of endemic Caesalpinioideae in Central Africa, a contribution to the evaluation of actual protected areas in the region. *International Journal of Biodiversity and Conservation,* 4: 118–129. https://doi.org/10.5897/IJBC11.150.

Ng, W. T., Silva, A. C. O., Rima, P., Atzberger, C. & Immitzer, M. 2018. Ensemble approach for potential habitat mapping of invasive *Prosopis* spp. in Turkana, Kenya. *Ecology and Evolution,* 8(23): 11921-11931. https://doi.org/10.1002/ece3.4649.

Nursamsi, I., Partasasmita, R., Cundaningsih, N. & Ramadhani, H. S. 2018. Modeling the predicted suitable habitat distribution of Javan hawk eagle *Nisaetus bartelsi* in the Java Island, Indonesia. *Biodiversistas,* 19(4): 1539-1551. https://doi.org/10.13057/biodiv/d190447.

Oney, B., Reineking, B., O'Neill, G. & Kreyling, J. 2013. Intraspecific variation buffers projected climate change impacts on *Pinus contorta. Ecology and Evolution,* 3: 437– 449. https://doi.org/10.1002/ece3.426.

Oppel, S., Iankov, P., Mumun, S., Gerdzhikov, G., Iliev, M., Isfendiyaroglu, S., Yeniyurt, C. & Tabur, E. 2014. Identification of the best sites around the gulf of Iskenderun, Turkey, for monitoring the autumn migration of Egyptian Vultures and other diurnal raptors. *Sandgrouse,* 36: 240–249.

Ovadia, Y., Fertig, E., Ren, J., Nado, Z., Sculley, D., Nowozin, S., Dillon, J. V., Lakshminarayanan, B. & Snoek, J. 2019. *Can You Trust Your Model's Uncertainty? Evaluating Predictive Uncertainty Under Dataset Shift. 33rd Conference on Neural Information Processing Systems* (NeurIPS 2019), Vancouver, Canada. Accessed 24 September, 2019. https://arxiv.org/pdf/1906.02530.pdf.

Parmesan, C. 2006. Ecological and evolutionary responses to recent climate change. *Annual Review of Ecology, Evolution, and Systematics,* 37(1): 637–669. https://doi.org/10.1146/annurev.ecolsys.37.091305.110100.

Pearson, R. G. & Dawson, T. P. 2003. Predicting the impacts of climate change on the distribution of species: Are bioclimate envelope models useful? *Global Ecology and Biogeography,* 12(5): 361–371. https://doi.org/10.1046/j.1466-822X.2003.00042.x.

Pearson, R. G., Raxworthy, C. J., Nakamura, M. & Peterson, A. T. 2007. Predicting species distributions from small numbers of occurrence records: a test case using cryptic geckos in Madagascar. *Journal of Biogeography,* 34: 102–117. https://doi.org/10.1046/j.1466-822X.2003.00042.x.

Pearson, R. G., Thuiller, W., Araujo, M. B., Marttinen, P., Brotons, L., McClean, C., Miles, L., Seguardo, P., Dawson, T. P. & Lees. D. C. 2006. Model-based uncertainty in species range prediction. *Journal of Biogeography,* 33: 1704–1711. https://doi.org/10.1111/j.1365-2699.2006.01460.x.

Pereira, R. S. 2002. Desktop GARP. Kansas: University of Kansas Center for Research. Available from http://www.nhm.ku.edu/desktopgarp/index.html.

Persechino, A., Mignot, J., Swingedouw, D., Labetoulle, S. & Guilyardi, E. 2013. Decadal predictability of the Atlantic meridional overturning circulation and climate in the IPSL-CM5A-LR model. *Climate Dynamics,* 40: 2359–2380. https://doi.org/10.1007/s00382-012-1466-1.

Peterson, A., Papeş, M. & Soberon, J. 2008. Rethinking receiver operating characteristic analysis applications in ecological niche modeling. *Ecological Modelling* 213(1): 63– 72. https://doi.org/10.1016/j.ecolmodel.2007.11.008.

Peterson, A. T. 2003a. Predicting the geography of species' invasions via ecological niche modeling. *Quarterly Review of Biology,* 78: 419-433. https://doi.org/10.1086/378926.

Peterson, A. T. 2003b. Projected climate change effects on Rocky Mountain and Great Plains birds: generalities of biodiversity consequences. *Global Change Biology,* 9: 647–655. https://doi.org/10.1046/j.1365-2486.2003.00616.x.

Peterson, A. T. 2006. Uses and requirements of ecological niche models and related distributional models. *Biodiversity Informatics,* 3: 59–72. https://doi.org/10.17161/bi.v3i0.29.

Peterson, A. T., Soberón, J., Pearson, R. G., Anderson, R. P., Martínez-Meyer, E., Nakamura, M. & Araújo, M. B. 2011. *Ecological Niches and Geographic Distributions, Monographs in Population Biology* 49. Princeton, NJ: Princeton University Press.

Peterson, A. T., Soberon, J. & Sanchez-Cordero, V. 1999. Conservatism of ecological niches in evolutionary time. *Science,* 285: 1265–1267. https://doi.org/10.1126/
science. 285.5431.1265.

Phillips, S. J., Anderson, R. P. & Schapire, R. E. 2006. Maximum entropy modeling of species geographic distributions. *Ecological Modelling,* 190: 231–259. https://doi.org/10.1016/ j.ecolmodel.2005.03.026.

Phillips, S. J., Dudík, M. & Schapire, R. E. 2004. A maximum entropy approach to species distribution modeling. In Greiner, R. & Schuurmans, D. (Eds.). *Proceedings of the 21st International Conference on Machine Learning.* 655–662. Alberta, Canada.

Phillips, S. J. & Dudik, M. 2008. Modeling of species distributions with Maxent: New extensions and a comprehensive evaluation. *Ecography,* 31: 161–175. https://doi.org/10.1111/j.0906-7590.2008.5203.x.

Phillips, T. J. & Gleckler, P. J. 2006. Evaluation of continental precipitation in 20th century climate simulations: the utility of multimodel statistics. *Water Resources Research,* 42: W03202. https://agupubs.onlinelibrary.wiley.com/doi/pdf/10.1029/2005WR004313.

Phipps, W. L., Diekmann, M., MacTavish, L. M., Mendelsohn, J. M., Naidoo, V., Wolter, K. & Yarnell, R. W. 2017. Due South: A first assessment of the potential impacts of climate change on Cape vulture occurrence. *Biological Conservation,* 210: 16–25. http://doi.org/10.1016/j.biocon.2017.03.028.

Pierce, D. W., Barnett, T. P., Santer, B. D. & Gleckler, P. J. 2009. Selecting global climate models for regional climate change studies. *Proceedings of the National Academy of Sciences,* 106: 8441–8446. https://doi.org/10.1073/pnas.090009 4106.

Pires, M. M., Perico, E., Renner, S. & Sahlen, G. 2018. Predicting the effects of future climate change on the distribution of an endemic damselfly (Odonata, Coenagrionidae) in subtropical South American grasslands. *Journal of Insect Conservation,* 22: 303-319. https://doi.org/10.1007/s10841-018-0063-y.

Prakash, V., Green, R. E., Pain, D. J., Ranade, S. P., Saravanan, S., Prakash, N., R. Venkitachalam, Cuthbert, R., Rahmani, A. R. & Cunningham, A. A. 2007. Recent changes in populations of resident *Gyps* vultures in India. *Journal of the Bombay*

Natural History Society, 104(2): 129-135. https://doi.org/10.1017/S0959270 917000545.

Preston, K., Rotenberry, J. T., Redak, R. A. & Allen, M. F. 2008. Habitat shifts of endangered species under altered climate conditions: importance of biotic interactions. *Global Change Biology,* 14: 2501–2515. https://doi.org/10.1111/ j.1365-2486.2008.01671.x.

Qin, A., Liu, B., Guo, Q., Bussmann, R. W., Ma, F., Jian, Z., Xu, G. & Pei, S. 2017. Maxent modeling for predicting impacts of climate change on the potential distribution of *Thuja sutchuenensis* Franch., an extremely endangered conifer from southwestern China. *Global Ecology and Conservation,* 10: 139–146. https://doi.org/10.1017/S00218596190006 86.

Rahbek, C., Gotelli, N. J., Colwell, R. K., Entsminger, G. L., Rangel, T. F. L. V. B. & Graves, G. R. 2007. Predicting continental-scale patterns of bird species richness with spatially explicit models. *Proceedings of Royal Society B-Biology Science,* 274(1607): 165-174. https://doi.org/10.1098/rspb.2006.3700.

Rangel, T. F. L. V. B., Diniz-Filho, J. A. F. & Bini, L. M. 2006. Towards an integrated computational tool for spatial analysis in macroecology and biogeography. *Global Ecology and Biogeography,* 15: 321–327. https://doi.org/10.1111/j.1466-822X.2006.00237.x.

Raxworthy, C. J., Ingram, C. M., Rabibisoa, N. & Pearson, R. G. 2007. Applications of Ecological Niche Modeling for Species Delimitation: A Review and Empirical Evaluation Using Day Geckos (*Phelsuma*) from Madagascar. *System Biology,* 56(6): 907–923. DOI: https://doi.org/10.1080/10635150701775111.

Ray, D., Behera, M. D. & Jacob, J. 2017. Evaluating Ecological Niche Models: A Comparison between Maxent and GARP for Predicting Distribution of *Hevea brasiliensis* in India. Proceedings of National Academy of Science, India, Sect. B *Biological Science,* 88(3). https://doi.org/10.1007/s40011-017-0869-5.

Ray, R., Gururaja, K. V. & Ramchandra, T. V. 2011. Predictive distribution modeling for rare Himalayan medicinal plant *Berberis aristata* DC. *Journal of Environmental Biology,* 32(6): 725-730.

Reichler, T. & Kim, J. 2008. How well do coupled models simulate today's climate? *Bulletin of the American Meteorological Society,* 89: 303–311. https://doi.org/10. 1175/BAMS-89-3- 303.

Rodríguez, J. P., Brotons, L., Bustamante, J. & Seoane, J. 2007. The application of predictive modelling of species distribution to biodiversity conservation. *Diversity and Distributions,* 13: 243–251. https://doi.org/10.1111/j.1472-4642.2007. 00356.x.

Rosenzweig, C., Karoly, D., Vicarelli, M., Neofotis, P., Wu, Q., Casassa, G., Menzel, A., Root, T. L., Estrella, N., Seguin, B., Tryjanowski, P., Liu, C., Rawlins, S. & Imeson, A. 2008. Attributing physical and biological impacts to anthropogenic climate change. *Nature,* 453(7193): 353–357. https://doi.org/10.1038/ nature06937.

Rustad, L., Campbell, J., Dukes, J. S., Huntington, T., Lambert, K. F., Mohan, J., & Rodenhouse, N. 2009. Changing Climate, Changing Forests: The Impacts of

Climate Change on Forests of the Northeastern United States and Eastern Canada. *Canadian Journal of Forest Research,* 39 (2): Retrieved from https://www.nrs.fs.fed.us/pubs/gtr/gtr_nrs99.pdf

Salas, E. A. L., Valdez, R., Michel, S. & Boykin, K. G. 2018. Habitat assessment of Marco Polo sheep (*Ovis ammon polii*) in eastern Tajikistan: modeling the effects of climate change. *Ecology and Evolution,* 8(10): 5124–5138. https://doi.org/10.1002/ece3.4103.

Sanchez-Cordero, V. & Martínez-Meyer, E. 2000. Museum specimen data predict crop damage by tropical rodents. *Proceedings of the National Academy of Sciences USA,* 97: 7074-7077. https://doi.org/10.1073/pnas.110489897.

Sanjerehei, M. M. & Rundel, P. W. 2017. The impact of climate change on habitat suitability for *Artemisia sieberi* and *Artemisia aucheri* (Asteraceae) – a modeling approach. *Polish Journal of Ecology,* 65: 97–109. https://doi.org/10.3161/15052249PJE2017.65.1.009.

Sargent, R. G. 2011. Verification and Validation of Simulation Models. In Jain, S., Creasey, R. R., Himmelspach, J., White, K. P. & M. Fu (Eds.), *Proceedings of the 2011 Winter Simulation Conference. IEEE.* Retrieved from https://ieeexplore.ieee.org/document/6147750.

Sarikaya, O., Karaceylan, I. B. & Sen, I. 2018. Maximum Entropy Modeling (Maxent) of Current and Future Distributions of *Ips mannsfeldi* (Wachtl, 1879) (Curculionidae: Scolytinae) in Turkey. *Applied Ecology and Environmental Research,* 16(3): 2527 - 2535. https://doi.org/10.15666/aeer/1603_25272535.

Schorr, G., Holstein, N., Pearman, P. B., Guisan, A. & Kadereit J. W. 2012. Integrating species distribution models (SDMs) and phylogeography for two species of Alpine Primula. *Ecology and Evolution,* 2(6): 1260–1277. https://doi.org/10.1002/ece3.100.

Scoccimarro, E., Gualdi, S., Bellucci, A., Sanna, A., Fogli, P. G., Manzini, E., Vichi, M., Oddo, P. & Navarra, A. 2011. Effects of tropical cyclones on ocean heat transport in a high-resolution coupled general circulation model. *Journal of Climate,* 24: 4368–4384. https://doi.org/10.1002/ece3.100.

Scott, J. M., Heglund, P. J. & Morrison, M. L. 2002. *Predicting species occurrences: issues of accuracy and scale*. Washington, D.C., Island.

Sekercioglu, C. H. 2007. Conservation ecology: area trumps mobility in fragment bird extinctions. *Current Biology* 17: R283–R286. https://doi.org/10.1016/j.cub.2007.02.019.

Sekercioglu, C. H., Schneider, S. H., Fay, J. P. & Loarie, S. R. 2008. Climate change, elevational range shifts, and bird extinctions. *Conservation Biology,* 22: 140–150. https://doi.org/10.1111/j.1523-1739.2007.00852.x.

Sekercioglu, C. H., Primack, R. B. & Wormworth, J. 2012. The effects of climate change on tropical birds. *Biological Conservation* 148: 1–18. https://doi.org/10.1016/j.biocon.2011.10. 019.

Setyawan, A. D., Supriatna, J., Nisyawati, Sutarno, Sugiyarto & Nursamsi, I. 2018. Predicting impacts of future climate change on the distribution of the widespread selaginellas (*Selaginella ciliaris* and *S. plana*) in Southeast Asia. *Biodiversitas,* 19(5): 1960-1977. https://doi.org/10.13057/biodiv/d190548.

Shabani, F., Kumar, L. & Esmaeili, A. 2015. A modelling implementation of climate change on biodegradation of Low-Density Polyethylene (LDPE) by *Aspergillus niger* in soil. *Global Ecology and Conservation,* 4: 388–398. https://doi.org/10.1016/j.gecco.2015.08.003.

Shabani, F., Kumar, L. & Taylor, S. 2012. Climate change impacts on the future distribution of date palms: a modeling exercise using CLIMEX. *PLoS ONE* 7(10): e48021. https://doi.org/10.1371/journal.pone.0048021.

Shabani, F., Kumar, L., Esmaeili, A. & Saremi, H. 2013. Climate change will lead to larger areas of Spain being conducive to date palm cultivation. *Journal of Food, Agriculture & Environment,* 11(3&4): 2441–2446. https://hdl.handle.net/1959.11/14065.

Shabani, F., Kumar, L. & Ahmadi, M. 2018a. Assessing Accuracy Methods of Species Distribution Models: AUC, Specificity, Sensitivity and the True Skill Statistic. *Global Journal of Human-Social Science: B Geography, Geo-Sciences, Environmental Science & Disaster Management* 18 (1): 1.0 - 2018. Accessed December 25, 2019. https://globaljournals.org/GJHSS_Volume18/2Assessing-Accuracy-Methods.pdf.

Shabani, F., Kumar, L., & al Shidi, R. H. S. 2018b. Impacts of climate change on infestations of Dubas bug (*Ommatissus lybicus* Bergevin) on date palms in Oman. *PeerJ,* 6: e5545. https://doi.org/10.7717/peerj.5545.

Sheehan, K. L., Esswein, S. T., Dorr, B. S., Yarrow, G. K. & Johnson, R. J. 2017. Using species distribution models to define nesting habitat of the eastern metapopulation of double-crested cormorants. *Ecology and Evolution,* 7: 409-418. https://doi.org/10.1002/ece3.2620.

Silvertown, J. 2004. Plant coexistence and the niche. *Trends in Ecology and Evolution,* 19: 605–611. https://doi.org/10.1016/j.tree.2004.09.003.

Sloan, S. & Sayer, J. A. 2015. Forest Resources Assessment of 2015 shows positive global trends but forest loss and degradation persist in poor tropical countries. *Forest Ecology and Management,* 352: 134–145. https://doi.org/10.1016/j.foreco.2015.06.013.

Soberon, J. & Peterson, A. T. 2005. Interpretation of models of fundamental ecological niches and species' distributional areas. *Biodiversity Informatics,* 2: 1–10. https://doi.org/10. 17161/bi.v2i0.4.

Songer, M., Delion, M., Biggs, A. & Huang Q. 2012. Modeling Impacts of Climate Change on Giant Panda Habitat. *International Journal of Ecology,* 2012: 108752, 12. https://doi.org/10.1155/2012/108752.

Sonti, S. H. 2015. Application of Geographic Information System (GIS) in Forest Management. *Journal of Geography and National Disaster,* 5: 145. https://doi.org/10.4172/2167-0587.1000145.

Sony, R. K., Sen, S., Kumar, S., Sen, M. & Jayahari, K. M. 2018. Niche models inform the effects of climate change on the endangered Nilgiri Tahr (*Nilgiritragus hylocrius*) populations in the southern Western Ghats, India. *Ecological Engineering,* 120: 355– 363. https://doi.org/10.1016/j.ecoleng.2018.06.017.

Sourabh, P., Thakur, J., Sharma, P., Uniyal, P. L. & Pandey, A. K. 2018. Habitat Distribution Modelling for Reintroduction of Endangered Medicinal Plants - *Ephedra gerardiana, Lilium polyphyllum, Crepidium acuminatum, Pittosporum eriocarpum* and *Skimmia anquetilia* in India. *International Journal of Ecology and Environmental Sciences,* 44(2): 207-216.

Stenseth, N. C., Ottersen, G., Hurrell, J. W. & Belgrano, A. 2005. *Marine Ecosystems and Climate Variation.* Oxford, Oxford University Press.

Stiels, D. & Schidelko, K. 2018. Modeling Avian Distributions and Niches: Insights into Invasions and Speciation in Birds. In Tietze, D. T. (Ed.), *Bird Species, Fascinating Life Sciences*. https://doi.org/10.1007/978-3-319-91689-7_9.

Stocker, T., Qin, D., Plattner, G., Tignor, M., Allen, S., Boschung, J., Nauels, A., Xia, Y., Bex, V. & Midgley, P. 2013. *Climate change 2013: the physical science basis.* Intergovernmental panel on climate change, working group I contribution to the IPCC fifth assessment report (AR5). New York.

Stockwell, D. R. B. and Peters, D. 1999. The GARP modelling system: Problems and solutions to automated spatial prediction. *International Journal of Geographical Information Science,* 13: 143–158. https://doi.org/10.1080/136588199241391.

Stockwell, D. R. B. & Peterson, A. T. 2002. Effects of sample size on accuracy of species distribution models. *Ecological Modelling,* 148: 1–13. https://doi.org/10.1016/S0304-3800(01)00388-X.

Stohlgren, T. J., Ma, P., Kumar, S., Rocca, M., Morisette, J. T., Jarnevich, C. S., & Benson, N. 2010. Ensemble habitat mapping of invasive plant species. *Risk Analysis,* 30: 224– 235. https://doi.org/10.1111/j.1539-6924.2009.01343.x.

Suarez-Seoane, S., Osborne, P. E. & Alonso, J. C. 2002. Large-scale habitat selection by agricultural steppe birds in Spain: identifying species-habitat responses using generalized additive models. *Journal of Applied Ecology,* 39: 755–771. https://doi.org/10.1046/j.1365-2664.2002.00751.x.

Summers, D. M., Bryan, B. A., Crossman, N. D. & Meyer, W. S. 2012. Species vulnerability to climate change: impacts on spatial conservation priorities and species representation. *Global Change Biology,* 18: 2335-2348. https://doi.org/10.1111/j.1365-2486.2012.02700.x.

Tebaldi, C. & Knutti, R. 2007. The use of the multi-model ensemble in probabilistic climate projections. *Philosophical Transactions of the Royal Society A: Mathematical, Physical and Engineering Sciences,* 365: 2053–2075. https://doi.org/10.1098/rsta.2007.2076.

Sutton, W. B., Barrett, K., Moody, A. T., Loftin, C. S., de Maynadier, P. G. & Nanjappa, P. 2015. Predicted changes in climatic niche and climate refugia of conservation priority salamander species in the northeastern United States. *Forests,* 6: 1-26. https://doi.org/10.3390/f6010001.

Swets, K. 1988. Measuring the accuracy of diagnostic systems. *Science,* 240: 1285–1293. https://doi.org/10.1126/science.3287615.

Teeffelen, A. J. A. V., Vos, C. C. & Opdam, P. 2012. Species in a dynamic world: Consequences of habitat network dynamics on conservation planning. *Biological Conservation,* 153: 239–253. https://doi.org/10.1016/j.biocon.2012.05.001.

Terribile, L. C., Olalla-Tarraga, M. A., Diniz-Filho, J. A. F. & Rodríguez, M. A. 2009. Ecological and evolutionary components of body size: geographic variation of venomous snakes at the global scale. *Biological Journal of the Linnean Society,* 98(1): 94-109. https://doi.org/10.1111/j.1095-8312.2009.01262.x.

Thackeray, S., Sparks, T., Frederiksen, M., Burthe, S., Bacon, P., Bell, J., Botham, M., Brereton, T., Bright, P. & Carvalho, L. 2010. Trophic level asynchrony in rates of phenological change for marine, freshwater and terrestrial environments. *Global Change Biology,* 16(12): 3304–3313. https://doi.org/10.1111/j.1365-2486.2010.02165.x.

Thomas, C. D., Cameron, A., Green, R. E., Bakkenes, M., Beaumont, L. J., Collingham, Y. C., Erasmus, B. F. N., De Siqueira, M. F., Grainger, A., Hannah, L., Hughes, L., Huntley, B., Van Jaarsveld, A. S., Midgley, G. F., Miles, L., Ortega-Huerta, M. A., Peterson, A. T., Phillips, O. L. & Williams, S. E. 2004. Extinction risk from climate change. *Nature,* 427: 145–148. http://dx.doi.org/10.1038/nature02121.

Thompson, C. M. & McGarigal, K. 2002. The influence of research scale on bald eagle habitat selection along the lower Hudson River, New York (USA). *Landscape Ecology,* 17: 569–586. https://doi.org/10.1023/A:1021501231182.

Thuiller, W. 2004. Patterns and uncertainties of species' range shifts under climate change. *Global Change Biology,* 10: 2020–2027. https://doi.org/10.1111/j.1365-2486.2004.00859.x.

Thuiller, W., Lavorel, S., Araujo, M. B., Sykes, M. T. & Prentice, I. C. 2005. Climate change threats to plant diversity in Europe. *Proceedings of National Academy of Sciences, USA,* 102: 8245–8250. https://doi.org/10.1073/pnas.0409902102.

Thuiller, W., Midgley, G. F., Rouget, M. & Cowling, R. M. 2006. Predicting patterns of plant species richness in megadiverse South Africa. *Ecography,* 29(5): 733–744. https://doi.org/10.1111/j.0906-7590.2006.04674.x.

Thuiller, W., Lafourcade, B. & Araujo, M. 2009a. *Mod Operating Manual for BIOMOD.* Laboratoire d'Écologie Alpine, Université Joseph Fourier, Grenoble, France.

Thuiller, W., Lafourcade, B., Engler, R., & Araujo, M. B. 2009b. BIOMOD: a platform for ensemble forecasting of species distributions. *Ecography,* 32: 369–373. https://doi.org/10.1111/j.1600-0587.2008.05742.x.

Thuiller, W. & Münkemüller, T. 2010. Habitat suitability modelling. In Møller, A. P., Fiedler, W., Berthold, P. (Eds.), *Effects of Climate Change on Birds.* New York. NY: Oxford University Press Inc.

Trautmann, S. 2018. Climate Change Impacts on Bird Species. In Tietze, D. T. (Ed.), *Bird Species, Fascinating Life Sciences.* https://doi.org/10.1007/978-3-319-91689-7_12.

Trisurat, Y., Shrestha, R. P. & Kjelgren, R. 2011. Plant species vulnerability to climate change in Peninsular Thailand. *Applied Geography,* 31: 1106–1114. https://digitalcommons.usu. edu/cwel_pubs/83.

Tuck. G., Glendining, M. J., Smith, P., House, J. I. & Wattenbach, M. 2006. The potential distribution of bioenergy crops in Europe under present and future climate. *Biomass and Bioenergy,* 30: 183–197. https://doi.org/10.1016/j.biombioe.2005.11.019.

Tyrberg, T. 2010. Avifaunal responses to warm climate: the message from last interglacial. *Records of the Australian Museum,* 62: 193–205.

UNCC. 2018. UN Climate Change News, 01 October 2018. *Wetlands Disappearing Three Times Faster than Forests.* Retrieved from https://unfccc.int/news/wetlands-disappearing-three-timesfaster-than-forests.

Upadhyay, M. 2009. *Making GIS Work in Forest Management.* Pokhara, Nepal: Institute of Forestry.

VanDerWal, J., Murphy, H., Kutt, A. S., Perkins, G., Bateman, B., Perry, J. & Reside A. 2013. Focus on poleward shifts in species' distribution underestimates the fingerprint of climate change. *Nature Climate Change* 3(3): 239–243. https://doi.org/10.1038/nclimate1688.

Venables, W. N. & Ripley, B. D. 2002. *Modern Applied Statistics with S,* 4th ed. Berlin: Springer.

Voous, K. H. 1960. *Atlas van de Europese vogels*. Amsterdam. NL: Elsevier. [*Atlas of European Birds*]

Vuuren, D. P. V., Edmonds, J., Kainuma, M., Kainuma, M., Riahi, K., Thomson, A., Hibbard, K., Hurtt, G. C., Kram, T., Krey, V., Lamarque, J. F., Masui, T., Meinshausen, M., Nakicenovic, N., Smith, S. J. & Rose, S. K. 2009. The representative concentration pathways: An overview. *Climate Change,* 109: 5–31. https://doi.org/10.1007/s10584-011-0148-z.

Walsh, R., Assis, A., Paula, A., Patton, J., Marroig, G., Dawson, T. & Lacey, E. 2016. Morphological and dietary responses of chipmunks to a century of climate change. *Global Change Biology,* 22(9): 3233–3252. https://doi.org/10.1111/gcb.13216.

Warren, D. L., Glor, R. E. & Turelli, M. 2008. Environmental niche equivalency versus conservatism: Quantitative approaches to niche evolution. *Evolution* (N.Y). 62: 2868– 2883. https://doi.org/10.1111/j.1558-5646.2008.00482.x.

Warren, R., VanDerWal, J., Price, J., Welbergen, J. A., Atkinson, I., Ramirez-Villegas, J., Osborn, T. J., Jarvis, A., Shoo, L. P., Williams, S. E. & Lowe, J. 2013. Quantifying the benefit of early climate change mitigation in avoiding biodiversity loss. *Nature Climate Change,* 3: 1–5. https://doi.org/10.1038/nclimate1887.

Watanabe, M., Suzuki, T., O'ishi, R., Komuro, Y., Watanabe, S., Emori, S., Takemura, T., Chikira, M., Ogura, T., Sekiguchi, M., Takata, K., Yamazaki, D., Yokohata, T., Nozawa, T., Hasumi, H., Tatebe, H. & Kimoto, M. 2010. Improved climate simulation by MIROC5: mean states, variability, and climate sensitivity. *Journal of Climate,* 23: 6312–6335. https://doi.org/10.1175/2010JCLI3679.1.

Weyant, J., Azar, C., Kainuma, M., Kejun, J., Nakicenovic, N., Shukla, P. R., La Rovere, E. Yohe, G. 2009. *Report of 2.6 versus 2.9 Watts/m2 RCPP Evaluation Panel*; Integrated Assessment Modeling Consortium: Geneva, Switzerland.

Whitton, F. J. S., Purvis, A., Orme, C. D. L. & Olalla-Tarraga, M. A. 2012. Understanding global patterns in amphibian geographic range size: does Rapoport rule? *Global Ecology & Biogeography,* 21(2); 179–190. https://doi.org/10.1111/j.1466-8238.2011.00660.x.

Wiens, J. A., Stralberg, D., Jongsomjit, D., Howell, C. A. & Snyder, M. A. 2009. Niches, models, and climate change: Assessing the assumptions and uncertainties. *Proceedings of National Academy of Sciences, USA.* 106(2): 19729-36. https://doi.org/10.1073/pnas.0901639106.

Wintle, B. A., McCarthy, M. A., Volinsky, C. T. & Kavanagh, R. P. 2003. The use of Bayesian model averaging to better represent uncertainty in ecological models. *Conservation Biology* 17: 1579–1590. https://doi.org/10.1111/j.1523-1739.2003.00614.x.

Wisz, M. S., Hijmans, R. J., Li, J., Peterson, A. T., Graham, C. H. & Guisan, A. 2008. Effects of sample size on the performance of species distribution models. *Diversity and Distributions* 14: 763–773. https://doi.org/10.1111/j.1472-4642.2008.00482.x.

Yang, X., Kushwaha, S. P. S., Saran, S., Xuc, J. & Roy, P. S. 2013. Maxent modeling for predicting the potential distribution of medicinal plant, *Justicia adhatoda* L. in Lesser Himalayan foothills. *Ecological Engineering,* 51: 83– 87. https://doi.org/10.1016/j.ecoleng.2012.12. 004.

Yee, T. W., Mitchell, N. D. 1991. Generalized additive-models in plant ecology. *Journal of Vegetation Science,* 2: 587–602. https://doi.org/10.2307/3236170.

Young, B. E., Franke, I., Hernandez, P. A., Herzog, S. K., Paniagua, L., Tovar, C. & Valqui, T. 2009. Using Spatial Models to Predict Areas of Endemism and Gaps in the Protection of Andean Slope Birds. *The Auk,* 126(3): 554-565. https://doi.org/10.1525/auk.2009.08155.

Zank, C., Becker, F. G., Abadie, M., Baldo, D., Maneyro, R. & Borges-Martins, M. 2014. Climate Change and the Distribution of Neotropical Red-Bellied Toads (*Melano phryniscus*, Anura, Amphibia): How to Prioritize Species and Populations? *PLoS ONE,* 9(4): e94625. https://doi.org/10.1371/journal.pone.0094625.

Zeng, Q., Zhang, Y., Sun, G., Duo, H., Wen, L. & Lei, G. 2015. Using Species Distribution Model to Estimate the Wintering Population Size of the Endangered Scaly-Sided Merganser in China. *PLoS ONE,* 10(2): e0117307. https://doi.org/10.1371/journal.pone.0117307.

Zhang, J., Jiang, F., Li, G., Qin, W., Li, S., Gao, H., Cai, Z., Lin, G. & Zhang, T. 2019a. Maxent modeling for predicting the spatial distribution of three raptors in the Sanjiangyuan National Park, China. *Ecology and Evolution.* https://doi.org/10.1002/ece3.5243.

Zhang, K., Zhang, Y. & Tao, J. 2019b. Predicting the Potential Distribution of *Paeonia veitchii* (Paeoniaceae) in China by Incorporating Climate Change into a Maxent Model. *Forests,* 10: 190. https://doi.org/10.3390/f10020190.

Zhang, K. L., Yao, L. J., Meng, J. S. & Tao, J. 2018. Maxent modeling for predicting the potential geographical distribution of two peony species under climate change. *Science of the Total Environment,* 634: 1326–1334. https://doi.org/10.1016/j.scitotenv.2018.04.112.

In: Critical Research Techniques in Animal …
Editors: Kaushalendra Kumar Jha et al.
ISBN: 978-1-53619-720-4

Chapter 5

A Symbiosis of Habitat Conservation and Eco-Tourism Development: A Case Study of the Valmiki Tiger Reserve, India

Debashis Debnath *** and *Jaynath Ram***
Sociology and Community Development Faculty,
Indian Institute of Forest Management, Bhopal, India

Abstract

The paper focuses on two important ecosystem related human activity systems: habitat conservation and eco-tourism. The former deals with the processes of preservation, protection and restoration of biodiversity, natural resources, ecosystem or environment for sustainable development. The latter is a conservation tool that could provide an economic alternative to more invasive and extractive activities, enabling greater public awareness of environmental issues. A symbiosis of these two systems makes the basis of environmental services towards sustainable development. To test the effectiveness of this integrated perspective, an empirical-based study was conducted in the Valmiki tiger reserve in Bihar, India through a participatory approach with the applications of sociological methods among the fringe villagers and the visited tourists with purposively random sampling. Despite constraints or hindrances by the technological influences, under the process of globalization, the study values the convergence of the integrated system and highlights various issues of awareness, education, knowledge, Naxalite (social unrest) movement in the local areas, etc. inside the fringe areas and simultaneously reveals the scopes of habitat conservation with people's ethos, religious values and ecotourism activities, promising the livelihood generation due to exosystemic balance.

---

* Corresponding Author's E-mail: ddebnathster@gmail.com.

INTRODUCTION

Habitat is the physical and biological setting in which organisms live and the other components of the environment are encountered (Krebs, 1985; Jones, 1987). Habitat conservation is an important process to preserve, protect and restore the biodiversity, ecosystems or natural resource sustainability for development (Environmental Défense Fund, 1999). Ecotourism is a tool to protect, conserve, restore and manage natural habitat and provides for beneficially socio-economic involvement of conservation areas, such as national parks, wildlife sanctuaries, biosphere reserves, and zoological and botanical gardens (Ceballos-Lascurain, 1987). It has recently emerged as a system for conserving the natural and cultural heritage of a nation that contributes to sustainable development (Stronza and Durham, 2008). Ecotourism is a new form of non-consumptive, educational, and romantic tourism that is adapted to relatively undisturbed and under-visited areas of immense natural beauty with popular cultural and historical importance. Ecotourism may thus be defined as a form of nature-based tourism, defined as "responsible travel to natural areas that conserves the environment and improves the well-being of local people" (TIES, 1990). Natural habitat conservation is essential for a profitable and sustainable tourism business, and the effective of products, money and people. The literature on nature tourism and the environmental impacts of the industry dates back to the late 1970s. Our understanding of these mechanisms, their ecological impacts and our capacity to manage tourism in protected areas lags behind the growth of tourism to protected areas. A rapid growth in nature tourism and tourism to protected areas has coincided with a shift in protected area management strategies towards integrated development.

Valmiki Tiger Reserve (VTR) includes both the Wildlife Sanctuary and Valmiki National Park (VNP) in Bihar, an eastern province of India, which is one of the earlier tiger reserves of the country, being established in 1994, and is spread over an area of 899.38 square kilometres (347.25 square miles), which is 17.4% of the total geographical area of the district West Champaran. In the VTR, ecotourism is based on natural fauna, flora (wildlife animals and forest) and religious indicators which is preferred by local as well as out station visitors. The park is situated at the bank of river Gandak in West Champaran district of North-western Bihar along the Indo-Nepal border. It is the only National park in Bihar. Valmiki Tiger Reserve (VTR) is a part of the Royal Chitwan Tiger Conservation Landscape (Dinerstein et al. 2006, cf from Kumar and Sinha 2016). In contiguity with Chitwan National Park and Parsa Wildlife Reserve of Nepal, it is one of the eleven Level 1 Tiger Conservation Units (TCUs) in the Indian subcontinent, where long term survival of tigers (*Panthera tigris*) is possible. Being an integral part of the TCU, the VTR is crucial for tiger conservation in the whole landscape (Wikramanayke et al. 1998, cf from Kumar and Sinha 2016). As of 2018, there were 47 tigers in the Reserve. Both the National Park and the Tiger reserves encounter problems such as forest degradation, soil erosion and poaching, hunting and human-wildlife conflict and its worldwide issue of the forest, wildlife animals and conflict by the Naxalite and common people (Kideghesho, et al., 2007). This tourism site is situated at the Indo- Nepal border, which influences Naxalism and increases illegal activities. Biodiversity losses are critical in all ecosystems. Many plant and animal species' populations are threatened in the reserve and park, due to shrinkage of forest and declination of geographical disturbance, human pressure. The conservation of natural resources and habitats has not been well developed. On the fringe areas and buffer zone of the park the "*Tharu*" tribal group is predominantly living, who are partly involved in the Naxalite activity. They also participate in illegal cutting, felling, and poaching. Due to their residential proximity to the border area of India and Nepal, these people may migrate to different countries whenever police action takes place (as reported on the 27 January 2019, Indian express and Google News).

Conservation, Tourism and Symbiosis

Ecotourism has become an important economic activity in natural areas around the world. It provides opportunities for visitors to experience powerful manifestations of nature and culture and to learn about the importance of biodiversity conservation and local cultures. At the same time, ecotourism generates income for conservation programs and economic benefits for communities living in rural and remote areas. Ecotourism is not merely an alternative to mass tourism, nor is it the only alternative, it offers sustainable opportunities for economic development to local people. There are some characteristic features describing the symbiosis of habitat conservation and ecotourism. First, ecotourism is primarily nature-based in terms of its attractions and products, with associated cultural influences – contemporary, historic and archaeological – constituting a significant ancillary component. Secondly, ecotourism fosters learning opportunities, partly through the provision of formal product interpretation, but also through the maintenance of conditions, that facilitates informal personal interactions with, and appreciation of the natural environment and associated cultural influences. Third and most contentiously, ecotourism should be managed so that it is conducive to sustainability (Fennell, 1999, 2002).

According to Ceballos-Lascurain (1987), ecotourism is an approach that acknowledges the links between the cultural landscape and the local communities which are directly connected with natural forest, wildlife animals and Non-Timber Forest Products. The socio-cultural life of local communities must be protected, due to the strong engagement between local people and their areas of residence inside or near the forest areas. Some authors have suggested that the term community-based ecotourism (CBET) ventures should be used to distinguish those initiatives which are environmentally sensitive, but which also aim to ensure that members of local communities have a high degree of control over the activities taking place, and a significant proportion of the benefits accrue to them (Liu, 1994 cf from Kumar and Sinha 2016; Ceballos-Lascurain, 1996).

Community-based ecotourism is a form of ecotourism where the local community has substantial control over, and involvement in, its development and management, and a major proportion of the benefits remain within the community (World Wildlife Fund International 2001). As such, this system may require substantial devolution of decision-making power to local social actors. This assumes that such local people are competent and knowledgeable, and their engagement with conservation management and tourism is beneficial for overall outcomes. It has been emphasized that community-based ecotourism should be seen and evaluated as just one tool in achieving this and its role may be to provide a more sustainable form of livelihood for local communities; encourage communities themselves to be more directly involved in conservation; and generate more goodwill towards, and local benefits from, conservation measures such as protected areas (World Wildlife Fund International, 2001).

The idea of tourism in any national park and wildlife sanctuary is based on habitat conservation and ecotourism and the reciprocal relationship between them. The prior planning, local involvement and control measures are perceived to increase ecotourism successfully (Kruger et al., 2005). Ecotourism contributes to nature conservation, socio-economic, environmental benefits, and even to the scientific and public consciousness towards the forests, wildlife species (Wearing and Neil, 1999). Ecotourism often prefers to no consumption of biodiversity and will protect for future preference. For habitat and wildlife conservation many organizations made and directly applied wildlife-based ecotourism to control poaching, hunting was cited to frequently deduce the demographic ratio of wilds (Altman et al., 2016).

In the Protected Area (PA) management, the conservation of flora and fauna is based on the policy-making, planning, and programming necessary for the survival and conservation of all bioresources with the management of keystone species and associate wildlife. The PA is based on an open-access regime with restrictions, so, there is a possibility that there will be damage the bioresources and wildlife, by the people living inside or in the vicinity of the PA. Therefore, in this system, there must be several ways to protect or save the wildlife and forest products from illegal-cutting or -felling of trees and killing of wildlife, by developing livelihood generation and ecosystem services. These stand out for both advanced conservation goals and improve the livelihoods of local people. Ecotourism facilitates the provision of an economic alternative, which promotes public awareness of environmental issues, and increase funds for conservation (Honey 1999, 2008). These benefits have facilitated a current large demand for the ecotourism sector and natural resources protection for sustainable development across the world in sustainable tourism policies. Ecotourism and nature conservation have a mutually beneficial relationship that is creating a sustainable partnership. The enhancement of the ecotourism sector activates competition from the protection of quality and quantity of natural habitat and increasing value of these natural resources and ecotourism industry to drive socio-economic benefits. That could be a useful way of encouraging new forms of business, increasing employment and promoting the conservation of landscapes (Castellani and Sala, 2010). Such a system placing as conservation of biodiversity and major components of ecotourism including local people of the PA, tourists, eco-infrastructure and PAs. Thus, it provides benefits socially, economically, managerially and politically (Hvenegaard, 1994).

The framework presented in Figure 1 shows that the core of sustainable development depends upon the ecotourism development and has a significant role supported by the fringe villagers of the protected area to give input in the habitat conservation through three interfaces viz cultural, ecosystem services and development planning and programs.

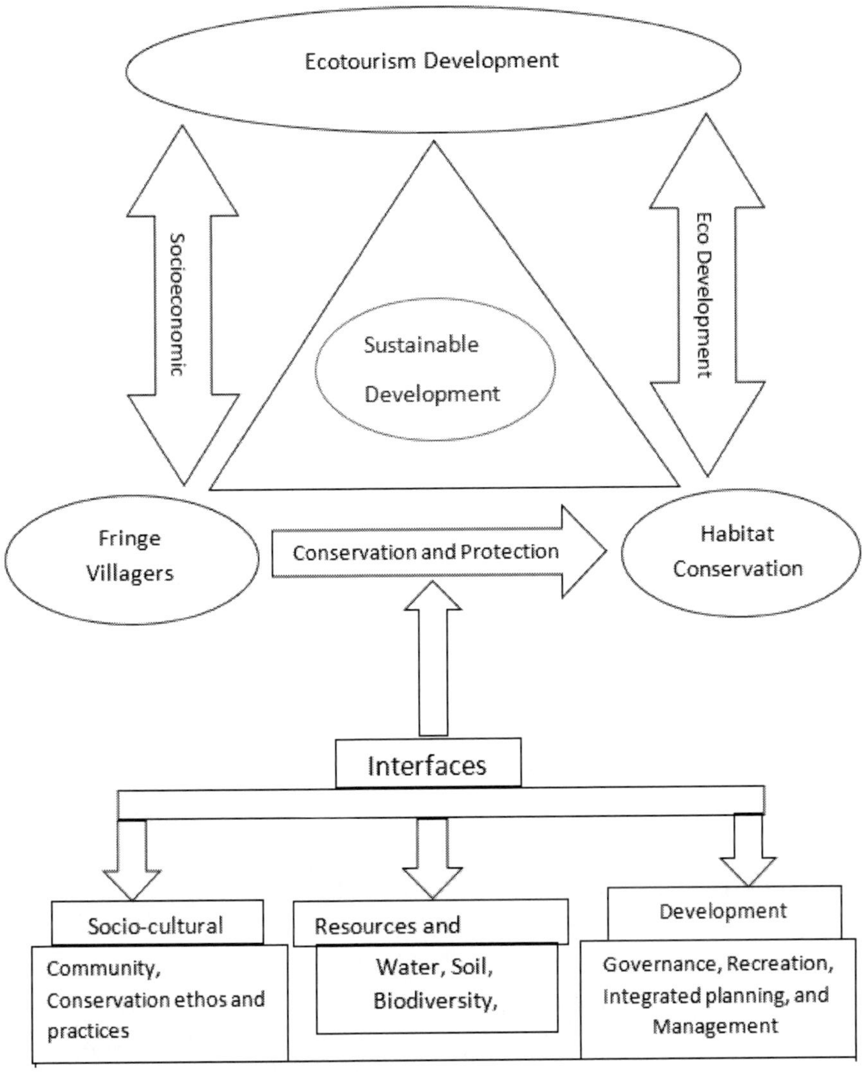

Figure 1. Framework model of sustainable conservation and development of ecotourism with livelihood opportunity.

Objectives of the Study

As already discussed, ecotourism is a strategy based on biodiversity conservation and livelihood generation of the communities with their participation towards sustainable development. The global trend towards "sustainable tourism" to denote the adoption of the principles of sustainable development by the tourism sector, towards creating a balance between social, environmental, cultural and economic contexts. The notion of sustainable tourism, hence, encompasses three dimensions, namely environmental conservation, economic efficiency and social equity.

The key issue is the development of an international network of protected areas, using 'ecotourism' approaches designed to stimulate positive socio-economic change within local communities as fringe villagers to maintain and improve the ecological health of protected areas. The objectives of this study are the following:

(i) To discuss the potentiality of conservation resources having important flora and fauna for commercial, aesthetic and medicinal value in the habitat and socio-cultural features of the fringe villagers at VNP.
(ii) To assess the potentiality of ecotourism and livelihood generation opportunities in the buffer zone and fringe villagers' areas around the park.
(iii) To study the ecotourism activities and interrelationship of habitat conservation.

Approaches and Methods

Study Site

The pristine forest and wilderness of VTR (Figure 2) is an excellent example of the Himalayan Terai landscape. This large, protected area has crisscrossing and meandering rivers, streams and rivulets, man-made canals; swamps and grasslands and identified with a rich variety of fauna and flora and natural beauty, which are maintained for recreational, educational and scientific use, where indigenous species of plant and wild animals protected in the park (Singh 2007). There are different types of major habitats viz. terrestrial and aquatic (grassland, forest cover, estuary, nest on the tree in VNP. Despite this, there are several pitfalls of habitat conservation and management in the park (Ram, 2018). Important factors include activities such as poaching, hunting, gathering, fuelwood collection, tree cutting or felling, and consequent soil erosion, and forest degradation, caused by common people and Naxalites.

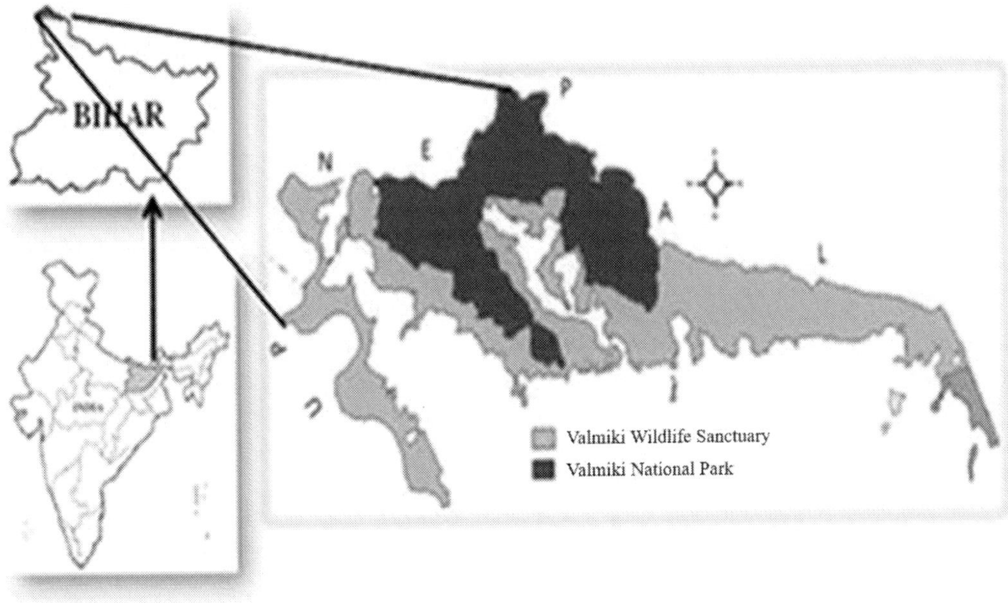

Figure 2. Location and map of study area (Valmiki Tiger Reserve). The area differentiated in dark green is National Park and light green is Wildlife Sanctuary.

STUDY METHODS

Selection of Villages

To generate qualitative and quantitative data, village was taken as a sampling unit. Altogether, four villages in two clusters were selected for quantitative and qualitative data to explore eco-tourism potential of the area. The villages were selected in two clusters; a. near Valmikinagar, where tourism activities are prominent, and b. in Done valley, which is a remote area with scope of nature and culture-based tourism. These villages are Bhediyari, Valmiki Nagar, Kotraha and Lakshmipur. Mostly they are "*Tharu*" community-based villages and close to Valmiki Nagar, where tourism activities are sound. Seventy households were purposively random sampled in these villages to collect data and information through schedules and questionnaire survey for habitat conservation and protection of fauna and flora. In addition to these, thirty-four respondents (26 tourists, 5 IFS trainee, 2 ACF trainee, 1 High Court Chief Judge, and 2 DFO) were selected in the protected areas.

Tools and Techniques

Various tools and techniques were used for data collection in the VTR for habitat conservation, species composition, species diversity and protection and management of biodiversity. The techniques included questionnaires, both structured and semi-structured, schedules, interviews with tourists, forest guards, tourist guides, and fringe villagers, stakeholders and managers of the ecotourism department, etc. for qualitative and quantitative data collection. The participatory tools were transect or forest walks and focus group discussion (FGD) in the villages with the residents keeping in view the structured questions with the main focus on the assessment of community perceptions, attitude, understanding and knowledge on habitat conservation and ecotourism (Ram, 2018).

A modified framework was developed on the criteria and indicators for showing the inter-relationship between habitat conservation and ecotourism, based on the approach of Castellani and Sala (2010), as given in Table 1, below.

Table 1. The framework used for assessing habitat conservation and ecotourism in Valmiki national park

Criteria	Indicators
1. Natural resources attraction	(i) Landscape, bhabar (Dun sal forest) & tarai tracts
	(ii) Forests, woodlands
	(iii)) biodiversity, fauna & flora
	(iv) wildlife animals
	(v) ecosystem
	(vi) water bodies, rivers
	(vii) soil
	(viii) grassland
	(ix) wetland
2. Diverse ecotourism activities (socio-culture activities)	(i) natural/environmental tourism
3. Cultural Diversities	(i) socio-cultural tourism and heritage preference
	(ii) adventure tourism
	(iii) some weeds grass (use in hut making and thatching
Others (research, education)	(i) Scientific research,
	(ii) Visitor's education
	(iii) Wildlife photography
	(iv) social awareness

(Source: Ram, 2018)

Observations and Results

Relevant data collected on the landscape, habitat conservation, fauna, flora, social, cultural and religious aspects, ecotourism activities, livelihood, etc. were analysed and presented under following sub-heads. This section has been supplemented with data from secondary sources. Some of the ecotourism related photographs from VTR are presented as Figure 3, 4 and 5).

Habitat Conservation

The local inhabitants of the PA living in the vicinity of VTR, that is in the buffer area, is carrying out the eco-development programmes and are the potential human resources for ecotourism activities. The landscape of VTR encompasses foothills ranges of Himalayan Shivalik with a mosaic of the cliffs, ridges, gorges, hills, streams and valleys, dense forests, open woodlands, grasslands, swamps and riverine fringe. Two rivers flow through the Park - Gandak and Masan river. River Gandak forms the western boundary of the Valmiki wildlife sanctuary. It enters India at Valmiki Nagar, where two rivulets Sonha and Pachnad join it forming a holy confluence '*Triveni*' (means tri-junction), which divide India and Nepal, also knowns as *Narayani*. Triveni has immense value from religious point of view. The Gandak river with the Himalayan range in the distance offers a great view and is a paradise for wild-life photographers. The River Gandak flows with different tributaries viz. Manor, Shapes, Kosil, Haha-Masan, Singha, Shanghai, Bandai, and Rau. This waterbody network plays significant role in vegetation growth and animal life as well as in the lives of local residents.

Fauna

VTR and VNP have diverse habitat spots to shelter wild animals, according to the Zoological Survey of India (ZSI) and World-Wide Fund for Nature (World Wildlife Fund International 2001). The charismatic tiger has good population in the reserve and its survey is conducted every year from November to January. Valmiki wildlife sanctuary is important for the conservation of habitat and wild animals. The major wildlife species found in the forest of VTR are the Bengal tiger (*Panthera tigris tigris*, Linnaeus, 1758), Indian rhinoceros (*Rhinoceros unicornis*, Linnaeus, 1758), Black bear (*Ursus thibetanus*, G. Cuvier, 1823), Indian sloth bear (*Melursus ursinus*, Shaw, 1791), otter (*Lutrogale perspicillata*, Geoffroy Saint Hilaire, 1826), Indian leopard (*Panthera pardus fusca*, Meyer, 1794), Indian wild dog or dhole (*Cuon alpinus alpinus*, Pallas, 1811), Indian bison (Gaur) (*Bos gaurus*, Smith, 1827), wild boar (*Sus scrofa*, Linnaeus, 1758), etc. There are several species of deer and antelopes found in VTR, which form an important part of the prey base for carnivores. These include the barking deer (Genus *Muntiacus*, Rafinesque, 1815), spotted deer or Chital (*Axis axis*, Erxleben, 1777), Indian hog deer (*Hyelaphus porcinus*, Zimmermann, 1780), Sambar (*Rusa unicolor*, Kerr, 1792), and Nilgai or blue bull (*Boselaphus tragocamelus*, Pallas, 1766).

Photo credit: Radhika Jha.

Figure 3. Some images from the study area. Entrance to the ecotourism zone for restricted visit (above) and popular Valmiki crossing with road sign leading towards Revered Valmiki Ashram (below).

Photo credit: Radhika Jha.

Figure 4. Images of the Study Area. Moist deciduous forest (above) and tiger pray base (below) of Valmiki Tiger Reserve.

Other species sighted in VTR include the striped hyena (*Hyaena hyaena*, Linnaeus, 1758), wild cat (*Felis* sp., Linnaeus, 1758), fishing cat (*Prionailurus viverrinus*, Bennett, 1833), langur (Subfamily Colobinae, Jerdon, 1867) flying squirrel (*Pteromyini*, Brandt, 1855), clouded leopard (*Neofelis nebulosa*, Griffith, 1821) and mongoose (*Herpestes edwardsi*, É. Geoffroy Saint-Hilaire, 1818). The tiger population in the Reserve was 10 in 2010, 22 in 2013 and 47 in 2018. Some new species have been introduced into the sanctuary, these including 53 species of mammals, 10 species of amphibians, 27 species of reptiles and 75 species of insects (Chakraborty et al., 1998, cf. from Kumar and Sinha 2016). Other species, such as the Indian flying fox (*Pteropus medius*, Temminck, 1825) and some bats (Order *Chiroptera*, Blumenbach, 1779) have been sighted in the Madanpur forest block on the main road from Madanpur to Valmiki Nagar. Primates such as Rhesus macaque (*Macaca mulatta*, Zimmermann, 1780) and common langur (Genus *Semnopithecus*, Desmarest, 1822), are also found in the sanctuary (Ram, 2018).

The reptiles which are commonly found in VTR are pythons (Family Pythonidae, Fitzinger, 1826), king cobra (*Ophiophagus hannah*, Cantor, 1836), common krait (*Bungarus caeruleus*, Schneider, 1801), banded kraits (*Bungarus fasciatus*, Schneider, 1801) and domuha snakes (sand boas) (Subfamily *Erycinae*, Bonaparte, 1831). Aquatic reptiles recorded in this area include gharial or fish – eating crocodiles (*Gavialis gangeticus*, Gmelin, 1789) and monitor lizards (Genus *Varanus*, Merrem, 1820) among others. Crocodiles are fairly common near Belahwa village, which is adjacent to the VTR. Gharials are found in River Gandak, as would be expected for a river linked to the Ganges. At present 241 bird species have been reported from VTR. Common bird species of VTR include the Nepal khalij pheasant (*Lophura leucomelanos*, Latham, 1790), three-toed quail (Family *Phasianidae*, Horsfield, 1821), paradise flycatcher (*Terpsiphone paradisi*, Linnaeus, 1758), grey shrike (*Lanius excubitor lahtora*, Linnaeus, 1758), green willow warbler (*Phylloscopus nitidus*, Blyth, 1843), tree pipit (*Anthus trivialis*, Linnaeus, 1758), white eye warbler (*Schistolais leontica*, Bates, 1930), green barbet (*Stactolaema olivacea*, Shelley, 1880), waders (Order *Charadriiformes*, Huxley, 1867), plovers (Subfamily *Charadriinae*, Leach, 1820), snipes (Family *Scolopacidae*, Rafinesque, 1815), ibises (Subfamily Threskiornithinae, Poche, 1904), storks (Family Ciconiidae, J. E. Gray, 1840), pitta (Genera *Pitta*, Vieillot, 1816, *Hydrornis*, Blyth, 1843), pied hornbill (*Anthracoceros albirostris*, Shaw & Nodder, 1807), White-eared night heron (*Gorsachius magnificus*, Ogilvie-Grant, 1899), emerald dove (*Chalcophaps indica*, Linnaeus, 1758). Pigeons (Family *Columbidae*, Leach, 1820) occurring in VTR include five types of green pigeons and purple wood pigeon. In the night several owls, owlets, nightjars, etc. can be easily sighted. The jungle of VTR abounds in various kinds of moths, caterpillars and butterflies (Order *Lepidoptera*, Linnaeus, 1758). Common butterflies like common mormon, great mormon, glassy tiger, great egg fly, club beak, grey pansy, lime butterfly, common crow can be easily sighted. Invertebrates mostly insects, constitute about 75 species. Ray (1998, cited in Kumar & Sinha, 2016) has given a description of 10 species of amphibians in Valmiki Tiger Reserve. Sur et al. (1998, cf. from Kumar and Sinha 2016) recorded 27 species of reptiles while Biswas and Bal (1998, cf. from Kumar and Sinha 2016) reported 35 species of hemipterans in the Reserve.

Photo credit: Radhika Jha.

Figure 5. Wild animal conservation awareness programme displaying importance of flagship wild animal Indian Bison. Vernacular language is used copiously for raising awareness among local visitors.

Flora

As per Champion and Seth (1968) classification, there are seven commonly known forest types found in VTR, which makes VTR a special destination for nature lovers.

- Bhabar – Dun Sal Forest
- Dry Siwalik Sal Forest
- West Gangetic Moist Mixed Deciduous Forest
- Khair – Sissoo Forest
- Cane Brakes
- Eastern Wet Alluvial Grassland
- Barringtonia Swamp Forest

Due to diverse topographical and edaphic factors, the reserve harbours varied vegetation types. The Botanical Survey of India has categorized seven vegetation types within the limits of the sanctuary and the national park:

- Moist mixed deciduous
- Open – land vegetation
- Sub-mountainous semi-evergreen formation
- Freshwater swamps
- Riparian fringes
- Wetlands
- Alluvial grasslands and high hill savannah

The sanctuary has mixed deciduous vegetation in plain areas. The sites are covered up with Rohini (*Mallotus phillipensis*), Sal (*Shorea robusta*) and Sihor (*Tribulus asper*), another natural forest of plantation Shisham (*Dalbergia sissoo*) Semal (*Salmalia malabarica*), Teak (*Tectona grandis*), Bamboo, Karam (*Adina cordifolia*), Asan (*Terminalia tomentosa*), Bahera (*Terminalia balerica*), Asidh (*Lagestromia parviflora*), Satsal (*Dalbergia latifolia*), etc. Singh (2007) also gave a general account of the vegetation pattern of the VNP. About 700 species belonging to 459 genera and 117 families were reported. Apart from general vegetation, endemic, rare, threatened plants had been identified. Singh (2007) has also reported plants having high economic potential, useful as edible (*Curculigo orchioides, Dioscorea alata, Cassia tora, Dendrocalamus strictus,* etc.), timber (*Anogeissus latifolia, Gmelina arborea, Terminalia alata, Toona ciliata,* etc.), dyes (*Acacia catechu, Indigofera astragalina, Wendlandia tinctoria, Woodfordia fruticosa,* etc.), gums and resins (*Anogeissus latifolia, Boswellia serrata, Sterculia urens,* etc.), medicinal (*Abrus precatorius, Andrographis paniculata, Elephantopus scaber, Helicteres isora,* etc.). Singh (2007) also identified some of the common causes for degradation of the natural vegetation in the park and suggested conservation strategies for the same. The observations and results in the present investigation are similar as reported by the above researcher.

Grasslands

Grasslands represent important habitat comprising several species of grasses and sedges like, munj (*Saccharum munja*), Kans (*Saccharum spontaneum*), Elephant grass (*Typha elephantina*), Narkat (*Phragmites karka*), *Vitiveria zizanioides*, *Imperata cylindrica*, Choranth (*Heteropogon contortus*), Sabai (*Eulaliopsis binata*), etc. occupying approx. 5% of the whole geographical area of the reserve. These grasses and sedges control soil erosion in VTR and VNP other than providing fodder to wild animals. The grasslands are spread over 44 km in reserve. These grasslands control and check wildlife animals in the core and buffer zones from out grazing of domestic animals. Due to anthropogenic pressure, there is degradation in the grasslands in the park. A total of 114 species of graminoids (sedges and grasses) were listed in the national park. Climbers are also common in the VTR. The common species being Mahulan (*Bauhinia vahlii*) Mahai (*Butea parviflora*), Panilat (*Vitis repanda*), Ramdatwan (*Smilax parviflora*), and Arar (*Acacia pinnata*). There are several medicinal species found in VTR. Some of them are Satawar (*Asparagus racemosus*), Safed Musli (*Chlorophytum borivilianum*), Dudhkoraiya (*Holarrahena antidysenterica*), Amla (*Emblica officinalis*), Piper (*Piper longum*), etc. In the recent years, grassland management has been brought under effect aiming at wildlife conservation, especially tiger conservation in the VTR.

Socio-Cultural Features

The people of the local areas have cultural traditions and historical background with various unique ethos and cultural tradition. *Tharu* people themselves say that they are the original inhabitants of the forests. They lived in the forests over hundreds of years practicing short fallow shifting cultivation with an understanding of the ecological process. They are embedded with conservation ethos of environment management. The scheduled tribes other than *Tharu* in the Valmiki landscape are collectively called Dhangar – which means retained labourer in Oraon dialect. Dhangar comprises four tribes: Oraon, Munda, Lohra and Bhuiya. Each Dhangar tribe has its own dialect and they observe their traditional festivals. Currently, their population is around 0.5 lakh. The tribal peoples were brought to the area from the Chhotanagpur Hills (Hills south to the Gangetic Plains) as laborers by the British. There are several theories on migration of *Tharus* to the region. The tribes in the area are not forest dwelling and their main occupation is agriculture. These tribal communities often present traditional myths and folktales in close association with natural setting. Under the cultural practices and conservation ethos, they manage the environmental resources and natural habitat. It also makes inextricably close association and intricate understanding of the ecological process. Based on the field experience with these communities, it is revealing the ways they perceive nature and their surrounding with socio-religious beliefs and cultural structures. The communities around the reserve are traditionally dependent on the forests for irrigation, firewood, thatching materials, timber, vegetables and medicinal plants.

Sacred Place (Land), Animals and Plants

These tribal communities, through their conservation ethos and cultural practices protect and preserve several trees like *Ficus religiosa*, *Ficus benghalensis* and *Mangifera indica*, etc., which they also worship, during the *Ram Navami* festival. Besides, they use part of the tree such as roots, flowers, trees and tubers, etc. as ritual articles in the offering. They also worship land and water for the ritual practices during their agricultural festivals or forest rites. These act as conservation practices of natural resources and their protection. It has its own values and importance to socio-cultural, scientific, environmental and economic aspects and so on. It goes to their sacred beliefs and environmental conservation as embedded in their ethos and practices.

Ecotourism Activities

Ecotourism comprises the tour in the natural setting, local people's involvement, education imparted to the tourists and environmental sustainability. In this connection the following ecotourism activities at VTR are listed:

1. Tour activities are nature-based in the park comprising a day trip to a special habitat area with wildlife components.
2. Specialised wildlife tours including safari tours, bird watching, an elephant ride and whale watching tours are conducted.
3. Staying in the park area is facilitated with good accommodation for tourists e.g., resort hotels, farm-stays, etc.
4. For the popularity of the park the tourists are presented with entertainment packages at the guest house and rafting at the Gandaki river and the jungle camp (situated on the bank of Gandak river). Managed locational attractions featuring a natural aggregation of wildlife (e.g., Bird breeding colonies, fish aggregation areas, migratory pathways for birds and mammals including waders and whales, overwintering insect aggregations, glow worms in caves).
5. The tiger reserve is the most religious tourism site where Jatashakar Temple, Nardevi Temple, Valmiki hermitage (most famous religious epic writer Sage Maharshi Valmiki), *Mai* (the goddess) Asthan close to Madanpur, *Jatashankar, Nardevi* and *Kaleshwar* are the temples near Valmiki Nagar. Sofa temple is located near Manguraha.
6. Nepal is near to India by road and by boat. So, the neighbouring inhabitants prefer to visit Valmiki tiger reserve. It is an attraction to close by Nepali tourists. Similarly, Indian tourists are also attracted to Nepal's tourism activities like *Triveni, Barah Ghambha, Shish Mahal* even to Royal National Park Chitwan. It is a holy place of the Hindus. The name is derived from the fact that three rivers Gandak, Sonha and Pachanad merge here soon after they emerge from the hills. Triveni is also believed to be the site at which the fight commenced between the Lords of Forest and Water (the Elephant and Crocodile). *Triveni Magh Mela* (which is a holy dip on an auspicious day, Magh Purnima (the full moon day in the month Magh, a month of local calendar in January-February) at the confluence of three rivers. This religious mela (fair) attracts Indian as well as Nepali tourists and local people during the mela. However, a good number of local pilgrims visit temples and sites of mythological importance. Similarly, during 10 days of *Chaitra Navami* and *Durga Puja* the temples inside the reserve are visited by thousands of devotees.

7. Due to *Magh mela,* fringe villagers and other stakeholders (hotels, guest houses, shopkeepers) benefit from different resource generation for livelihood during this period.
8. Nature-based tours that include wildlife (e.g., National Park tour with game drive, regional protected area tour, day trip to specific habitat areas with wildlife component).

Throughout this religious *mela* (fair), many people visit the VTR. The advantage of the tourism department is revenue gain but disadvantage of *mela* (fair) is environmental pollution on human and animals due to heaps of garbage, plastic bottles, poly bags, throughout Valmiki Nagar, mela (fair) place and on both sides of the road.

In the inquiry of the importance of the VTR, 63.8% of the fringe villagers feel the importance of protected areas, 18.1% have no information about it and the rest 18.1% people are fully aware of ecotourism and its benefits. When they were asked about the impacts of ecotourism, 63.8% know it, but 18.1% do not know about it and 18.1% have clear knowledge about its impacts. On the questions of conservation, 54.5% are aware of conservation, 9.09% have little knowledge about conservation and its benefits, 36.41% do not know about it. This indicates that the local people have to be educated through different awareness programme like the importance of ecotourism, how it could help for employment opportunities and help in the livelihood of local people. How it helps conserve the flora and fauna which is highly required for the bright future of their children.

Linking Ecotourism with Livelihoods

By creating economic incentives for impoverished villagers or their communities, ecotourism is thought to encourage local guardianship of indigenous resources. When asked about the livelihood generation after ecotourism started in the VTR, most respondents agree with livelihood opportunities for the fringe villagers in the reserve. Overall, 70 respondents were interacted, of which 80% (56 respondents) preferred the opportunity to generate livelihoods for local people. 4.28% (3 respondents) preferred conservation, 4.28% (3 respondents) preferred business, 10% (7 respondents) preferred ecotourism activities and 1.42% (1 respondent) preferred to alternative livelihoods like driving, services to hotels, resorts, etc.

According to recorded data that showed total of 27,657 tourists visited in six-year. They contributed, by generating rupees 1,825,453 as revenues through ecotourism activities. Total 11.94% of the local community received livelihood through ecotourism in a different field in the tiger reserve. A total of 20% of the local community remain satisfied and informed that livelihood opportunity increased by ecotourism. About 48% villagers opined that ecotourism activities in their area would help increasing income to the villagers, while 23% said that it would lead to improvement in infrastructure facilities. Some of the respondent also felt that ecotourism would make their village popular.

DISCUSSION

Assessment of Potentiality for Habitat Conservation

The facts above indicate that the VTR, including the VNP and wild-life sanctuary, are locations that receive protection because of their recognized natural, ecological or cultural values. This has the high potentiality of natural setting comprising cliffs, foothills, ridges, rivers, tributaries, waterbodies, streams, dense forests, open woodlands, grasslands, swamps and riverine fringe, that provides a good-view of natural attractions or products, with associated cultural influences – contemporary, historic and archaeological for the tourists for photography. As protected areas are intertwined with cultural artifacts, and their story is entwined with that of human civilization in these protecting places and resources as indigenous communities guarding sacred sites with cultural traditions and heritage. Thus, the habitat conservation is also constituted with cultural diversities of many indigenous people, on the economic or material benefits from nature and consequently humans began to appreciate the value of nature itself and the need to protect it.

Inter-Relationship of Habitat Conservation with Ecotourism

The whole fringe villagers of the tiger reserve are engaged in agriculture, which is the main source of livelihood in the protected area. The Bettiah of West Champaran district, where VTR & VNP are situated, have natural beauty and religious centres that attract and motivate the tourists. This place gradually is growing as an ecotourism centre and advantage of the forest as well as the local community to generate livelihood opportunities. Enlightened external and local people feel that there are negative impacts on the forests as well as on the livelihoods of local residents since the forest area has been declared as PA. Some believe the conservation of the flora and fauna an immediate necessity, but they are few, especially those who are members of the Eco-development committee. This indicates that there is a need of raising awareness among people concerning PA and only then the aim of ecotourism can be achieved.

Assessment of Potentiality of Ecotourism for Livelihood Generation

Today economy of the world has been transformed into the process of globalization with the removal of economic boundary among the nations at the global levels and making "outsourcing and off-shoring" in the productions as an effect of the crisis in capitalists and labour crisis. This economy has been based on production quality, branding multinational and so on. These issues have influenced conflicts and competing actions, enabling and constraining sustainable development. Hence the approach for the process of conservation has come into the consequence to the various efforts of biodiversity conservation & management, livelihood generation of the poor people and sustainable development to combat the climate variability in the forest ecotourism. Ecotourism is such an effort based on biodiversity conservation, livelihood generation of the communities with their participation for sustainable development. Protected areas have good scopes and opportunities to generate livelihoods. These can be initiated based on tour activities comprising day trips to special habitat areas with wildlife components, with special sighting in the wildlife tours including safari, bird watching, elephant riding, etc. along with entertainment packages and visits different fairs distributed throughout the year.

In this regard, the oft-quoted notion that 'wildlife stays because wildlife pays' is important because it suggests, in a more broadly based context, that resources must continue to be made available for those who make their livelihood in and around protected areas – spaces which remain as living, working landscapes for the support of populations in a sustainable fashion (Munro & Willison, 1998).

CONCLUSION

India's protected areas (PAs) comprised important strongholds for biodiversity harbouring a variety of species of flora and fauna, which include ecological, socio-economic, ecotourism and socio-cultural importance. These are major issues and challenges due to lack of awareness, education, knowledge, ecotourism development (strongly with well facilities), sociocultural development, the participation of fringe villagers in protection, preservation and conservation to support habitat conservation. Ecotourism, as alternative tourism, involves visiting natural areas to learn, to study, or to carry out activities environmentally and friendly. It focuses primarily on experiencing and learning about nature, its landscape, flora, fauna and their habitats as well as cultural art and craft from the locality. Ecotourism and habitat conservation generate livelihood and employment for tourist guides, guards, drivers, eco-friends, supporting staff of forest, etc. in developing ecotourism site and forest conservation. This is a better supporting work to nature and PAs that reduce poverty and raise the socioeconomical status of fringe villagers. Conservation and sustainable use of biodiversity are prerequisites for the long-term continuation of forest ecosystem services, and sustainable forest management (SFM) criteria.

Awareness among the tourists along with local residents concerning flora and fauna through environmental education should be provided so that the people could learn more and more about the importance of biodiversity conservation. The park authority has planning and management to conservation and protection, which is a challenge for forest staff without the participation of the local community. In the tiger reserve and buffer areas, local tribal community, the *Tharu*, have a significant role in the participation for protection with their indigenous knowledge on the park ecosystem in collaboration with the forest department and fringe villagers.

The following issues need to be solved by the forest department in collaboration with fringe villagers (see also the following framework, Figure 6 for a summary of future possibilities):

1. Less development of infrastructure at ecotourism site (e. g. roads, guesthouses, parks).
2. Minimum participation of local community in ecotourism at VTR.
3. Less social and mutual relationship between forest staffs and fringe villagers with the forest understanding.
4. Minimal livelihood generation throughout ecotourism in the tiger reserve.
5. There is a lack of proper management of cleanliness, canteen, the road inside of the park, guide, safari-visit.

There is lack of main participation of fringe villagers in habitat protection. But forest staffs try to motivate and increase awareness to local people with collaboration of local NGO. Overall observation and finding of research are necessary to generate livelihood and conserve habitat in a "win-win" situation with sustainable development. In the VTR ecotourism has the potentiality to generate employment in a different part of the PAs, which conserves natural resources and enhances the socio-economic status of fringe villagers in the buffer zone as well.

In that context, it may be said, nature tourism or wildlife tourism in VTR is in infancy. Flow of tourists in the reserve is meagre. Lack of infrastructure for tourist stay and poor condition of roads in the past might be the reason for this. In November 2012, the twelve-bed hotel Valmiki Vihar was renovated and opened for tourists, which has helped attracting local tourists as well as those from other states. The place was infamous for anti-social activities till early 2000 and law-and-order situation was not up to the mark. Due to this reason outsiders did not prefer to visit the place. Due to low abundance and considerable human disturbance, the sightings of wildlife are poor, especially those of tigers. This is another reason why the reserve is not much popular among wildlife tourists (see the framework, Figure 6) which also summarises the present study has been developed as the future perspective of sustainable ecotourism development in VTR (Ram, 2018).

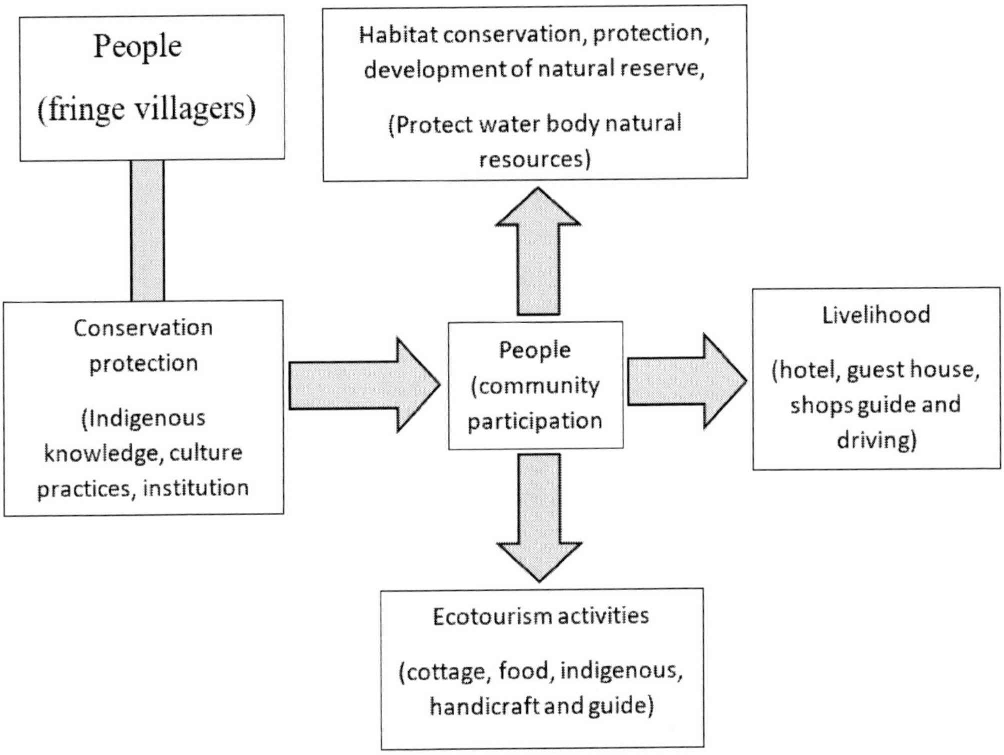

Figure 6. Structural framework people's involvement for habitat conservation and livelihood generation.

Acknowledgments

We are thankful to Ms Radhika Jha, PhD. Research scholar, Lucknow University for providing some photographs of Valmiki Tiger Reserve. Contribution of K. K. Jha, Ex-Professor, IIFM, Bhopal, India and Michael O. Campbell, Simon Frazer University, Canada is also duly acknowledged in providing valuable inputs for improvement of the text.

References

Altmann, M. C. G. (2016). *Wildlife-based ecotourism as sustainable conservation strategy: ecological and management indicators of conservation impact*. Thesis submitted for Master of Arts, Prescott College, Arizona, USA in Environmental Studies: Conservation Science and Policy. December 2016. Adopted from: https://www.researchgate.net/publication/311355842.

Castellani V. & Sala, S. (2010). Sustainable performance index for tourism policy development. *Tourism Management*, *31*(6), 871-880.

Ceballos-Lascurain, H. (1987). The future of 'ecotourism'. *Mexico Journal*, 13–14.
Champion, H. G. & Seth, S. K. (1968). *A Revised Survey of Forest Types of India.* New Delhi: Manager of Publications, Governemnt of India Press.
Environmental Défense Fund. (1999). *New Red List Reveals European habitats under threat.* IUCN.
Fennell, D. (1999). *Ecotourism: An Introduction.* New York: Routledge.
Fennell, D. (2002). *Ecotourism Programme Planning.* Wallingford: CABI.
Honey, M. (1999). *Ecotourism and Sustainable Development: Who Owns paradise?* (Second Edition in 2008). Washington: Island Press.
Hvenegaard, G. T. (1994). A Status Report and Conceptual Framework. *The Journal of Tourism Studies*, *5*(2), 24. University of Alberta.
International Ecotourism Society TIES. (1990). *Guidelines.* Washington, DC: International Ecotourism Society.
Jones, G. E. (1987). *The Conservation of Ecosystems and Species.* New York: Croom Helm.
Kideghesho, J. R. (2007). Factors influencing conservation attitudes of local people in Western Serengeti, Tanzania. *Biodiversity and Conservation*, *16*(7), 2213 – 2230.
Krebs, C. J. (1985). *Ecology: Experimental Analysis of Distribution and Abundance.* 3rd Ed. New York: Harper & Row.
Kruger Oval. (2005). Biodiversity conservation, affluence, and poverty: Mismatched costs and benefits and efforts to remedy them. *Ambio*, *21*(3), 237–243.
Kumar, R. & Sinha, S. (2016). *Planning and Implementation of Ecotourism in Valmiki Tiger Reserve.* LAP LAMBERT Academic Publishing.
Lubeck, L. (1990). East African Safari Tourism: the environmental role of tour operators, travel agents and tourists. In Kusler, J. (Ed.), *Ecotourism and Resource Conservation: collection of Papers*, Madison: Omnipress.
Ram, J. N. (2018). *Role of Fringe Villagers in Habitat Conservation & Ecotourism in Valmiki National Park Bihar, India.* M. Phil Dissertation. Indian Institute of Forest Management, Bhopal (Memeo).
Singh, R. K. (2007). Floristic wealth of Valmiki National Park, Bihar- An overview. *Indian Journal of Forestry*, *30*(2), 159-166
Stronza, A. & Durham, W. H. (Eds.). (2008). *Ecotourism and Conservation in the Americas.* CABI.
Wearing, S. & Neil, J. (1999). *Ecotourism: Impacts, Potentials and Possibilities.* Oxford: Butterworth-Heinemann.
World Wildlife Fund International (2001). *Guidelines for community-based ecotourism development.* WWF International.

In: Critical Research Techniques in Animal …
Editors: Kaushalendra Kumar Jha et al.
ISBN: 978-1-53619-720-4

Chapter 6

TECHNOLOGY IN GOOD GOVERNANCE: A CASE OF FRA AND GEOINFORMATICS

Arvind Kumar Jha*
Former Principal Chief Conservator of Forests,
Social Forestry, Maharashtra, India

ABSTRACT

The policies governing the forests of India have moved over time from the format in 1952 that allowed its management from the revenue point of view to that declared in 1988 which prioritized environmental stability and maintenance of ecological balance and made derivation of direct economic benefits subordinated to this aim. It however maintained that rights and concessions enjoyed by the tribals and other poor people living in and near forests should be fully protected and their domestic requirements should be the first charge on forest produce. The rights and concessions, which were related to the carrying capacity of forests had to be primarily for the *bona fide* use of the communities. The Scheduled Tribes and Other Traditional Forest Dwellers (Recognition of Forest Rights) Act, 2006 (Forest Rights Act or FRA) legislated by the Ministry of Tribal Affairs (MoTA), Government of India (GoI) was an attempt to recognize forest rights in favour of a well specified target population while simultaneously providing them an opportunity to participate in the strengthening of the conservation regime of forests and maintenance of ecological balance. The implementation of the FRA had to be fair, objective and transparent, considering the invaluable role of forests in terms of provision of human life-supporting eco-services, and the possibility of misuse of provisions of such an Act by people with vested interests, encroachers, rural elite, and ineligible claimants could not be ruled out. Hence, the use of geoinformatics for decision making in this regard was a unique initiative taken in the state of Maharashtra in India. This chapter examines the direct impact on governance and the related spin-offs.

---

* Corresponding Author's E-mail: dr_arvindjha@yahoo.com.

INTRODUCTION

Good governance is an essential tool for translating policy imperatives into sustainable and equitable development on the ground. Development in developing and some middle-income nations is however commonly marred by one of the most serious concerns in the society viz. prevailing corruption in the governing system. Ansari (2010) and Chaudhry (2010) reiterated that large scale corruption penetrated all levels of governance and become a major threat to development. The legal and administrative steps taken so far to improve ethical norms in the legislature, executive and judiciary have proved inadequate. Corruption has introduced malpractices of different forms in governance which has questioned productivity and delivery of services. In today's time E-Governance provides a certain roadmap for efficient delivery of services at the door step. This plays a vital role, while reducing corruption from our society. This also provides better relations between the government and customers, the government and business operatives and the government and government services with better utilisation of resources. It is widely successful in India using private public partnership models (Dhumal, 2015).

On the other hand, Information and Communication Technology (ICT) has played a key role in development and economic growth of the developing countries of the World. Political, cultural, socio-economic developmental and behavioural decisions today rest on the ability to access, gather, analyze and utilize information and knowledge. ICT which covers geoinformatics, needs to focus on 6Cs, such as computers, communication, connectivity, cyber laws, cost and common-sense to emerge as an effective and well governed country in 21st century (Al-Shalabi 2005, Chandra and Malaya 2011). However, good governance is concerned with the deliverance of sound economic and developmental policies and services in an effective manner to the people. The sound policies meant to secure justice, empowerment, employment, efficient delivery of services, eradication of poverty and making a healthy social setup are essential components that should be implemented through good governance (Grindle 2012; Chhikara 2015).

The Scheduled Tribes and Other Traditional Forest Dwellers (Recognition of Forest Rights) Act, 2006 (FRA), is considered an important landmark in the history of forest resource use and management in India. The Act aims at recognizing traditional rights of well defined categories of forest dwellers on the one hand, and maintaining the ecological balance on the other with a view to provide sustainable livelihood options to forest dwelling scheduled tribes (FDSTs) and other traditional forest dwellers (OTFDs), including those who were forced to relocate without receiving their legal entitlement to rehabilitation before 13 December, 2005.

The FRA also provides forest right of land to those evicted due to State development interventions without having been given land compensation, and where the land acquired by the Government has not been used within five years of the said acquisition for the purpose for which it was acquired. Thus, the underlying objective of the Act has been to strike a balance between the potentially conflicting interests of the forest dwelling communities and conservation of forests and wildlife resources. Instant text is confined to good governance through technology intervention in implementation of FRA, a legal tool brought into existence to empower the STs and OTFDs. It discusses merits and challenges of FRA, and success and limitation of Technology intervention. Nevertheless, provisions of some more forest related Acts and Official reports relevant to the context are extracted and mentioned in the text as and when required.

The Forest Rights Act 2006: Merits and Challenges

The Scheduled Tribes and Other Traditional Forest Dwellers (Recognition of Forest Rights) Act, 2006 was legislated by the Government of India (GoI) with an impressive preamble that reads as "An Act to recognize and vest the forest rights and occupation in forest land in forest dwelling Scheduled Tribes and other traditional forest dwellers who have been residing in such forests for generations but whose rights could not be recorded ...". The objective was also to empower such individuals and their communities to work in the interest of sustainability of the resource base on which their forest rights were contingent. It said: "... the recognised rights of the forest dwelling Scheduled Tribes and other traditional forest dwellers include the responsibilities and authority for sustainable use, conservation of biodiversity and maintenance of ecological balance and thereby strengthening the conservation regime of the forests while ensuring livelihood and food security of the forest dwellings Scheduled Tribes and other traditional forest dwellers". The Act, as an important step towards the above promised to address the issue of "the long standing insecurity of tenurial and access rights of forest dwelling Scheduled Tribes and other traditional forest dwellers including those who were forced to relocate their dwelling due to State development interventions". Rules under FRA were notified on 1st January 2008 in the name of Scheduled Tribes and Other Traditional Forest Dwellers (Recognition of Forest Rights) Rules (FR Rules) and were later amended in 2012.

A careful consideration of the above as well as specific provisions of the FRA and Rules made there under brings out the following significant aspects:

- The target populations covered as beneficiaries under this Act are basically of two categories of people that, by their definition itself, fall in the lowest rung of society by virtue of the fact that they primarily reside in and depend on the forests or forest lands for *bona fide* livelihood needs. They are called FDSTs and OTFDs.
- There are thirteen types of Forest Rights (FRs) mentioned in Section 3(1) and one in Section 4(8). For each type of Forest Right, the Act indicates specific target resources along with conditions on the fulfilment of which FRs are to be recognized.

- There are two broad categories of FRs viz. individual and community forest rights recognized under the FRA and the process for the same includes determination of the nature and extent of the same. Thus, in each case the 'Title' on Forest Right has to specify both nature and extent which, quite obviously, is an essential precondition for the sustained exercise of forest rights.
- The section 5 of the Act contains duties of holders of FRs which prescribes that holders of any forest right, *Gram Sabha* viz. the village assembly consisting of all adult members of the village, and village level institutions in areas where there are holders of any forest right are empowered to protect the wild life, forest and biodiversity, catchments area, water sources, and other ecological sensitive areas; and also to preserve the habitat of FDSTs and OTFDs from destructive forces affecting their cultural and natural heritage. Thus, the law makers have ensured through this Act that responsibility and authority are inextricably linked to FRs so far as the FR holders are concerned. All other agencies, including *Gram Sabha* (constitutional body at village level), are supposed to help and support these two categories of people in that by integrating their roles or interfacing them with the already existing systems and mechanisms operationalized through other relevant Acts and institutional apparatuses of governance.
- The Act specifies very precise conditions regarding forest land and also Community Forest Resources for being considered as eligible under the Act. These must be fulfilled before the same can be considered fit for recognition of FRs on them. Any forest land or forest resource that does not fulfil those conditions cannot be considered under the Act for recognition of FRs. Two simple examples make this fact clear: (a) A forest land may be considered for the right of cultivation only if is self-cultivated for livelihood, is occupied before the 13th day of December 2005, and is also under occupation on the date of commencement of this Act; and (b) The FR under Section 3(1)(i) of the FRA; which is the right to protect, regenerate or conserve or manage any community forest resource; can be recognized only if the area under claim is proved to have been traditionally protected and conserved by the claimant community for sustainable use and also if, as required under section 4(3), the FDSTs and/or OTFDs community who are claimants had the said area under their occupation before the 13th of December 2005.
- The FRA also specifies that, save as otherwise provided in FRA itself and in the Panchayats (Extension to the Scheduled Areas) Act 1996 (PESA), it's provisions are in addition to and not in derogation of the provisions of any other law for the time being in force.

An objective assessment of the FRA and Rules made thereunder clearly reveals that the law makers had, through this act, aimed to uplift two neatly defined categories of populations/communities i.e., FDSTs and OTFDs (Jha 2009). Similarly, they had been cautious enough about the safety of forest resource also when they; for example, through the important condition (as specified in Section 3(1)(i) of the Act); legislated to give FR under 3(1)(i) to only such FDST and/or OTFD communities who occupied the forest land under claim and also possessed past traditional experience of protecting and conserving community forest resource for sustainable use.

FRA IMPLEMENTATION CORE: DEMOCRATIZATION

The FRA was a unique initiative in the furtherance of democratization movement in India wherein the process deployed in the 'quasi-judicial function' of taking decision about recognition of FR in favour of legally eligible people; and as a corollary, for the rejection of ineligible cases; has been designed to be participatory in nature. The *Gram Sabha* (Village assembly constituted of all adult members of the village) was designated as the authority to initiate the process for determining the nature and extent of individual as well as community forest rights that could be given to the FDSTs and OTFDs. The Forest Rights Committee (FRC) which had no mandatory representation of any Government functionary in it was given the task of verifying, on behalf of the *Gram Sabha*, the nature and extent of FRs claimed by the applicants. Later, the *Gram Sabha* had to pass resolution on the recommendation of this Village Level Committee and forward the same to the Sub Divisional Level Committee (SDLC) for examining the resolutions and the maps of the *Gram Sabha* to ascertain the veracity of the claims.

The SDLC having one member each from the Forest Department and Tribal/Social Welfare Department has Sub Divisional Officer (belonging to the Revenue Department) as the Chairman while the member from the Tribal Department works as member secretary. It also has three members from members of the Block or *Tehsil* level *Panchayats* (legal body) out of the total of six and works to forward the claims with the draft record of proposed forest rights through the Sub-Divisional Officer for final decision to the District Level Committee (DLC) which again has three members of the district *Panchayat* as its members out of its total number of six members. The DLC is chaired by the Collector (member of the Revenue Department) and has member secretary from the Tribal/Social Welfare Department while the district level Forest Officer is a member. Even the State Level Monitoring Committee (SLMC) headed by the Chief Secretary of the State, which is supposed to monitor the process of recognition, verification and vesting of FRs has three Scheduled Tribes member/members of the Tribes Advisory Council as its members.

It is clear that the direct involvement of Government functionaries as representatives of Revenue, Forest, and Tribal/Social Welfare Departments begins in the scheme of things from SDLC upwards. As such, it had become imperative that at the most crucial stage of scrutiny i.e., at the village level, the Government provided a system that could enable the Forest Rights Committees to examine the FR claims with objectivity. It was also important that the *Gram Sabha*s were duly oriented so that they, although using the parameter of 'majority' for arriving at final decisions, could actually 'weigh' the evidences and take decisions considering merits of the case. Additionally, with the increased probability of grabbing of benefits by the ineligible people due to swollen number of claims forcing the decision makers overlook technical aspects and side track procedural requirements due to target oriented implementation, a verifiable and auditable robust system for the whole exercise in general was necessary to be attempted. After all, any lapses in accomplishing the task of determination of rights through such decentralized process did have an inherent potential of adversely impacting the invaluable ecological asset of the whole nation in addition to the sustainability of exercise of forest rights by the FDSTs and OTFDs themselves.

FRA VIOLATION: IMPLICATIONS

Forest Right claims can be rejected either on account of ineligibility of claimants or due to inadmissibility of claims or both. All forest lands and forest resources on which claims of forest rights are rejected or are wrongly recognized are nothing but encroachments of forest land and resources. Encroachment on a government land or public property is fundamentally an 'unlawful gaining upon the right or possession of another'. This has been dealt with by Honourable Supreme Court of India (SC) in their order dated 19 September 2018 in the case *Janabai* vs. Additional Commissioner and Others [Special Leave Petition (civil) 24212 of 2017]. To be precise, every case of recognition of forest rights by DLC in violation of FRA provisions is the example of an illegal act and allowing them to continue amounts to persistence of state supported 'encroachment' on the rights of other stakeholders.

Policy and Legal Perspective

The Section 13 of FRA mentions: 'Save as otherwise provided in this Act and the Provisions of the PESA, the provisions of FRA shall be in addition to and not in derogation of the provisions of any other law for the time being in force'. Exceptions are indicated in respect of the application of provisions of the Forest (Conservation) Act 1980 (FCA) for the facilities mentioned in Section 3(2) subject to the conditions specified therein and also for grant of forest rights as per Section 4(7) of the FRA. In this regard, the following points emerge:

(i) Erroneous recognition of Individual Forest Rights (IFRs) amounts to violation of various provisions of the FRA, FCA, and Indian Forest Act 1927 (IFA); and in case of Protected Areas (PAs), that of the Wild life Protection Act 1972 (WPA) too. As regards Community Forest Rights (CFRs), such action violates the FRA, FCA, IFA, Biological Diversity Act 2002 (BDA), PESA in scheduled areas, and also the WPA in case of PAs. In case of PAs, it is simultaneously violative of the SC orders in Interlocutory Application (I.A.) No. 548 on 14-02-2000 and also that in I.A. No. 707 passed on 18-02-2002 (Jha 2019).

(ii) The orders recognizing forest rights erroneously amount to violation of the Doctrine of Public Trust in relation to the use of public lands and natural resources.

(iii) Erroneous recognitions and their recurrence due to poor monitoring reveal the failure of the State and statutory authorities to perform their constitutional functions in relation to Articles 48 A and 51(A)(g) of the Indian Constitution and also their inability to anticipate, prevent and address the causes of such acts that would eventually lead to environmental degradation.

(iv) Unlawful recognitions impact adversely the 'right to life' of one and all.

(v) Illegal recognitions amount to gross injustice towards every citizen of the country as a result of misappropriation of public property.

(vi) The principal aim of Indian Forest Policy 1988 has been fixed to 'ensure environmental stability and maintenance of ecological balance including atmospheric equilibrium which are vital for sustenance of all lifeforms, human, animal and plant'. It declared that the derivation of direct economic benefit must be subordinated to this principal aim. Any illegally granted FR is fundamentally against this aim.

Social and Governance Perspective

(i) Illegalities in the recognition of forest rights under FRA cause irreversible damage to the expected social impact of FRA's intent so clearly expressed through it's preamble that envisions sustainable use and conservation of biodiversity and also strengthening of the conservation regime of the forests by and for the benefit of the target population viz. FDSTs and OTFDs.

(ii) Recognition of FRs on Forest lands and forest resources to ineligible people is to the gross disadvantage of the genuine FDSTs and OTFDs.

(iii) Illegal recognitions amount to gross injustice towards every citizen of the country on account of damage to services that are related to people's livelihoods and economy.

(iv) Illegal recognitions of FRs irrevocably damage the rights of future generations and intergenerational equity.

(v) Illegal recognitions, their proliferation and continuance of encroachments critically hurt the confidence of the law-abiding common citizens in governance.

(vi) Unlawful recognitions and enjoyment of benefits on their basis by the ineligible claimants throw to irrelevance the basic efforts behind bringing in this legislation and also that of spreading all over the nation the framework of its implementation at the cost of public money.

Ecological and Forest Management Perspective

(i) Considering the excessive degree of fragmentation and increased ingress of ineligible human population into the habitat of wildlife due to the flawed and illegal implementation of FRA, the existence of total gene pool including wildlife is endangered.

(ii) The principles of management of forests on landscape basis and ecological considerations get completely compromised and replaced by management of mosaic of Community Forest Resource area patches based on their services required for local level requirements articulated through the resolutions of individual *Gram Sabhas*. It is worth noting that the CFR areas can range from a few hectares to just few hundred hectares depending on the size of the CFRs of villages. It happens because the definition of village defined in the Section 2(p) of FRA has been adopted from the PESA which specifies that the village could be as small as a hamlet.

(iii) With larger and larger populations moving into and exploiting the forest resources, the above situation with micro plans implemented primarily for serving the village economy in a politicized environment has the potential of damaging inextricably the long term existence of biomes and forest related ecosystem services.

(iv) Illegal recognition of Right on Community Forest Resources under FRA has the potential of having serious adverse implications for the overall environment, watersheds, wildlife habitats, and also climate change. It amounts to outsourcing illegally to innumerable villages the crucial job of protection and maintenance of the ecological health of the country.

Fulfilment of Good Governance's Criteria: Impending Challenges

Good Governance requires a humanistic approach; i.e., it should be participatory, consensus oriented, transparent, responsive, equitable and inclusive, effective and efficient, accountable, and should follow the rule of law. The implementation process of FRA and its Rules intrinsically contains in itself some broad elements like participatory decision making, transparency, inclusiveness, and responsiveness. FRA's effective implementation in the field, however, required that all the features of good governance be factored into its implementation strategy duly considering not only the unique socio-political and cultural milieu prevalent in the villages but also the impression created about the Act by the leaders and opinion makers. This was also important because of the fact that the FRA was enacted in 2006 while its implementation started in 2008 only after the Forest Rights Rules were notified.

Every legal instrument having a welfare ethos has a specific target group defined under its provisions and it is obvious that any grant of benefit to ineligible persons or communities is outrightly illegal and an injustice to the genuine beneficiaries intended to be serviced by the Act. Although FRA was no exception, most of the local level politicians, non-forester bureaucrats, and opinion makers avoided to appreciate and propagate the crucial fact that FRA had just two sets of populations as its target group and they are well defined under section 2 of the Act as FDSTs and OTFDs. They did not communicate correctly to the masses that the intention of the parliament was actually to recognize the individual as well as community rights of the above mentioned two specific sets of people that existed at the lowest rung of even the rural and tribal society and also uplift and empower them with a view to ensure that the vision behind the FRA as enunciated in its preamble could be achieved. The FRA was being projected to be the panacea for all tribals and forest dwellers of the country (Haaque, 2019). In the process they also ignored the fact that in the name of removal of 'historical injustice' meted out to the FDSTs and OTFDs in the past, the hurried implementation of the Act without appropriate scrutiny had the potential of causing 'historical injustice' to the forest resources as well as to all the populations that depended on them for ecosystem services. Well orchestrated publicity about the FRA by naming it as '*patta* (land ownership) granting Act' tended to encourage new encroachments on the forest land by all sections of rural society.

Against the above backdrop, despite the FRA giving predominance to *Gram Sabha* as the key player in the villages, the effect of *de-facto* overlaying of the philosophy behind political party based structures over people's psyche already divided on caste, religion, and community lines worked to influence processes as well as decisions at the *Gram Sabha* level. The Forest Department's predicament was increasingly becoming visible in their reaction to the actions on the part of people who were emboldened to encroach due to the Section 4(5) of the Act that said: "Save as otherwise provided, no member of a FDSTs or OTFDs shall be evicted or removed from forest land under his occupation till the recognition and verification procedure is complete". People of all categories in large numbers had started adopting a strategy wherein they made fresh encroachments overnight, filed a claim with the *Gram Sabha* and just relaxed asking the forest administration to wait for the final decision on their claim. It was actually a matter of exception where some non-government organizations, for example, in the Yawal area of Jalgaon District, stood with the Forest Department against this modus operandi. The forest administration already branded and widely publicized as being anti-tribal for every question asked by them on validity of claim or evidence was isolated and often cornered by the fellow committee members themselves. Most of the senior level offices, in the above background, avoided taking any active part in the implementation saying that it was an Act by the Tribal Department and its implementation was their responsibility. All this, unfortunately, was happening ignoring the basic fact the sustainability of forest rights of the FDSTs and OTFDs itself depended on the safety and sustainability of forest resource.

Political leaders were openly making promises to the people about grant of *pattas* which, in common parlance means 'ownership' of land, whereas the Act talked of ownership of Forest Rights of habitation and self-cultivation on forest land keeping the legal status of the land unchanged. To make things worse in the field, this was being done without examining in detail the eligibility of beneficiaries under the Act. The Government of India wanted the FRs to be recognized at a fast pace and accordingly targets were set for the implementing agencies (GOI Committee Report: Manthan 2010, Page 47). Regular review was taken by the Central Government and representatives of Ministry of Tribal Affairs (MoTA) were assigned the task of visiting the States and take review of the progress. Narrative of the Act, in the socio-political domain was fast getting transformed from being a welfare instrument for the people of two communities to that of a wider and popular political agenda of forest land distribution. The bureaucracy had to perform as per targets, the states often competed with each other in terms of number of FR cases recognized, and circumvention of rules and procedure at different levels, it appeared, was not going to remain an unlikely occurrence. It may not perhaps be out of place here to mention that it was perhaps this aspect which the MoTA attempted to cover when it amended the FR Rules in 2012 and notified the proviso in Rule 12(10) as 'no recommendation or rejection of claims shall be merely on any technical or procedural grounds'.

Further, to quote just one example, notwithstanding the fact the FR Rule itself prescribed something as subjective and vague as the 'statement of elders other than claimants, reduced in writing' as an admissible evidence for recognition and vesting of forest rights; the guidelines issued, from time to time, by the MoTA kept interpreting definitions in a way so as to relax various criteria and conditions thereby creating an atmosphere wherein the job of intrusion of ineligible people in the scheme of things could become easier with time. It was anybody's guess that the challenges in implementation of FRA would go up as days passed. The fact remains that from the beginning itself, the honest implementer had to save himself also from the likely outcomes of faulty implementation making it critically important for him to create and firmly place a well thought out and widely acceptable implementation system that met the requirement of good governance.

FRA Implementation: Challenges and Responses

Considering the background as well as implications mentioned above, the need of establishing a verifiable and fool proof system on the ground was just non-negotiable. The idea was to create a system for implementing the Act that conformed to the principles of good governance and was characterized accordingly by the qualities of being participatory, consensus oriented, transparent, responsive, equitable and inclusive, effective and efficient, and accountable.

As a matter of fact, in many cases especially in far flung areas, documentary evidence pertaining to self-cultivation or habitation on or before 13 December, 2005 in the forest land claimed for self-cultivation or habitation was not available. Identifying forest lands occupied by claimants but not recorded as such by the Forest Department was a challenge. A major challenge for the implementers was also to identify and reject cases in which claimants had encroached after the cut-off date of 13th December 2005 that was prescribed by the Act as one of the parameters for the claim of self-cultivation and habitation to qualify as a forest right under FRA. The FR Rules came into existence on the 1st January 2008. The task was thus to find out in 2008 whether self-cultivation or use of the forest land for habitation was in existence as on 13th December 2005 on the area under FR claim. Similarly, with the impression of *'Patta'* distribution legislation allowing FR claim to the extent of 4 hectares' gaining ground, many people had started expanding their encroached areas after 2005. Such cases had just a part of their claimed area as admissible for recognition of forest right under FRA. Identifying, after 2008, such 'parts' of area under claim and rejecting them was not easy.

The process of recognition had to strictly follow the rule of law. As per FRA, the FRs had to be examined, determined and recognized with due mention of 'nature' and 'extent' and this primary task was the responsibility of the Forest Rights Committee which had to be a group of ten to fifteen villagers. In this regard timely measurement of the claimed land was a real challenge. The FR Rule 4(3) prescribed that necessary assistance be provided to the *Gram Sabha* by the authorities in the State. State's Directorate of Land Records (DILR) had neither the staff strength nor the equipment to complete the measurement of forest lands under FR claim at the pace at which the GoI expected the recognition of forest rights to be completed. Considering the fact that there were an estimated number of more than 0.3 million claim cases expected to be filed, the charges required to be paid to them for the measurement was also a major issue. Several Non-Government Organizations (NGOs) were against assigning the task of measurement of forest lands under FR claims to the Maharashtra Forest Department. Measuring forest land under claim in rural and far flung tribal areas within the targeted time needed an urgent solution.

Considering the ground level reality in the tribal and rural areas in general and in the areas where tribal people were in smaller numbers, making sure that the claim recognition as well as exercise of recognized FRs including performance of duties prescribed in Section 5 of the Act for the Forest Right holders was participatory and consensus oriented was an uphill task. This required repeated organization of well focused orientation trainings along with frequent provision of updated knowledge about the provisions of FRA, FR Rules as well as guidelines that descended from the MoTA. It was important to bring all the members of the *Gram Sabhas*, SDLCs, as also DLCs on the same platform so far as the objective, content, implementation process, and expected outcomes of FRA's implementation was concerned. A feedback from some of the FRCs reflected that, as they were living in the same village, they were afraid of rejecting the claim cases of the rural elite and powerful. Accordingly, though they were declared as 'public servant' for the purpose of FRA implementation, it was essential to immunize and make them confident on the ground in this regard. For the implementer of the Act all above came up as preconditions for aligning the implementation system with the desired criteria of legal soundness, equitability and inclusiveness.

Transparency required that the information pertaining to the implementation process, claims, claims verification, claims recognition, etc. are made quickly available to all the concerned stakeholders/departments/authorities for helping decision making and establishment of faith in the system. Considering the fact that the spread of implementation system had to cover more than 14000 villages, and simultaneously required to establish flow of communication with 110 SDLC offices and 28 DLC offices where FR claims needed processing, this aspect required an innovative approach. Appropriate responsiveness and accountability on the part of the authorities, as also effectiveness and efficiency of the whole process depended strongly on the choice of this approach.

GEOINFORMATICS IN FRA IMPLEMENTATION

Considering all the factors and the issues prevalent in the field, it was decided to go for a geoinformatics related technological intervention in the implementation of FRA which could be 'objective' in its approach. A 13-digit alpha numeric ID (Identification code) was to be allotted to every claimant with the first two digits indicating his District, next two indicating Taluka, next five indicating Village, next three his initials and the last one to accommodate claimants having same initials. With this ID fed into the GPS (Garmin 76CXS) machine, polygon of the forest land under FR claim was to be recorded and overlaid on Cartosat-1 satellite images of two time series using an indigenously developed software. One could thereby examine visually the extent as well as type of land use as evidenced on the satellite image pertaining to the relevant points of time.

This system's output could be seen as polygon on the relevant forest land overlaid simultaneously on the satellite images of two time series appearing side by side on the computer screen. This was appropriately integrated into the decision-making process adopted by the DLCs. The National Committee on Forest Rights Act, in its report of December 2010, had referred to the satellite imagery based claim verification system in Maharashtra and noted: "The Maharashtra procedure of measuring all claimed areas by GPS at the time of submission of the claim itself appears to be the best procedure, provided it is accompanied by training for FRC members to use GPS instruments, and more importantly, public demonstration of the accuracy and transparency of this method to villagers". (Government of India Committee Report: Manthan 2010). A summary of the actions taken in the State of Maharashtra by the Tribal Research and Training Institute (TRTI), Pune for operationalizing an approach adhering to good governance principles is presented in Tables 1 and 2.

Table 1. Summary-State of Maharashtra : Actions by the Tribal Research and Training Institute (TRTI), Pune

S. No.	Requirement	Actions taken by TRTI to meet the requirement
1.	Follow the rule of law	The Act and Rules translated into vernacular language, major tribal dialects with available copies. Local language film, on implementation process screened in all villages. Training for FRCs, Gram Sabha members, SDLC, DLC members. (A total of about 0.1 million persons trained) The fear in the minds of the village level authorities was removed by giving them a Proforma based Claim verification document that could also save them from being accused as subjective. Detailed guidelines for DLCs and SDLCs. Geoinformatics based system deployed to view the land use position on a past date viz., 13-12-2005
2.	Participatory and Consensus oriented	Proforma based Claim verification system containing the ethos of participatory orientation and consensus in decision making provided to FRCs and *Gram Sabhas.* Detailed guidelines issued to the DLCs and SDLCs. Increased understanding for Forest Rights Committee members to objective and fearless the in decision-making environment. The villagers were also trained to use GPS instruments for measuring the land on which FR claims were made, increasing implementation, and cooperation. Geoinformatics based software, with satellite images of the relevant periods (2005-06 and 2007-08) provided to the DLCs in order to reduce conflicts make the final decision-making consensus oriented as far as possible.
3.	Transparent	Development of dynamic web site of TRTI to manage information regarding implementation on day to day basis. A well-designed Management Information System capturing the progress of implementation placed on the website. All GPS polygons of forest land under claim (with unique 13-digit ID for each claimant) uploaded on the TRTI website and opened for viewing and superimposing on the satellite images using an in-house developed software (TRTI-VGIS) using specified passwords.
4.	Responsive	A monitoring team set up at TRTI office for random check of claim cases using geoinformatics. A web based short messaging service (SMS) alert system put in place for sending responses/urgent messages/ instructions/feedback to the DLCs and below.
		All responses to field level queries and minutes including those of State Level Monitoring Committee meetings uploaded regularly on the website.

Table 2. Summary State of Maharashtra Actions by the Tribal Research and Training Institute (TRTI), Pune

S. No.	Requirement	Actions taken by TRTI to meet the requirement
5.	Equitable and inclusive	Implementation framework provided by the FRA and Rules itself was followed to ensure reasonable compliance in this regard. 'TRTI-VGIS' software developed at TRTI, Pune was installed at all District Committees and it used CARTOSAT 1 Satellite data of 2005-06 and 2007-08 with superimposed GPS polygon measurement. This system with its unique feature of using evidence created by the 'eye in the sky' was propagated to ensure unbiased handling of claims. Additionally, a number of villagers as also members of NGOs desirous of handling the GPS machine were trained and even the claimants were allowed to handle GPS instrument to improve the sense and degree of inclusiveness.
6.	Effective and efficient	The system's effectiveness and efficiency were ensured by taking action as per the points mentioned above.
7.	Accountable	The possibility of verification of any action at any time in future itself makes the decision maker feel accountable. Verifiability of the decision as well as decision making process was made possible by the TRTI by putting in place a robust proforma based implementation and monitoring system coupled with the availability of objectively verifiable evidence at any point in time.

Technology Adopted

Considering the beneficiary group, the involvement and role of institutions like Forest Rights Committee and *Gram Sabha* as also the role of Tribal Development Department of the State of Maharashtra as the nodal agency in the state, it was decided to adopt the simplest, user friendly, and objective-oriented technological intervention. The details are as under:

System Architecture and Outputs

The nature and extent being the most important parameters in recognition as well as exercise of Forest Rights, the claim verification by FRCs was designed to begin with the acceptance of the claim form from the claimant and doing first the GPS measurement of the forest land under claim to determine its extent. The GPS machine with the 13-digit code entered in it was used to create the polygon duly linked exclusively and permanently with the name of the claimant. The GPS machine was taken to the office of the Sub Divisional Officer who was heading the SDLC. Here the GPS data was uploaded for immediate transmission to the server. All DLCs could thereafter download the polygon from the server, overlay the same on the satellite images by using the TRTI-VGIS software and analyse the case for scrutiny of its land use as on 13 December, 2005 and also for land use change, if any, after that date (OneWorld Foundation, India 2011).

The system architecture is shown in the Figure 1. At the Sub Divisional Office, however, the 'print' option yielded a measurement report of the forest land in a standard template as shown in Figure 2. The details contained in this document helped the implementers not only in assessing nature of claim and the 'extent' of area under claim along with its locational parameters but also provided important check points to monitor the timely processing of cases through use of time stamps, for example, of measurement of land and uploading of GPX file on the system.

Table 3. Technology Adoption

Hardware	**Garmin 76 CSx GPS with accuracy of +/- 2.5 m with 20 tracks storage capacity**
Softwares	Garmin MapSource TRTI-VGIS PostgreSQL – POSTGIS database
Satellite Images	Sensor: IRS Cartosat -1 Type: Panchromatic (B/W) Spatial Resolution: 2.5 m Swath: 23 X 23 km
	Best scale to view: 1:5000

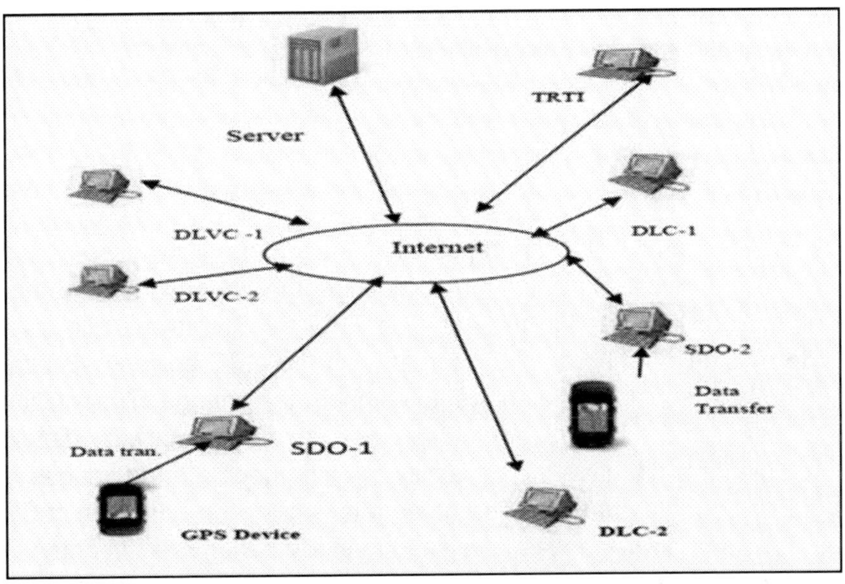

Figure 1. Schematic representation of system architecture used for capturing and analyzing the geo-spatial data overlaid on satellite image. Representative images of Server and computer systems are located at different stations. Mobile GPS machines are used for data transfer. Two way and unidirectional arrows indicate data transfer routes.

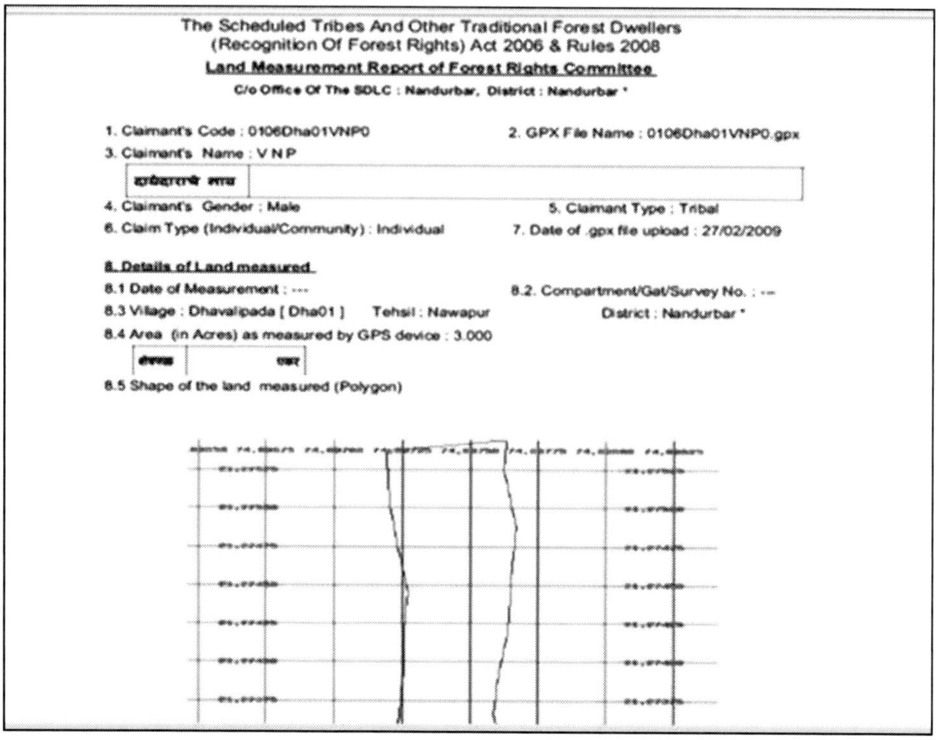
The Scheduled Tribes And Other Traditional Forest Dwellers
(Recognition Of Forest Rights) Act 2006 & Rules 2008
Land Measurement Report of Forest Rights Committee
C/o Office Of The SDLC : Nandurbar, District : Nandurbar *

1. Claimant's Code : 0106Dha01VNP0
2. GPX File Name : 0106Dha01VNP0.gpx
3. Claimant's Name : V N P
4. Claimant's Gender : Male
5. Claimant Type : Tribal
6. Claim Type (Individual/Community) : Individual
7. Date of .gpx file upload : 27/02/2009

8. Details of Land measured
8.1 Date of Measurement : ---
8.2. Compartment/Gat/Survey No. : ---
8.3 Village : Dhavalipada [Dha01] Tehsil : Nawapur District : Nandurbar *
8.4 Area (in Acres) as measured by GPS device : 3.000
8.5 Shape of the land measured (Polygon)

Figure 2. Diagrammatic representation of standardized template showing details of the claimant, type of claim, and extent of area under FR claim along with its GPS polygon on the earth's geographic coordinate system. Template shows details of the claimant, type of claim, and extent of area under FR claim along with its GPS polygon on the earth's geographic coordinate system.

TRTI-VGIS Software

The GPS polygons overlaid on the satellite images of the two time-series (2005-06 and 2007-08) provided critical information about the land use around the date of 13 December, 2005 indicated in the Act. The TRTI-VGIS software showed the two images side by side on the computer screen with the GPS polygon simultaneously overlaid on both. This helped the analyst to compare the change in vegetation and/or land use as seen on the images of the two time-series. This software developed on open-source platform did have the capability of using google earth images also and was useful if the google images of the dates relevant for the purposes of FRA were available. The same software was later used to study the encroachments done after 2005 by using images of 2011-2012 also (Maharashtra Forest Department Study 2012). Some examples of detection of cases in which forest lands were wrongly claimed under forest right of cultivation under FRA are shown in the Figure 3 and Figure 4 below:

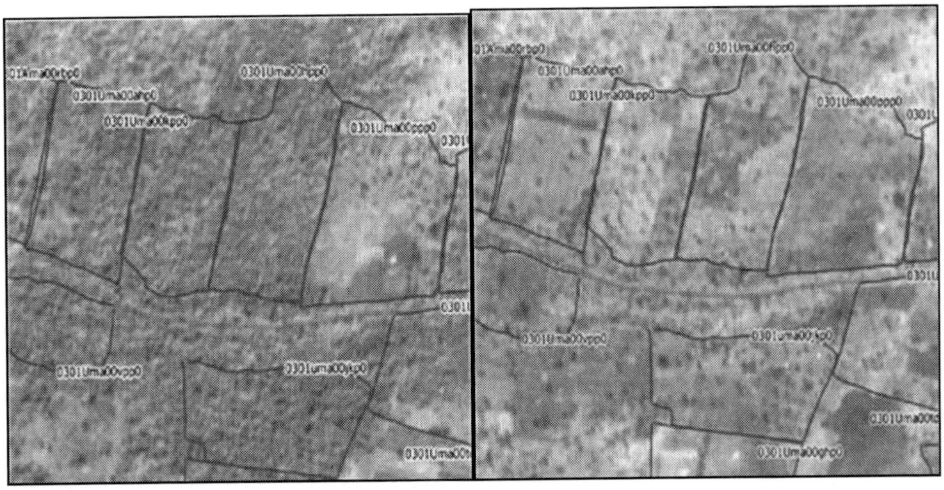

Figure 3. Land use change (Forest/tree cover to agriculture). Software output shows that area with vegetation cover in November 2005 were claimed (left). December 2008 imagery shows cultivation after tree cutting (right). Please note the alpha numeric codes associated with each polygon having different landuse signature used for settling the claims.

Figure 4. Image of a parcel in Gadchiroli District, Maharashtra with claims in polygons. This shows evident detection of wrongly claimed areas especially on the fringes of forest (apparent unmistakable presence of trees in forest area).

Effectiveness of the System

The strength of geoinformatics system as well as Maharashtra's experience was duly incorporated in the study report of the Joint Committee of Ministry of Environment and Forests and Ministry of Tribal Affairs - the National Committee on Forest Rights Act (Government of India 2010, Pages 60-64). A separate study, however, was conducted by the Forest Department in 2011-12 to analyse the finally granted Individual Forest Rights cases under Section 3(1)(a) for cultivation. A total of 35,044 cases (spread in 1514 villages) were analyzed which covered 602.17 km^2 (as per GPS and MapSource software). Most of the cases were from the Dhule (11,577), Nandurbar (11854), and Gondia (6014) districts of Maharashtra having large number of claims (Maharashtra Forest Department Study 2012).

On the claimed area front it was found that 287 cases were with duplicate polygons meaning thereby that the same land was under the claim of two persons, and in 210 cases the areas under claim by two persons had more than 50% overlap. There were 1466 cases where the claim was on more than the limit of 4 ha (0.004 km^2) prescribed by the Act, FRA. Further, it was found that 29,671 (85%) out of 35,044 cases were fully under agriculture in 2005 meaning thereby that the cases were fit for grant of rights from the point of view of land use, area wise it covered 46,969 ha (78%) and thus were correctly recognized. Amongst the wrongly granted cases, 7% were of completely barren land, 2% had full tree cover on them and 7% showed combination of tree cover, agriculture and barren as per signatures on the satellite images of year 2005-06. Needless to say, all cases that were wrongly granted violated not only the FCA and IFA, but also WLPA and the orders of the Honorable Supreme Court of India in case the forest land in question was located in the PAs. The result of the study was sent to all Government Departments with a view to bring in further improvement in the implementation of the Act.

Technological Intervention: A Win-Win Game

The system of implementation of FRA utilizing the above technological innovation worked with reasonably good results for quite some time despite the pressures created by and on behalf of the ineligible people from time to time in their attempt to get undue benefits under the Act. In 0.204 million cases the creation of GPS polygon as the first step towards fixing the extent of the claim was followed. Till May 2019, the number of titles of Individual Forest Rights and Community Forest Rights distributed in the State are 0.165 million and 7084, respectively.

Planned intervention using modern technology, thus, ensured that a transparent and evidence-based objective scrutiny as well as monitoring system was in place. The decision-making authorities (Committees) could work fearlessly, avoid political as well as public pressure by citing the result on the grounds of satellite image-based analysis. The confidence level of Committees went up and they could deliver quick and fair decisions granting forest rights to eligible claimants. Ineligible claims on account of land use became easier to reject. New offenders were discouraged after knowing that the technology based on geoinformatics could detect ineligible claims and also encroachments done after 2005. Accordingly, forest land was saved from encroachers (Jha 2018). Wildlife sanctuaries like Yawal got due attention in so far as getting them free from encroachments was concerned (Jha & Chaudhary, 2019).

There was reduction in conflicts at all levels including in the villages in cases where the same piece of land was being claimed by two persons. Inter departmental coordination on Act implementation became relatively smooth (especially amongst the Revenue, Forest and Tribal Development Departments) and the forest department did not have to fight against wrong claims made on the basis of false evidences of cultivation as on 13 December, 2005. The claimant ID came up as a potential link for developing an on-line system to benefit the forest right holders with developmental inputs. Online data base could be utilized to bring Forest management and forest rights' sustainability issues on GIS frame work. With all the above, coupled with the objectivity and transparency introduced by the technological intervention improved the confidence of people in governance.

LIMITATIONS WITH TECHNOLOGY IMPLEMENTATION

The technological intervention in the form of web-based verification system based on geoinformatics did bring in discipline, objectivity, transparency, and accountability in dealing with the claim cases under FRA but there were limitations too. The limitations were faced on three counts viz. administrative, socio-political and technological. The administrative limitations included the complete lack of any precedence in the Tribal Department of use of GIS related technology and its application. Further, The Tribal Department, although made the nodal agency for implementation of FRA and its members appointed as member secretaries of the SDLCs and DLCs possessed hardly any experience of handling the exercise of forest rights or about maintaining a non-exploitative relationship between tribal and rural populations with the forest resources, which, as a matter of practice was handled traditionally by the Forest Department. In such a situation the introduction of the concept and practical aspects of Geoinformatics for ensuring strict determination of 'nature' and 'extent' of forest rights was often seen by them as well as the Revenue Department as causing hindrance to implementation of FRA. Resultantly, dilution of the evidential importance of satellite based proof was produced by the MoTA through amendment to the Forest Rights Rules in 2012 which said that the satellite imagery and other uses of technology may supplement other form of evidence and shall not be treated as a replacement. Further guidelines by MoTA vide its letter dated 27 July 2015 said: "Use of any technology, such as, satellite imagery, should be used to supplement evidences tendered by a claimant for consideration of the claim and not to replace other evidences submitted by him in support of his claim as the only form of evidence. The effect in the field was obvious and the use of geoinformatics suffered a blow.

The socio-political forces imposed limitations not just because of the large number of quite vocal and influential claimants who belonged to the ineligible category and saw the geoinformatics based system as working against their interest but due also to the atmosphere created as a result of actions of MoTA that promoted circumvention of satellite based evidence that, undoubtably, was the best objective proof of genuineness of claim pertaining to a backdated situation. The Forest Department possessed considerable experience of using GIS based technology in its management practices. Strong objections were raised quite vociferously by some NGOs advocating against involvement of forestry personnel in the work of GPS measurement of forest lands under FR claim. This was done to propagate a feeling that the forest department was highly biased in favour of forests and would work against the interest of the claimants. Although one positive fallout of this was capacity building of even the local villagers in handling of GPS instrument, the atmosphere caused by such forces created doubts on the sincerity of a wing of Government, misled the people on the objectivity of such technology, and eventually also affected the continuance of this technology based system in scrutiny of FR claims.

On the technological front, the major limitation was non availability of satellite images for some areas for the relevant time period. The resolution of Cartosat-1 was relatively low especially for claim cases involving smaller areas. The difference between fallow land (one/two year) and that under cultivation could only be assessed based on boundary marks as the signature of cultivation was difficult to be discerned in some satellite images. The availability of high-resolution color imageries such as Quick Bird/IKONOS/WorldView-2 would have been a better option but they were expensive.

CONCLUSION

Use of geoinformatics in the implementation of FRA had been an initiative which was of one of its kind in the Tribal Development Department of Maharashtra. That was an attempt, rightly so, to provide well defined forest rights, both in terms of nature as well as extent to the genuine claimants under the Act. The innovation in implementation simultaneously attempted to reject the claims of ineligible people and in effect, at the same time, protect the forest resources in the interest of the genuine forest right holders. The report of the National Committee on Forest Rights Act had lauded it as the best procedure. More than 0.1 million people were trained in Maharashtra on the implementation of FRA with the new technology. The task of taking it forward was tough not only because of the difficulties in bringing in a new system into the work culture of bureaucracy but also because the ineligible claimants attempting to grab the benefits of FRA were far too many and also had the capability to manipulate systems to their advantage.

The satellite-based system favoured poorest of the poor by providing a non-challengeable objective evidence for his habitation/cultivation of forest land; the GPS machine provided him with an instant map of the piece of forest land occupied by him which, as if by magic, was created before his own eyes without him parting with any of his money or material for the job; and the proforma based claim verification system linked with his thirteen digit code protected him from the exploitative web of the authorities and rural elite so far as proving the genuineness of his eligibility and link with the piece of forest land claimed for forest right under FRA was concerned. In contrast, the same instruments mentioned above worked to sieve out the ineligible claimants establishing the faith of genuine right holders in governance.

A case study "Geoinformatics for Forest Rights, Maharashtra, India" on this initiative found a place in the 'ICT Facilitated Access to Information Innovations: A Compendium of Case Studies from South Asia' brought out by OneWorld Foundation, India in 2011 with support from World Bank Institute and Canadian International Development Agency. It was acknowledged as Best Practice by Department of administrative reforms & public grievance, GoI and finds a place on its website on 'Governance Knowledge Center'. It received certificate of recognition from eWorld Forum in 2011 as Best Public Choice ICT initiative for the underserved Community. It also received India's eGovernance National Award (Silver) in 2011.

The hiccups did begin with the ineligible claimants, apparently the rural elite and powerful duly backed by some local organizations, opposing the processes in the scrutiny system using geoinformatics in the name of causing delays. This was followed by the State Level Monitoring Committee asking the Committees to grant FRs without measuring the areas under FR's claim in a bid to apparently accelerate the pace of finalization of claim cases, and also due, in some areas, to gradual avoidance of creation of verifiable evidences to serve perhaps craftily the interest of rural elite and powerful people. The main blow to the use of satellite image based system, however, came when MoTA amended the FR Rules in 2012 to include Rule 12A(11)(explanation 2) which mentioned 'The satellite imagery and other uses of technology may supplement other form of evidence and shall not be treated as a replacement'. To make things worse for use of geoinformatics that had, like a third umpire, the unique capability of weeding out inadmissible claims, MoTA issued a letter to States on 12 July, 2012 which said: "Use of any technology, such as, satellite imagery, should be used to supplement evidences tendered by a claimant for consideration of the claim and not to replace other evidences submitted by him in support of his claim as the only form of evidence."

More than ten years have passed since the FRA, with a noble idea of supporting the lives of specifically defined target groups of FDSTs and OTFDs as contained in its preamble, started getting implemented in India. The amendments in the Rules, modification in the guidelines, increase in clamour for re-opening just the rejected cases and not asking for cancellation of wrongly recognized ones, pouring in of new claims even now by the rich and powerful due to relaxation of eligibility criteria, frequent protests in favour of grant of forest rights to all, and complete politicization of the Act as an instrument for garnering votes are surely going to harm not only the future of the genuine FDST and OTFDs but also the forest resources on which their forest rights are dependent.

It is high time the MoTA examines the impact of the amendment to FR Rules done in 2012 regarding use of satellite-based evidence on the recognition of FRs and accordingly revisits its position on the matter. It needs to implement its own directives regarding geo-referencing contained in the letter dated 27 July 2015 addressed to the Chief Secretaries of all States for examining the cases cleared without the use of satellite image based evidence also and simultaneously take corrective steps to stop misuse of the FRA. Needless to say, the use of geoinformatics with high resolution satellite images in FRA's implementation in the whole country will not only help to ensure grant of admissible forest rights to the genuine claimants and reduce environmental conservation related tensions arising due to illegal grant of forest rights but will also reinforce, in the overall interest of one and all, the faith of people in governance.

References

Al-Shalabi, H. (2005). Role of information technology for good governance and society. *Journal of social sciences*, *1*(2), 81-83.

Ansari, H. (2010). Corruption a threat to national security. *The Tribune*, English Daily published from Chandigarh, India, April 10.

Chaudhry, P. (2010). Black Money and Corruption in India: Necessary Evils to Survive?: A Commentary. In: *Developing Economies: Innovation, Investment…* ISBN: 978-161122-541-9. Editor: Joanne M. Carcillo. 2010 Nova Science Publishers, Inc. New York.

Chhikara, B. S. (2015). Good Governance: Role of science, technology and innovations. *Integrated Research Advances*, *2*(1), 22-30.

Dhumal, P. S. (2015). E-governance-an application of information and Communication technology. *International Journal of Engineering Research and General Science*, Volume *3*, Issue 2, 648-652. http://oaji.net/articles/2015/786-1431083092.pdf.

Goswami, V. (2014). Geo-informatics for Forest Rights: Harnessing technology for better livelihood support and access to forest rights. *Case Studies on e-Governance in India – 2013 – 2014. Ministry of Communications and Information Technology, New Delhi.*

Government of India. (2010). *Manthan: Report by National Committee on Forest Rights Act*, December 2010.

Grindle, M. (2012). Good governance: The inflation of an idea. *Planning ideas that matter., 25982.*

Haaque, T. (2019). *Securing the Forest Rights of Tribals*. Retrieved from https://www.thehindubusinessline.com/opinion/securing-the-forest-rights-of-tribals/ article26449073.ece.

Jha, A. K. (2009). *Forest Rights Act-A Conservation Tool*, *The Indian Forester*, Vol. *135*, No.11, November 2009.

Jha, A. K. (2018). Saving Yawal Wildlife Sanctuary and the like: Is *Loco parentis* an answer?, *Vana Premi*, Sept 2018.

Jha, A. K. (2019). 'Moving from Breaking up or clearing of Forest Land' to 'Occupation': Understanding the term 'encroachment' and beyond; *Vana Premi*, July 2019.

Jha, K. K. & Chaudhary, P. (2019). Unravelling the complexities of Protected Area Management in Two Developing Countries: Issues of Human Displacement and Wildlife Conflict. In Campbell, M. (Eds), *Geomatics and Conservation Biology.* New York, NY: Nova Science Publishers, Inc. New York.

Lee, J. I. & Wolf, S. A. (2008). Critical Assessment of implementation of the Forest Rights Act of India, *Land Use Policy*, Vol. *79*, December 2018 (834-844).

Maharashtra Forest Department Study. (2012). *Study of Changes in Vegetative Cover and Land Use in Forest Land with Claims and Forest Rights Finally Recognized Under the Forest Rights Act 2006.*

In: Critical Research Techniques in Animal …
Editors: Kaushalendra Kumar Jha et al.
ISBN: 978-1-53619-720-4

Chapter 7

Similar Yet Different Rice-Farming Ecosystems: UNESCO Associated Cases of Two Culturally Rich Asian Nations

Kaushalendra Kumar Jha***
Indian Institute of Forest Management, Bhopal, India
Department of Forest, Environment and Climate Change,
Uttar Pradesh, Lucknow, India

Abstract

Human cultures have affected the landcover of many landscapes, this having important impacts on plant, animal and habitat ecology. This chapter explains the process of landcover change in two landscapes in Indonesia and India. The *Subak* landscape has been recognized as a cultural heritage site by UNESCO in Indonesia, on account of a legacy and tradition carried over more than hundreds of years (spanning 9 centuries). The *Apatani* cultural landscape of India is also listed as a proposed cultural heritage site, on similar lines as the Indonesian example. Although there are some dissimilarities in the form of cultural faith, landscape topography crop composition, natural resources, etc. the major commonality between these two landscapes is terraced rice cultivation with an indigenous irrigation system developed and perfected over hundreds of years by the local inhabitants. However, these cultivation systems include several biophysical components interwoven with the cultural fabric and embedded in a human ecosystem. Such components discussed in the text include people and their culture, farming and irrigation systems, agriculture model and biodiversity, constraints and threats, etc. Their traditional heritage-based knowledge helped them live through the ages in sufficiency and independency.

* Corresponding Author's E-mail: jhakk1959@gmail.com.

INTRODUCTION

The cultural landscape in every corner of the world is a result of interactions between humans and nature. Diverse people living in different environments and with different cultures have developed strategies to survive in their landscapes, creating numerous forms of such landscapes in their efforts to sustain their communities (Luchman et al., 2009; Yetki, 2017). Two prominent examples, the *Apatani* landscape and the *Subak* landscape, from Asia with UNESCO as a common link between them are known for their uniqueness in maintaining their own world independently on a sustainable basis. A prominent component of these landscapes is traditional and irrigated paddy rice cultivation. The ecosystem of these landscapes is interwoven with different biotic and abiotic components like agriculture crops, trees, domestic and wild animals, spring water, etc. around the skilled human beings who manage the resources guided by spiritual and religious belief.

The *Apatani* cultural landscape of Arunachal Pradesh (AP) is on the tentative list of UNESCO cultural heritage sites while the *Subak* system in Bali's cultural landscape is an already recognized site since 2012 (https://whc.unesco.org). Separated by a distance of > 6500 km, located in two democratic countries of diversified peoples in southern Asia, the former site is in India whereas the latter in Indonesia (Figure 1). The two different people, the *Apatani* and the *Balinese*, from different biogeographic zones, have different culture and religious faith but a common goal of sustainable livelihood through the judicious use of natural resources. In brief, the two ecosystems are similar yet different in several aspects as discussed in the narratives that follow.

UNESCO heritage criteria related to the listing of these two systems have the following narratives, extracted from https://whc.unesco.org. The *Apatani* system is listed as a proposed site under criteria iii and v for qualification (https://whc.unesco.org/en/tentativelists/5893/).

(i) Exceptional testimony to the cultural traditions of the *Apatani* tribe responsible for maintaining the landscape more or less in the same state for centuries together
(ii) Strong traditional institutions, cultural practices and spiritual beliefs which guided the *Apatani* in their characteristic wet rice cultivation and management of other natural resources
(iii) Outstanding traditional human settlement in an upland area after an incessant struggle for survival in the most adverse environment
(iv) Need to protect from the threat of globalization, ideological onslaught from outside the area, and subsequent changing values in the society

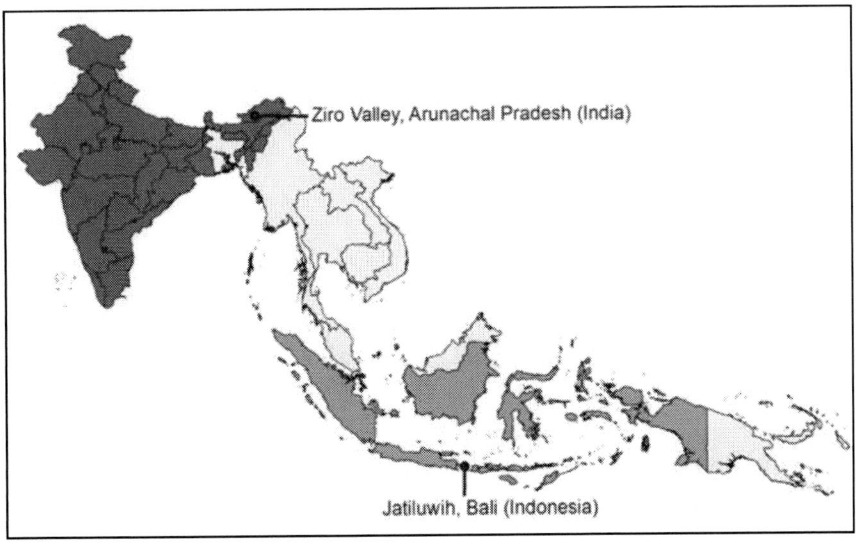

Figure 1. Location of two heritage sites, Ziro (*Apatani*) valley and Jatiluwih (*Subak*) tableau. These two sites are distantly located and separated by marine and large land mass indicating development of rice farming ecosystem independently.

The *Subak* system has been listed as recognised under criteria iii, v and vi (https://whc.unesco.org/en/list/1194/)

(i) The traditional philosophical concept of *Tri Hita Karana*, since the 12th century
(ii) Water management of the *Subak* landscape under the auspices of congregations of water temples, for a harmonious relationship with the natural and spiritual world
(iii) An intricate series of rituals, offerings and artistic performances
(iv) *Subak* as a democratic and an egalitarian system
(v) System based on ancient religious traditions, including *Saivasiddhanta* and *Samkhyā* Hinduism, Vajrayana Buddhism and Austronesian cosmology

However, the surroundings of these two indigenous populations, *Subaki* and *Apatani*, are very good examples of unique human habitation ecosystem units. This chapter intends to present the concept of human (habitat) ecology built on spatial and temporal interrelationship among different environmental components enveloping the two ancient and unique human landscapes. Human activities and interaction with nature involves very strong traditional ethnic knowledge and indigenous mechanism of sustainable production of food, wood and healthcare products. The process involved in performing such task incorporates durable socio-political structure, religious or spiritual faith, cultural activities and above all harmonious living of the communities. This conceptual model of human ecology is depicted in Figure 2.

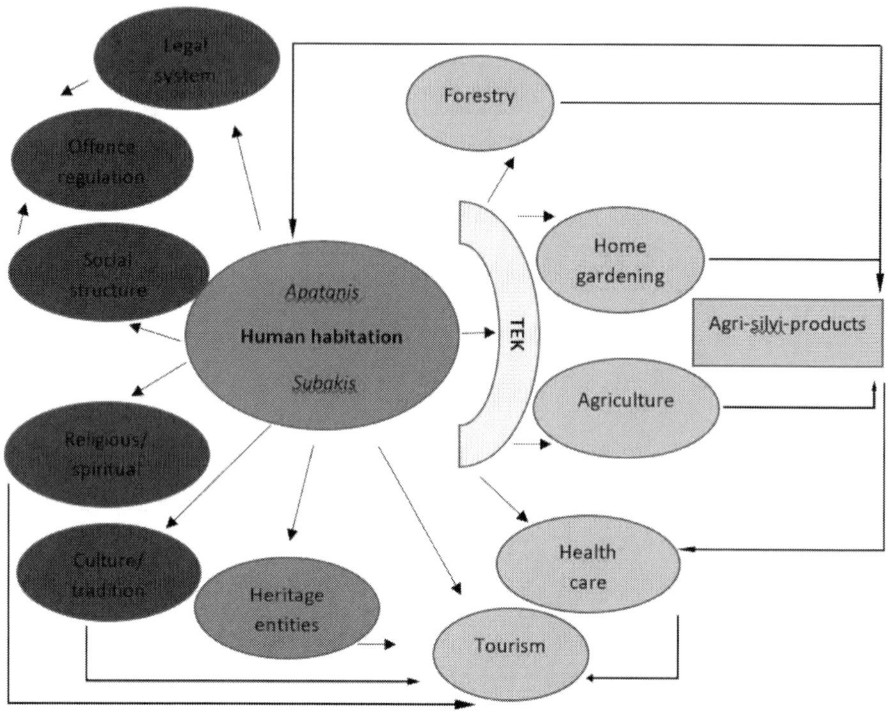

Figure 2. Flow chart showing selected social, natural and built environment components of human ecology. Purple, blue and green boxes, respectively, represent built environment, social aspects and economic activities mostly nature based and using Traditional Ecological Knowledge (TEK). Arrows are unidirectional connecters.

METHODOLOGY

The information contained in the text is based on observation, semi-structured interviews, informal talks, questionnaire surveys, literature consultation and hermeneutics (interpretive analyses). *Subak* related facts are based on field observation, some informal talks, interpretation and literature survey only. Apart from these methods, questionnaire surveys and semi-structured interviews were conducted in *Apatani* related data collection. Electronic mediums like academia.edu, researchgate.net and scholar.google.com was used for literature searches on *Subak* and *Apatani* heritage sites, terrace farming, irrigation system, culture and religion, biodiversity and ecotourism. These keywords returned many results but only the most appropriate ones were selected for consultation. The effort was to eliminate unreliable literature and concentrate on peer-reviewed articles and conference papers and book chapters. Selection was narrowed down to 88 references listed at the end of the text. Some cross references were consulted using google.com. Facts from some local language papers, quoted in other reprints have been referred to in order to cement the knowledge gap.

Heritage Landscape

Ziro valley and *Subak* (Jatiluwih, etc.) tableau, the two heritage landscapes have different physiographic features and climatic conditions. The former has flat habitation and agriculture fields with a gentle slope giving rise to paddy terraces at the elevation of 1525m asl. The paddy fields are devoid of trees, but bamboo plantations and pine orchards are close to the habitation which is generally comprised of tightly grouped houses in the foothills. The lofty forests are mainly on the hills. Contrastingly, Jatiliwih is composed of mountainous steep slopes converted to paddy terraces bearing a resemblance to multistoried building piled up one above another at an altitude of approximately 700m asl (Windari & Hadi, 2017). The houses and temples are generally in loose groups. The trees are interspersed in the cultivated fields and are concentrated along the vertical banks of stream ravinous slopes. The forests are distant from the hamlets or wards (*banjars*).

The *Apatani* heritage site confined to Ziro valley is surrounded by hills. It has an area of more than 1058 km^2 of which 33 km^2 is under cultivation. A comparatively smaller area is under settlement and the rest is covered by forests and plantations. This valley is endowed with a temperate climate with the summer temperatures ranging from12°C to 25°C, and in winter from 0.6°C to 15°C. The average annual rainfall of the area is 1200 mm. Frost is common in the winter season. Bali is just south of the equator and has a tropical climate which is hot all year round. It has a rainy season from November to March and a relatively dry season from April to October with some showers. Temperatures are high and uniform throughout the year (October to March: around 27°C; June to September: around 26°C). Humidity is high during the rainy season. In the interior of Bali island, the weather on the uplands is more likely to be cloudy and rainy. There are a surprising number of microclimates in Bali, some of them of historical significance (Pringle, 2004). There are five recognized *Subak* landscapes in Bali of which Jatiluwih is the largest *Subak* (3.03 km^2; Salamanca et al., 2015) and is often considered a representative site. A network of five rice terraces and their water temples covers 195 km^2 (Regan, 2013).

People and Culture

Demographically India is dominated by Hindus (79.8%), while Indonesia is contrastingly a Muslim majority (86%) state. However, the Bali province of Indonesia has a Hindu majority dominance (>90%; Sordalen, 2011). *Apatani*, the most secluded aboriginals of India in the northeastern most corner of the country, do not practice any traditional religion (Hinduism, Islam, Sikhism, Christianity, Buddhism and Jainism). They have spiritual faith in the Sun and the Moon (*Donyi-Polo*). Nevertheless, external cultural influence can be seen as a small population of this tribe has adopted Christianity as well as Islam in the past few decades.

The *Subak* (a Balinese organization or irrigation association, which is both civic and professional; Pringle, 2004) people (hereafter *Subaki*) from Bali practice Hinduism following *Tri Hita Karana* philosophy (*Parahyangan*, *Palemahan and Pawongan;* Lestari et al., 2017). This means that the *Subaki* respect humans, worship nature and revere god. Bali's official religion is Hindu, but it's far too animistic to be considered in the same vein as Indian Hinduism. The Balinese worship the trinity of *Brahma*, *Shiva* and *Vishnu*, three aspects of the one (invisible) god, *Sanghyang Widi*, as well as the *dewa* (ancestral gods) and village founders. They also worship gods of earth, fire, water and mountains; gods of fertility, rice, technology and books; and demons who inhabit the world underneath the ocean. They share the Indian Hindu belief in karma and reincarnation, but much less emphasis is attached to other Indian customs. There is no 'untouchable caste', arranged marriages are very rare, and there are no child marriages (www.lonelyplanet.com).

Origin and Migration

Historically, the original place of the *Apatani* was Lipyo Lembyan, which may be termed as modern-day Mongolia. Later, this group of people migrated to a place called Miido Lembyan (probably Tibet) and further migrated to Nyme Lembyan (probably the area of present AP, India and China). Finally, they reached a place called Biirii, presently known as Ziro, where they settled permanently. There are several folktales related to the settlement in the Ziro valley (Farooqy, 2017). One logical story narrates that the *Nyishi*, a dominant and aggressive tribe of the AP mountains forced the *Apatani* to settle into the valley in order to stay together and evade attacks and primarily protect their beautiful women.

The original races of Indonesia, similar to the Austromelanesoid and the Mongoloid, mixed with the migratory group from south of Asia which crossed the South Asia Sea ca. 300 BC as per Putra (2017). Later, bigger race migrations continued from all parts of Asia over a long period of time and formed the recent Indonesian nation. However, the model of the village community has been identified as taken from the model of Doab, a village community model developed by the Aryans on the banks of rivers of the Ganga and Yamuna (Putra, 2017). Thus, Balinese Hindu link and *Subaki* cultural similarity could be traced to Indian Hindu practices, traditions and customs.

Heritage Homes

Most Indigenous *Bali Aga* (mountain Balinese) villages use linear housing patterns in accordance with traditional Balinese philosophies (*Tri Hita Karana*, *Tri Angga* and *Tri Mandala*; Yudantini & Jones, 2015a; b). Traditional houses in Jatiluwih village, Tabanan have a specific spatial configuration that is the local identity of the area. Moreover, with the development of tourism in this village, traditional houses are not only a source of pride, but they also serve as an object of commodification (Dwijendra et al., 2017). However, the *Subaki* or the Balinese have separate houses for families. The materials used are locally available stones, wood and other plant products (Yudantini, 2018). Analogous to the human body, the housing compound has a head (the family temple with its ancestral shrine), arms (the sleeping and living areas), legs and feet (the kitchen and rice-storage building), and even an anus (the garbage pit or pigsty). There may be an area outside the house compound where fruit trees grow and/or pigs are reared.

The Balinese house looks inward – the outside is simply a high wall. Inside there is a garden and a separate small building or *bale* for each activity – one for cooking, one for washing and the toilet, and separate buildings for each 'bedroom.' In Bali's mild tropical climate, people live outside, so the 'living room' and 'dining room' are open verandah areas, looking out into the garden. The whole complex is oriented on the mountains – sea (*kaja–kelod*) axis. There are several variations on the typical family compound (www.lonelyplanet.com). For example, the entrance is commonly on the sunset side (*kuah*), rather away from the mountains and towards the sea (*kelod*) side, but *never* on the sunrise (*kangin*) side or in the direction of the mountains (*kaja*).

Contrastingly, *Apatani* houses are made predominantly of bamboos, pine wood and cane. Individual houses are joined together in a row to keep them close knit. A big rectangular hall on a raised platform is divided into different areas for living space in front of a fireplace or kitchen area and a washing area to the rear. The area on the longer sides of the hall, adjacent to kitchen, is used for sleeping. In some houses, bedrooms are separate but attached to the main hall. The kitchen heat is also used to dry the firewood kept above, hanging from the ceiling along with other household materials, including pig-meat. The basement space is used for pig rearing and sometimes domestic poultry. Open front and back verandah are used for sitting out or craft work (Sundriyal et al., 2002). There may be several reasons for such house design, such as colder climate, security from outsiders, construction material availability, etc.

Religious Practices

The series of religious rituals in *Subaki* is based on the Hindu religion in Bali, worshiping the Almighty as the preserver of life (*Lord Vishnu*) and the Goddess of fertility (*Dewi Sri*) to bless prosperity and abundant agriculture (Pitana, 1999 in Norken, 2016). The *Subaki* conduct several joint ceremonies: starting with signaling water diversion from source to irrigation canal (*Mapag Toya* or *Mendak Toya*) performed at *Ulun Empelan* Temple and ending with an event called the *Pakelem* ceremony at the *Ulun Danu* Temple. In addition to these joint ceremonies, individual ceremonies are also conducted, like channeling water from a canal to a rice field (*ngendagin*) and offerings (*mantenin*) to the rice stored in a barn (*lumbung*). A ritual by a landowner is performed at the temple (*ulun carik* or *sanggah catu* or *sanggah pengalapan*) located at the upstream of the field and at the entrance of a water intake. During these religious rituals the procession type and magnitude slightly vary between one *Subak* to another (Norken et al., 2015; Pitana, 1993; Sushila, 1987 in Norken, 2016).

As spiritual ceremonies in the traditional farming period, there are many prayers (*sembahyang*) and ceremonies (*upacara*) performed along the process of farming. These happen during seed sowing (*upacara ngawiwit*), start of planting (*upacara mamula*), or for warding off of pest (*upacara neduh*), maintaining water purity (*upacara mapag toya*), worshiping the god of vegetation (upacara tumpek bubuh), rejecting pests and evil spirits (*upacara penangkluk merana / mapakelem*), full moon praying (*sembahyang purnama*), zeromoon praying (*sembahyang tilem*), embracing harvest time (*upacara merebu*), five yearly local ceremony (*upacara mesabeh*) and ten yearly offering ceremony at *Payangan Gianyar* (*nyelung*) (Arif, 2015).

The *Apatani* did not develop temple or idol worship like the Hindus but have evolved some special festivals being celebrated in the region and reflect the unique and rich culture. They also, like the *Subaki*, celebrate festivals at community as well as individual level. *Yapung*, *Myokung*, etc. are performed collectively at the village level for the well-being of the human kind. However, they have three major festivals viz., *Murung*, *Myoko* and *Dree*. *Murung* festival is mainly celebrated by individuals for the well-being, prosperity and happiness of the concerned family and lineage. *Myoko* is celebrated to welcome the ancestors (*Ato Dihu* and *Ato Raru*) for the welfare of humankind. *Dree* is celebrated with great pomp and show for better cultivation, protection from storms, pests, natural calamities, bumper harvest and for the well-being of all plants, animals and mankind.

Four major agricultural rituals associated with *Apatani* agrarian culture for bumper harvests of agriculture and allied field crops are *Kide Midu*, *Dree Sodu*, *Yapung Mudu* and *Tamu Padu*. During *Kide Midu*, mother earth is prayed by offering chicks and eggs for protection against pest and other soil borne diseases, so that plant and vegetables grow healthy for better harvest. A taboo of not entering the paddy field and vegetable garden follows to avoid a bad omen coming to the offender. *Dree Sodu* is celebrated to please the god by sacrificing *Mithun*, cow, goat, chicken and eggs for a bumper harvest of paddies and other agriculture produce. Entry in the paddy field is prohibited for 7 days and the vegetable garden and forest for 3 days so that the agriculture god will be pleased and prevent and control the agriculture pests and other diseases effectively. *Yapung Mudu* is the worship of mother sky, to persuade her not to bring torrential rain, hail stone and to mitigate thundering during the peak period of agricultural fruit bearing stage, before the harvest of the paddies. As another taboo, work is restricted on the agriculture fields, vegetable gardens and forests for 3 to 7 days. *Tamu Padu* is performed occasionally, if there is any sudden change of climate leading to adverse effects on agriculture. Priests chant to god to stop adverse conditions by offering chickens and eggs. After performing rites for the subsequent three days, any entry into paddy fields and vegetable gardens is prohibited (Tayo et al., 2017).

Some other rituals (*Myokung*, *Danyi*, *Tamu*, *Tazer Hulii*, *Korlang*, *Kharung*, *Roppi*, *Udeh Tenii*, *Udeh Uii*, *Moreh eha*, *Aji eha*, etc.) are also performed by individuals. *Myokung* is performed for the fertility of soil after seeds are sown in the field or nursery. All gods and goddess are invoked, and sacrificial rituals conducted. *Danyi* (for sun goddess) is performed mainly for the well-being of all human beings, plants and animals as without the rays of sun no living being can survive. *Yapung* (for the sky goddess) is done to appease the goddess for good weather and sufficient rain to enhance the good growth of plants and crops in the paddies. *Tamu* (god of insects and pests) is performed when crops are transplanted from nurseries to the field. The rituals are important for (Bamin & Gajurel, 2015).

In case of natural death, the *Tazer Hulii* ritual is performed wherein a chicken is sacrificed. Another ritual *Ropii (Talii Gontii)* is performed in case of un-natural death. During this ritual the priests and relatives or family members taking part in the death procession tuck in a piece of a plant (*Cyclosorus glandulosus Millo Tarih*) in their hair knot on the forehead (*Piiding*), and rucksack (*lecha*) so as to prevent themselves from the spirit who faced an unnatural death (Bamin and Gajurel 2015). At the end of the procession green twigs are tucked onto the roof of a symbolic sacred home erected within the habitation.

FARMING MODEL

As reviewed by Castonguay et al. (2016), the rice terraces support several ecosystem processes and functions, and supply various ecosystem services. They benefit not only the local population but also domestic and international tourists. Besides their role in the production of rice for local communities, such terraces are at the centre of a complex of agroecosystems. They supply additional provisioning ecosystem services, like fish and snails (Lu and Li 2006, Garaway et al. 2013), and forest goods, like fruits and medicinal plants (Hayama, 2003; Gu et al., 2012). They also provide important regulating services, like flood mitigation, water-resource conservation, erosion reduction, and heat mitigation (Agus et al., 2006; Burkhard et al., 2015). Additionally, cultural services provided by rice terraces include religious identity (Lansing, 1987; Lansing & Kremer, 1993), tourism (Lorenzen & Lorenzen, 2010; Truong et al., 2014), and national pride (Nozawa et al., 2008). Two such systems, the valley terraces of India and mountain terraces of Indonesia, are presented (Figures 3 and 4) and discussed below.

The *Apatani* valley rice cultivation has undergone a steady evolution and modification through concise management of land that led to well-drained terraces with perfect bunds (*Agher*). The cultivation field, based on culture practices, can be categorised into three types viz. *Jaibee-aji*, *Pitang-aji* and *Miding. Jaibee-aji* is the marshy agricultural field normally kept without watering during the fallow period.

Pitang-aji requires water supply during fallow period, at least for two months to maintain the productivity of the weed infestation. *Miding* is a small size land maintained for raising nursery wherein water is maintained round the year except for preparation and sowing of seeds. The *Apatani* fertilized their rice fields with domestic waste to improve crop productivity. These are done during the fallow period of November to February. The crop residue is recycled by burning and decomposition after harvesting.

The *Apatani* also developed an economically more viable model of rice cultivation by integrating it with fish culture using common carp (*Cyprinus carpio*, Linnaeus, 1758), grass carp (*Ctenopharyngodon Idella*, Valenciennes in Cuvier & Valenciennes, 1844)) and silver carp fish (*Hypophthalmichthys molitrix*, Valenciennes, 1844)). Integrating fish with rice cultivation assured increased economic productivity and year-round employment opportunities for farmers. The plots utilized for rice-cum-fish culture are mainly fed on organic manure from poultry, pigs and cattle, and plant waste like rice husk, local beer, ashes from household burn, and compost like decomposed straw and weeds.

Figure 3. Glimpses of some "Rice Terrace Farming" ecosystem components from Bali, Indonesia. Top left: Multi-storied mountain terraces (background) in Agri-silviculture system at Ubud, Bali. Top right: Rice terraces with high risers. Bottom left: Religious procession by Balinese people with well-decorated offerings in the campus of Tirta Empul temple. Bottom right: influx of international tourists in the foreground premises of Tirta Gangga temple.

Figure 4. Glimpses of the components of "Rice Terrace Farming" ecosystem in Ziro valley, India. Top left: Valley terraces with paddy fields on gentle slope of homegarden (Agroforestry) model. Top right: Close up of the wider terraces with low risers in between along with bamboo fencing for fish farming. Bottom left: Mithun (*Bos frontalis*), part of animal husbandry component of homegarden and prized possession of the wealthy *Apatani*. Bottom right: Pine plantation (background) next to the rice terraces in pre-sowing stage.

The *Subak* Jatiluwih farmers form several groups that are specifically engaged in livestock raising in addition to rice farming. Two of the three groups (*Jungan Kelompok*) are engaged in the field of farming and cattle ranching and the other group (*dukuh sari*) is involved in the field of pig farms (Tirtawati et al., 2019). The *Subaki* had also developed traditional methods of wet rice farming over a long period of time by creating terraces and irrigation channels around them. This farming required manure and other organic matter to maintain productivity. Along with this, natural nutrients generated by the rich stew of the paddy fields created a stable system in which soil fertility did not decline over time.

Rice cultivation has shaped the social landscape in Bali. The intricate organisation necessary for growing rice is a large factor in the strength of community life. Rice cultivation has also changed the environmental landscape. Terraced rice fields cascade down hillsides like giant steps, in shades of gold, brown and green (www.lonelyplanet.com). However, the transformation in paddy culture in *Subak*/Bali/Indonesia started during the early 1970s with the introduction of high yielding miracle rice developed to ensure self-sufficiency by the International Rice Research Institute as a part of the Green Revolution. Along with very high production, this brought in problems of pest and disease, expensive culture due to regular application of chemical fertilizer and pesticide, all in addition to the risk of secured yield. The new variety also turned out to be pasty and insipid resulting in coerced acceptance as compared to traditional strains. Moreover, the *Subaki* were also forced to buckle under impositions from the authority for changing their age-old cultural practice of organic farming. Though the overall green revolution practices are still in place today, the government has dropped the emphasis on them after receiving complaints from farmers, social scientists and agricultural technicians (Pringle, 2004).

Irrigation System

Subak, an irrigation association is among the most important Balinese organisations. This association manages the downstream end of this system, coordinating not only the use of water, but the planting and harvesting cycles which are extremely important to ensure that individual farmers get the water when they need it and that the agricultural cycle is timed to minimise pest infestation. Due to steep intervening ravines, water is tapped high on the mountain slopes to reach lower elevations. Such tapped water passes through the canals, ditches, tunnels and terraces, which organise the flow of water to farmers. Each *Subak* is agricultural as well as religious in nature, and has its own temple dedicated to the goddess of rice (*Dewi Sri*). In addition, it has its own irrigation (water) temple (Pringle, 2004). The water temples are necessary to coordinate cropping patterns so that there is enough irrigation water for everyone and to reduce pests by coordinating fallow periods (Sepe, 2000).

The land used for rice cultivation has been irrigated efficiently through a well-managed irrigation system since time immemorial by the *Apatani.* They have developed a scientific method of irrigation through indigenous techniques with available local materials in the valley. This system of irrigation has been worked and perfected through community (*Bogo*, group of farmers) involvement on equitable sharing of water resources. Every stream that flows from the surrounding hills (*boring*) is tapped soon after it emerges from the forest, canalized at the periphery of the valley and diverted towards paddy fields by a network of primary, secondary and tertiary channels. The surplus water is drained back to the main canal without outflow of any organic matter or soil erosion. The agricultural fields have been made along elevational gradients. At higher elevations in the valley, fields are fitted with small diameter bamboo pipes, where the volume of water intake is less. Larger diameter pine pipes are used at the lower ends of the valley where the volume of water is more. Currently, large plastic and iron pipes are replacing pine pipes as the plastic and iron pipes give more durability and longevity in maintenance. These irrigation systems are managed by the traditional farmers' groups led by *Bogo Ahtoh*. The vision of these groups is reflected in the management and sharing of water in the community, which recognizes that water is a common concern that binds the group. Each plot owner is bound to provide equal outflow of water to the neighbouring plots.

The traditional village council (*Bulyang*) ensures compliance of such regulations. Each year canal repair is done through the collective participation of one person from each household (Dollo, 2007; Kacha, 2018).

IRRIGATION STRUCTURE

The *Subak* irrigation system consists of the simple gravity-irrigation principle, with the hydraulic structures consisting a diversion/intake (*buka or bungas*), or weir (*empelan*). It has primary canals (*telabah gede*), secondary canals (*telabah pemaron*), tertiary canals (*telabah cerik*) and quarternary canals (*telabah pengalapan*) to distribute the water to several owners of paddy fields. There is primary storage (*tembuku aya*), secondary storage (*tembuku pemaron*), tertiary storage (*tembuku cerik*), and other storage to distribute water among several owners of paddy fields (*tembuku penyahcah*). There may be tunnels (*aungan*) to break through hills. The structure at the end of a tunnel is called *kibul*, drain (*pengutangan*), and the receiving water body (*pangkung*) (Yekti et al., 2013 in Yekti, 2017).

The Ziro valley uses a well-managed water supply canal system, which is managed by the cooperative efforts of all the beneficiaries under the supervision of selected person(s) (*Bogo-Ahtoh*). They maintain the main canal (0.8 to 1.5 m in width and 0.60 to 1.2 m in depth) and the feeder canal (0.48 to 0.85 m in width and 0.45 to 0.75 m in depth). The feeder canals or pipes are branched to feed many terraces by blocking or opening the connecting pipes. The water in each feeder canal or pipe is regulated by wooden planks or stone to check the erosion. The flow of water from one field to another is maintained through a ditch (*Muhgo*) on bund (a structure designed to prevent flooding or breaches in construction) and two outlet pipes. The ditch is for outflow of excess water and maintenance of a desired depth. Two outlet pipes are placed to check overflow (upper one) and to drain water completely (lower one). Generally, the bunds are made up of soil, and supported by split bamboo and wooden pieces at the base if there is chance of erosion due to splashing of water. Initially, 1-2 cm water is maintained just after plantation in the field and then increased gradually with rice plant growth and size of fish using inlet pipe. During agri-pisciculture activities water is drained out using outlet pipes from time to time, which coincides with harvesting of fish and weeding (Dollo et al., 2009).

BIODIVERSITY

Both regions support high plant and animal diversity, but the current chapter restricts itself to biodiversity known and used by the two different peoples, *Apatani* and *Subaki*. They developed their traditional knowledge for the use of resources for life sustenance and livelihood support. Not only did they practice agriculture for simple living, but they mastered knowledge about ethnobotany (Srivastava et al.,

2010; Sujarwo et al., 2014; Caneva et al., 2017) as well. Animal-based resources were also used efficiently.

Other than grains and seeds (rice, millet, corn), the *Apatani* grew leafy vegetables, fruits, rhizomes and tubers. Sixteen indigenous land races of rice and 4 of millets were used (Dollo et al., 2009). Several varieties of fish (*tali*, *papi*, *ngilyang*, *tabu,* etc. Nimchow et al., 2010) were reared in multiple cycles and consumed locally throughout the year in dry and fresh form (Kumar & Ramakrishnan, 1990). The *Apatani* traditional knowledge regarding non-agricultural crops is very strong. They found more than 100 plant species to be used as Nontimber Forest Products (NTFPs) for various purposes, like medicine, edibles in the form of fruits and vegetables and construction material like canes (Kala, 2005; Srivastava et al., 2010). Although a majority of NTFPs were growing in community forests, as many as 32 NTFPs were grown in homegardens for quick access and additional yield on a frequent basis (Jha, 2015). Goat, fowl and pig raising are very common among the *Apatani* (Dutta et al., 2012). Different varieties of bamboos, canes and pines are grown and harvested for house construction.

Under *Subak* and *Abian* (farmer's organisation performing agriculture on dry land) management, agriculture has provided enough food and become the major means of income for the people's livelihoods in the region. The species conserved by them include Balinese rice (*Beras bali*), upland crops, fruit trees, flowers, vegetables and livestock, as well as precious genetic resources from specific crops and livestock such as snake fruit (*salak bali*, *Salacca zalacca* (Gaertn.) Voss), Balinese corn (*jagung seraya*), *kelapa dalam* variety of coconut, cashew nut, and Balinese cattle (*Sapi bali*) through agricultural production (MHDCRI, 2015). Home gardening is also done by the indigenous Balinese for cultivation of ethnobotanical species (Sujarwo & Caneva, 2015).

Both the *Apatani* and the *Subaki* developed traditional knowledge about the medicinal use of plants for treatment of various ailments and carried forward through traditional healers and elders. They discovered medicinal components in different plant parts, like roots, stem, leaves, fruits, flowers and whole plants. Literature suggests that the *Apatani* used as few as 105 plants (Kala, 2005; Srivastava et al., 2010, etc.) for treating several ailments like, gastric disorder, constipation, jaundice, diarrhoea, tuberculosis, leprosy, dysentery, conjunctivitis, wound healing, loose motion, stomach ache, headache, hypertension, throat pain, tooth ache, deworming, cough and cold, etc. Important plants used for healing preparation are *Ageratum conyzoides* (L. 1753), *Centella asiatica* ((L.) Urban), *Clerodendrum colebrookianum* (Walp.), *Houttuynia cordata* (Thunb.), *Oroxylum indicum* ((L.) Benth. ex Kurz), *Spilanthes paniculata* (*Acmella oleracea*, (L.) R.K. Jansen), *Paederia foetida* (L.), common guava *Psidium guajava* (L.), and ginger *Zingiber officinale* (Roscoe), etc. (Khongsai et al., 2011; Yakang *et al*., 2013; Bhuyan 2015, Tilling et al., 2015; Perme et al., 2015).

"*Usada Bali*" is the traditional system of medicine practiced by the *Subaki* which recognises more than 450 plants out of which the most important 50 plants are listed by Sutomo (2007). A few described in detail are *Zingiber oficinale*, *Acorus calamus*

(L., 1753), *Alstonia scholaris* ((L.) R.Br.), *Acacia catechu* (or *Senegalia catechu* (L.f.) P.J.H. Hurter & Mabb.), etc. This *usada* knowledge originated from India and spread out to Bali in the 5th century (Sutomo, 2007). Traditional medical material is used by the *balian* (healer) to cure various diseases in the form of *loloh* (herbal drink), *boreh* (powder), *tutuh* (drop), *sumbuh* (spurt), and *oles* (rub oil). It is said that there are only nine special quality plants that have been clinically tested, namely: *Indian bay leaf (Syzygium polyanthum*, (Wight) Walp.), green chiretta (*Andrographis paniculata* (Burm.f.) Nees), turmeric (*Curcuma longa*, L.), red ginger (*Zingiber officinale* rovb. var. rubra), bay cedar (*Guazuma ulmifolia* Lam.), Java ginger (*Curcuma zanthorrhiza*, Roxb.), common guava (*Psidium guajava*, L.), Javanese long pepper (*Piper retrofractum*, Vahl), and cheese fruit (*Morinda citrifolia*, L.) (Rasna & Tantra 2017). However, Hobart (1990) gives an account that a famous Javanese sage, Mpu Kuturan, after arriving in Bali meditated and acquired knowledge of the pharmacopoeia. He narrates about a few medicinal plants like, *Ficus benjamina* (L. 1767), *Sida rhombifolia* (L.), *Erythrina lithosperma* ((Hassk.) Merr.), *Sterculia foetida* (L.), etc. out of minimum 150 useful plants of medicinal value.

FARMING SUSTAINABILITY

Organic farming has been attracting greater attention as a result of the growing preference of farmers and consumers for safety and health benefits of organic products. The government also promotes organic products through a sustainable agricultural production system in balance with the environment (i.e., organic farming) from a production system that was becoming increasingly dependent on chemical fertilizers (Syuaib, 2009; Willer & Kichler, 2011; Shiotsu et al., 2015). The *Subaki* and the *Apatani* were already doing organic or sustainable farming but it was marginally impacted by Green Revolution initiatives.

The *Subak* system has comprehensive principles of agriculture in a spiritual sense concerning the maintenance of soil, water, plants, and animals. Farmers show gratitude to soil, water, sunlight, plants, animals and others. They perform a ceremony (*otonan*) for plants and animals every 6 months. Technology applied in traditional farming systems was of a higher level. They cultivated *Crotalaria juncea*, a kind of legume (*orok-orok*) approximately for four weeks. They used organic fertilizer (manure, pigs, chickens, goats, buffalo, household waste and kitchen ash) while practicing intercropping and crop rotation. Farmers always used home grown quality seeds and diversified food cropping. They did not use toxic materials (chemicals or synthetic materials), but always adhered to the principle of ecology (Arif, 2015).

Weeds are the major problem of rice farming in *Apatani* fields, but this is converted into opportunity by doing weeding (*Ahru-hodo*) four times in a year and composting that for organic farming. The rainwater drainage passing through the elevated houses carries human and animal waste along with garbage and is directed in such a way that it reaches the paddy field to enrich the soil. Paddy fields are also enriched by loads of silts, loams and humus coming through irrigation channel (Farooqy et al., 2017).

Table 1. Structure and function of *Apatani* farmers' organisation. Asterisks indicate gender differentiation

SN	Name and description of the groups	Task of the group
1.	*Bogo**: A farmer group sharing the common water sources. Group size is between 3-600 households depending on village size.	Construction and maintenance of water supply system and regulation of the efficient sharing of water among the group.
2.	*Aji Lenda*: A group which has their fields in the same area. Group size is 50-350 households.	Construction and maintenance of footpaths to allow access to and from fields.
3.	*Sulu Sikhii**: A group, which has their fields in the same area. Group size is 50-350 households.	Construction and maintenance of fencing to protect the agricultural fields from domestic and wild animals.
4.	*Tanser Patang***: Groups organised during field preparation and weeding. Group size is 5-15 households.	Field and nursery preparation, seed sowing, and weeding.
5.	*Konchi Patang**:* This group works in the morning between 5 am to 8 am. Group size is 5-10 households.	Field preparation, transplantation and weeding.
6.	*Halying Patang**:* This group shares labour during seedling transplantation. Group size is 5-15 households.	Transplantation of seedlings, particularly paddy and millet.
7.	*Enthee Patang:* This group forms to share labour during crop harvesting. Group size is 8-12 households	Harvesting and carrying of harvests.
8	*Bijee Lenda*:* A group having bamboo garden in same locality. Group size is 70-300 households.	Construction and maintenance of footpath, a way for carrying bamboo, timber and fuelwood.

Source: Dollo (2007), *Exclusively males, **Exclusively females, Non-asterisk groups are mixed gender

Democratic and Egalitarian Social System

Democratic Representation

The simple form of *Subak* organisation consists of *subak* members (*Krama*) and its leader/chairman (*Pekaseh*) along with an information provider (*Kesinoman*) acting on rotation basis (Putra, 2017). A *Subak* may have a secretary (*Penyarikan*) and a treasurer (*Petengen*) in it (Lestari et al., 2017). There is a collective agreement that a *Subak* leader (*Pekaseh Subak*) should be someone from the same *Subak* and with a very good attitude: honest, wise, fair and humble. He is democratically elected directly by each member of the community (Arif, 2015). *Buliyang* (village council), administrative and judicial unit is represented by all the clans from *Apatani* villages. Such representatives are not ordinary people, but they are the man of wealth, statesmen, a man of ability, capability, and expert in tribal laws or customary laws (Koj 2018). The post of *Buliyang* is hereditary, where the father nominates his son for the post. Nevertheless, strong loyalty and trust within the clans ensures that every member of the society is represented through their respective clan-representative. Any decision made by the village council is considered as public opinion (Dubey, 1998 quoted in McDonald, 2017; Sohkhlet and Lalyang, 2013). Thus, *Buliyang* is considered an administration by the people, for the people and of the people (Koj, 2018).

Cooperative Functioning

The *Apatani* farmers work in groups and cooperate with each other. They have eight different types of informal organisations with assigned tasks and a group leader responsible for leading the activities. Among the listed groups in Table 1, the *Bogo* group responsible for construction and maintenance of water sources is considered the most important as there are limited water sources for irrigation in the valley, and good water management is essential for efficient production in the paddy-cum-fish system (Dollo et al., 2009).

SOCIOPOLITICAL STRUCTURE AND ADMINISTRATION

Indigenous Organisation

Indigenous governance structure of the *Subaki* and the *Apatani* is presented in Figures 5 and 6. *Subak* is an organisation that operates in a hamlet or village, sometimes overlapping more than one. A small *Subak* is led by only a chairman called *Kelian Subak* or *Pekaseh* and assisted by a secretary called *Penyarikan* or *Juru Tulis*. For a large *Subak* the *Prajuru* consists of *Pekaseh* (chairman / head), *Petajuh* (deputy chairman), *Penyarikan* or *Juru Tulis* (secretary), *Patengen* or *Juru Raksa* (treasurer), *Kasinoman* or *Juru Arah* (channel of information) and *Saya/Bhaga/Penyade* (special assistant or working groups relating to various types of activities). Further large *Subak* or multiple *Subak* which has the same water source is called *Subak Gede*. This is chaired by the *Pekaseh Gede* and a deputy of *Pekaseh Gede* (Sushila, 1987; Pitana, 1993; Sutawan 2008; Norken et al., 2015).

Other components are the same. In the 1980s *Subak* organization began to be formed covering a watershed called *Subak Agung* and headed by *Pekaseh Subak Agung*. *Subak* can also be divided by smaller portions called *tempek* and led by *kelihan tempek* under a *Pekaseh* (Pitana, 1993; Sutawan, 2008). Later there has been addition of a working group called *penyade* or *bhaga* which is responsible for the implementation of *Parahyangan*, *Pawongan*, and *Palemahan* aspects, tailored to the needs of each *Subak* (Norken et al., 2016; Sutama, 2016).

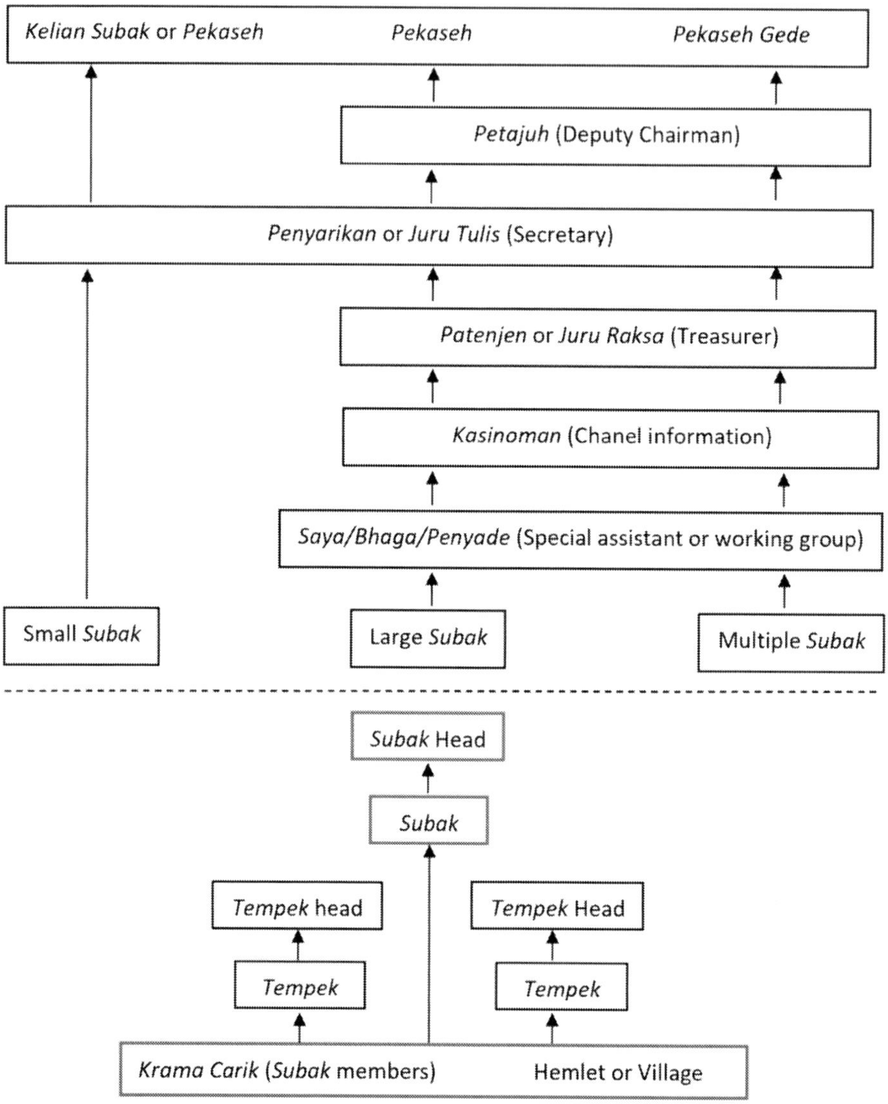

Source: Letsari et al., 2017; Norken et al., 2017.

Figure 5. Chart showing basic structure of Subak at the bottom (green boxes) where Tempek is subgroup of Subak. Structure above the dashed line is *Subak* of different sizes. Arrows mostly indicating hierarchical position but Secretary, Treasurer and *Kasinoman* have almost the same ranking but status superiority.

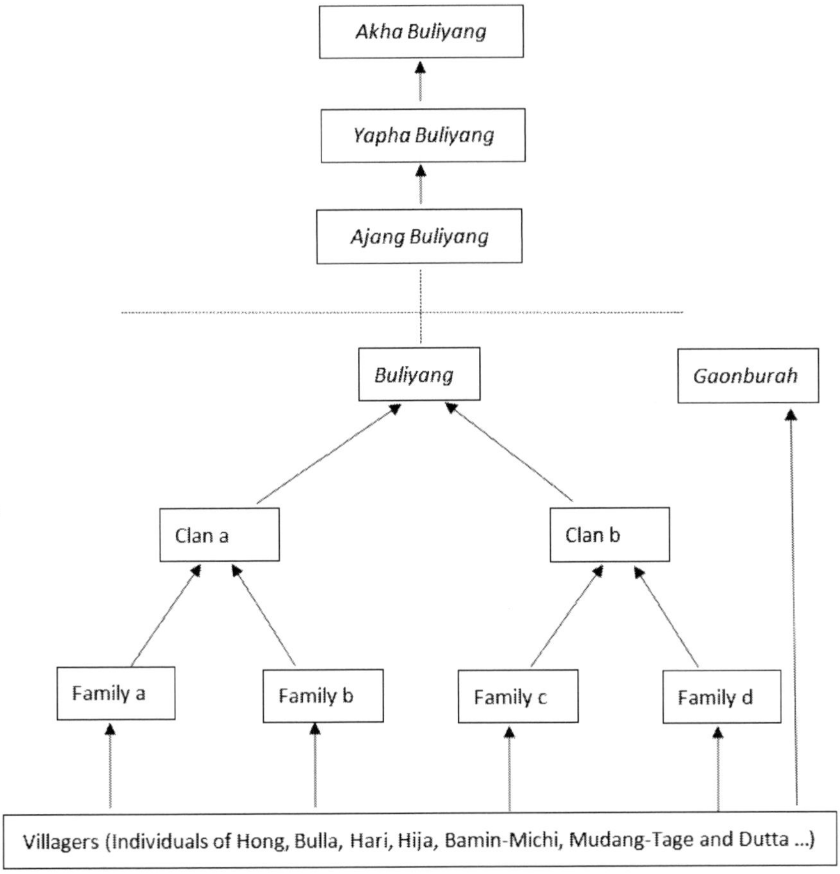

Source: Sohkhlet & Lalyang (1993); Dollo et al. (2009); McDonald (2017).

Figure 6. Chart showing sociopolitical structure in *Apatani* system. Arrows indicate hierarchic position.

There are seven historic villages (Hong, Bulla, Hari, Hija, Bamin-Michi, Mudang-Tage and Dutta) within the Apatani Valley (Srivastava et al., 2010; Dollo et al., 2009; Pant, 2000) established many years ago and each village is an independent political unit having a village council (Sohkhlet & Lalyang, 1993; Kani, 2012 in McDonald, 2017). A recent article puts the number of villages as thirty-five (Yakang et al., 2013) which includes some new villages like Biila, Kalung, Lempia, Posumla, Reru, Tajang, etc. The Apatani social structure is organized around clan membership. Historically, tribal laws were upheld and village administration accomplished through clan representatives known as buliyang (https://wwf.panda.org). Different levels of buliyang is also responsible for resolving the disputes.

Customary Laws

Apatani society is divided into a large number of clans *(nahi)* each of which reside together in a cluster in the village (*lemba*). The clan is a main social unit, which acts in solidarity. Earlier there were two distinct classes, the higher class (*mite*) and the lower class (*mura*), between which marriage was prohibited. In recent years, these distinctions have almost disappeared, along with the practice of child marriage, which was also common in the past. Governance is done through a system of village headmen (*gaonburas*), and a local village council (*buliyang*), which settles disputes at the village level according to customary law (Sundriyal et al., 2002).

The Traditional Village Council institution is of three tiers, namely *Akha Buliyang*, *Yapha Buliyang* and *Ajang Buliyang*. In each tier of council, there are one or two representatives from each clan. *Akha Buliyang* holds the highest authority on any decision making body. Role of *Yapha Buliyang* is that of mediator and interpreter in nature. *Ajang Buliyang* acts as messenger. The positions in *Buliyang* are hereditary or selected through democratic means. They are arbiters of tribal law and upholders of tribal justice, however, the individual *Buliyang* are primarily the spokesmen of their own clan or village (Dollo et al., 2009). They observe strict penal provisions mentioned in Table 2 to keep crime under check.

Table 2. *Apatani* Customary laws related to natural resource management

Offence	Customary law
Trespassing of paddy fields, and millet crops	The offender has to pay twice the cost of bamboo grove, kitchen garden or vegetable destroyed crops and also a fine of one full grown cow to the village authority.
Burning of house/ bamboo garden/ pine grove with ill-intention or willful causing of fire accidents.	The offender pays the cost of destroyed property and a fine of one full grown female mithun and one medium size mithun
Intentional destruction of boundary	The accused has to pay twice the cost of material damage and a fine of a full-grown cow to the village authority
Intentional destruction of fencing	Similar to above
Intentional destruction of water pipe in the paddy fields	Similar to above
Intentional destructionof ritual place	Similar to above

Source: Dutta et al. (2012).

Although the *Subak* has a territorial limit where the smallest civic entities are hamlets or wads (*banjars*). There are several *banjars* in a village, or *desa*, the next larger unit in the society. It often owns property, such as a meeting hall, market shed and gamelan orchestra. It may decide who can reside where. It has other rule-making authority which is to be taken seriously. If one doesn't attend *banjar* meeting, he faces a penalty. *Banjar* councils perform many important community functions, such as road upkeep, participation in major funerals and the organisation of quasi-police structures.

On similar lines as above, the *Subak* members or community (*Subaki*) is governed by customary rules called *awig-awig* (written legal codes or law book) *Subak*. *Awig-awig* is used to manage customary village organization and has a very important role in maintaining the natural (*palemahan*), social (*pawongan*), and spiritual environment (*parhyangan*) based on *Tri Hita Karana* (Astawa et al. 2018). Scope of *awig-awig* includes names and places, the basic principles, membership rules, the rules of the religious aspect, the rules of irrigation, setting fines, etc. (Norken et al., 2017). As a rule, *awig-awig* generally consists of a chapter called *Sarga* and section called *Palet*, and article called *Pawos*. The *awig-awig* is independently developed by the *Subak* community and adopted through a *Subak* meeting known as *paruman subak* (Putra, 2017). Most *subaks* possess *awig-awig* which details the rights and responsibilities of *Subak* members. This includes public obligations, guidelines concerning land and water use, legal transactions of land transfers, and collective religious ceremonies as well (Wiguna et al., 2005). The *Subaki* are very serious about attending the meetings of community members. The leader can decide the time of *ngayah* by himself if most of the members cannot attend a meeting to discuss the time of *ngayah* or other collective activities. In case of any violation, the offender is fined heavily. Community members who do not want to come *ngayah* without reason are fined an amount of 10 thousand rupiah which is to be submitted to the community meeting at *bale subak*. However, if he is unable to join the meeting due to a genuine reason, like illness or any other difficult condition then the fine imposed is only 5 thousand rupiah per day of *ngayah*. If a member of the community commits a crime against other members, he is excluded from the association in the community (Arif, 2015). There is also penalty provision for water theft, endogamous marriage and child marriage (Fox, 2012). Marriage among cousins, divorce, deserting after engagement, pawning and selling of rice fields and gardens or home yards, adoption of local tradition by immigrants are also regulated/prohibited by *awig-awig* with sanctions and penalties (Astawa et al., 2015 & 2018).

Heritage Ecotourism

The *Apatani* tribe inhabiting Ziro valley have been able to preserve much of their cultural heritage as embodied in certain rituals, uniqueness in paddy-cum-fish culture practices, tattooing systems, scenic or aesthetic beauty of land, rich flora and fauna, pleasant climatic conditions and hospitable nature of people. However, the lack of sufficient infrastructure such as the network of transport and communications and other tourism amenities and pathetic conditions of the existing facilities make this wonderful land less accessible to tourists. Tourism in the valley is relatively less developed as compared to other popular tourist destinations of the country (Thakuria, 2005). Therefore, it has a lower impact on the heritage ecosystem, which may be good for its continued preservation. But the *Apatani* society, with the onslaught of economic changes and modern value systems, is in transition. Modernization and the encounter with alternate cultures are radically impacting on their traditional culture and practices (Hilaly, 2018).

However, ecotourism can provide a positive impact on the preservation of the local environment and indigenous cultures since it is expected to be able to develop self-identity and pride among the local people. The community participation in eco-tourism has remarkably increased the sense of ownership of the local community members in protecting and conserving pristine local environment of the *Apatani* plateau. They are increasingly realizing that development and conservation is nature's gift to the people for their livelihoods and utilizing the resources as their valuable assets (Mize et al., 2016). Further, community-based ecotourism allows full involvement of the local community in planning, implementing and managing processes of ecotourism businesses and obtaining any benefits. There are some key aspects of success, such as support from government and community organizations to community formed committees or institutions for ecotourism management, local ownership to any possible facilities, homestay being the primary choice for accommodation, a local guide and management responsibility of local community (Budhi & Lestari, 2016).

The *Subak* empowerment model in Jatiluwih village has following features: (i) the model is conducted by *Subak* members using group approach. Education, training, group dynamics are used as strategies to improve group awareness, knowledge, skills and attitudes in order to have the ability to solve any problems faced by the community. The *Subak* members are provided foreign language training, entrepreneurship training and nature conservation training; (ii) there is a profit oriented new institute to support any activities conducted by the *Subak* members, such as establishment of a *Subak* cooperative under *Subak* organization management, as the place of economic activities so that the *Subak* members will be more economically empowered, in addition to its function as a social institution (Lestari et al., 2017).

THREATS TO THE HERITAGE

Both the heritage systems have faced almost similar impacts of globalization and economic changes including the green revolution. The green revolution, though it enhanced food sufficiency, changed the face of traditional farming. Local paddy variety (more acceptable to the people) was lost, soil pollution increased and insect-pest harmony with crops was disturbed. This shift in the agricultural system however seriously disturbed the traditional cycle of rice-farming that had previously been sustainable (Miura & Sarjana, 2016). Cultural services like, traditional ecological knowledge (TEK) and spiritual or religious experience and their resulting benefits, are gradually being lost, possibly due to commercialization of the traditional culture for tourism purposes (Bulilan, 2007). Spread of Christianity (Nozawa et al., 2008) and other religions and the high costs of holding non-negotiable rituals brings the same results, particularly in the case of *Subaki* (Salamanca et al., 2015). Decline in the TEK, especially in the *Apatani* valley, is due to the youth not taking interest in traditional activities as they are influenced by cosmopolitan culture resulting in adoption of modern lifestyle (Jha, 2015, 2016; Rechlin & Varuni, 2006). Also, the younger generation is moving out of their traditional homes in search of better education and/or employment (Farooqy, 2017).

Migration is already a common phenomenon in these two heritage sites and elsewhere which has impacts on farmers' decisions and land use (Mckay, 2003). Anticipated acceleration of migration could result in widespread abandonment of the terraces (Castonguay et al., 2016). Expansion of tourism at a faster pace, especially in the *Subak* landscape is taking its toll as the traditional land (rice field) is being diverted to nonagricultural uses year after year (Finlayson & Paramita, 2013; Pastika, 2014). Loss of water rights and the widespread adoption of labor-saving production technologies render the *Subak*'s ceremonial and community-oriented ties meaningless (Jha & Schoenfelder, 2011). The rapid development of (Bali) tourism has brought many changes in various aspects of community life resulting in severe threat to the sustainability of the *Subak* irrigation system (Tirtawati et al., 2019).

Though the above threats are genuine and impacting the systems currently, the cultural heritage still remains strong in both the places. Suartika (2018) suggested that less outside interference will guarantee that the authenticity of culture and environment will be maintained. However, minor modifications in heritage building suited to the need of tourism without damaging the originality and maintenance or renovation of original architecture may enhance tourism (Budihardjo et al., 2016). Many believe that conservation intervention will be able to regain the lost part of paradise (Jha, 2015; Castonguay et al., 2016, Farooqy, 2017, etc.). Nevertheless, Salamanca et al. (2015) advocated that the role of relevant government is crucial in empowering the farmers to oversee and manage the heritage they have built.

CONCLUSION

The ecosystem of heritage rice terrace farming in two different regions with varying and similar events or processes is presented in brief through a chart (Figure 7). Different components of the rice farming system have been practiced, developed and perfected by several generations of the local population without use of machines and western technologies. The cottage techniques (compost, ash, wooden tools, spades, etc.) have been involved in perfecting the irrigation system in undulating terrain. Local rice varieties have been developed and production has been maintained by the use of organic farming based on domestic and local products. There is involvement of cultural practices in safeguarding the crop rearing processes. All these have a positive impact on the maintenance and preservation of the heritage ecosystem. However, the green revolution has affected the age-old practice by interfering with traditional farming due to introduction of high yielding external rice varieties, chemical fertilizers and pesticides. Though the production was amply increased, local varieties resistant to pests started disappearing and people started missing non-sticky rice in their meals. People, especially the younger generation, started abandoning traditional farming and migrating towards the cities or industries in search of more lucrative jobs. Fast developing tourism also had a negative impact as the traditional land was being diverted to this industry.

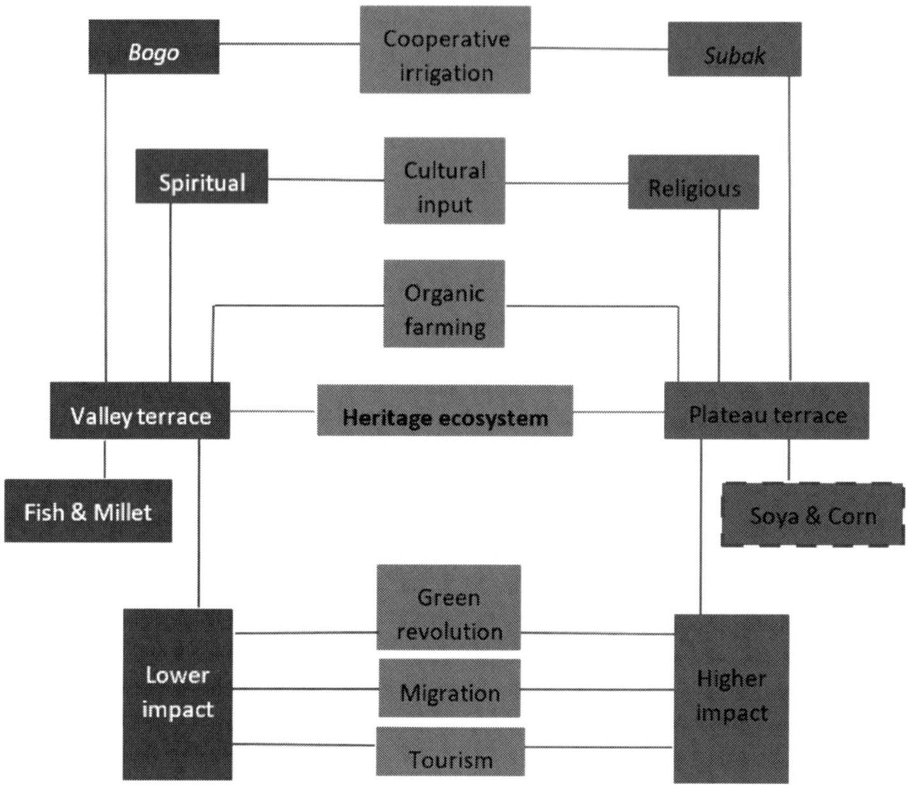

Figure 7. Functionality of "Heritage (Rice terracing) Farming System" in India and Indonesia. Red and maroon boxes are, respectively *Subaki* and *Apatani* practices with differences while green boxes are common events or similarity in the two systems. Porous box is the result of later modification. Lines are the event/object connecters.

Despite having above similarities both the distantly practiced systems also had dissimilarities. The Indian system has valley terraces surrounded by vegetated hillocks while the Indonesian system has mountain terraces with interspersed trees. Irrigation channels were trained on the mountain slope itself in case of the latter while it was done at foothills in the former. The job of creating and maintaining the channels was distributed to all the owners (*Subak* residents) and a group of persons (*Bogo*), respectively in the case of the *Subaki* and the *Apatani* system. Cultural aspects are embedded in the agriculture practices. While the *Subak* system is heavily loaded with the religious beliefs of Hinduism, the *Apatani* followed only spiritual practices like taboo and festivities. Although both practice the agroforestry model of livelihood sustenance, the *Subaki* have been growing only rice and doing animal husbandry along with use of some tree products. At the same time the *Apatani* have added fish rearing with rice paddy and animal husbandry along with home gardening (plantation of bamboo and pines). Unlike the use of single cropping systems by the *Subaki*, the *Apatani* grew millets and corn also. Although pigs and chickens were reared in the homestead, the *Apatani* never practiced ranching like their counterparts but raised *Mithun* (a bovine) in stray form in the forests. Modern development and globalization along with enhanced tourism impacted both traditional heritage ecosystems but the impact was much higher in the case of the Indonesian system than the Indian system. Nevertheless, whatever is lost from the heritage system, efforts are being put in to regain the original form.

Acknowledgments

A team of scholars lead by Dr Tripathi from NERIST, Itanagar is thanked for helping collect data and understanding the subject in Ziro valley. Dr Yuni from Udayana University, Bali and her colleagues including Shree Ketut also deserve appreciation for helping learn *Subak* system in the field especially in Ubud area and at Tirta temples. Radhika Jha constantly helped in improving the contents.

References

Agus, F., Irawan, I., Suganda, H., Wahyunto, W., Setiyanto, A. & Kundarto, M. (2006). Environmental multifunctionality of Indonesian agriculture. *Paddy and Water Environment*, *4*(4), 181-188. http://dx.doi.org/10.1007/s10333-006-0047-5.

Arif, A. S. (2015). *The Real Subak – Integrated Organic Farming as Tri Hita Karana Manifesto*. Bali ACLA 2015 International Symposium Agricultural Landscapes of Asia: learning, preserving, and redefining.

Astawa, I.G., Budiarsa, M., Mbete, A. & Simpen, I. W. (2019). The Interpretation of Ecological Preservation in the Awig-awig (Customary Law) Text of Tenganan Pegringsingan Village: Positive Discourse Analysis. *e-Journal of Linguistics*, *13*(1), 55—71.

Astawa, I. G., Budiarsa, M. & Simpen, I. W. (2018). Power and Ideology Revealed in The Awig-awig (Customary Law) Text of Tenganan Pegringsingan Village: Critical Ecolinguistic Study. *International Journal of Linguistics*, *10* (5), 45-60. https://doi.org/10.5296/ijl.v10i5.13656.

Bamin, Y. & Gajurel, P. R. (2015). Traditional Use and Conservation of Some Selected Plants Used in Festivals and Rituals in Apatani Plateau of Arunachal Pradesh, India. *International Journal of Conservation Science*, *6*(2), 189-200. Retrieved from http://ijcs.ro/public/IJCS-15-18_Bamin.pdf.

Bhuyan, M. (2015). Comparative Study of Ethnomedicine among the Tribes of North East India. *International Research Journal of Social Sciences*, *4*(2), 27-32. Retrieved from http://www.isca.in/IJSS/Archive/v4/i2/7.ISCA-IRJSS-2014-284.pdf.

Budhi, M. K. S. & Lestari, N. P. N. E. (2016). Community Based Ecotourism Development in Jatiluwih Village Bali. *International Journal of Applied Business and Economic Research*, *14*(3), 1865-1880.

Budihardjo, R., Tobing, R. R. & Antariksa. (2016). The impact of tourism towards sustainability and the change of architecture. *International Journal of Research in Engineering and Technology*, *5*(8), 269-274.

Bulilan, C. M. R. (2007). Experiencing cultural heritage and indigenous tourism in Banaue. *Philippine Quarterly of Culture and Society*, *35*(1), 100-128.

Burkhard, B., Müller, A., Müller, F., Grescho, V., Anh, Q., Arida, G., Bustamante, J. V., van Chien, H., Heong, K. L., Escalada, M., Marquez, L., Truong, D. T., Villareal, S. & Settele, J. (2015). Land cover-based ecosystem service assessment of irrigated rice cropping systems in southeast Asia—an explorative study. *Ecosystem Services*, *14*, 76-87. http://dx.doi.org/10.1016/ j.ecoser.2015.05.005.

Caneva, G., Traversetti, L., Sujarwo, W. & Zuccarello, V. (2017). Sharing Ethnobotanical Knowledge in Traditional Villages: Evidence of Food and Nutraceutical "Core Groups" in Bali, Indonesia. *Economic Botany*, *71*(4), 303–313. https://doi.org/10.1007/s12231-017-9395-x.

Castonguay, A. C., Burkhard, B., Muller, F., Horgan, F. G. & Settele, J. (2016). Resilience and adaptability of rice terrace social ecological systems: a case study of a local community's perception in Banaue, Philippines. *Ecology and Society*, *21*(2), 15. http://dx.doi.org/10.5751/ES-08348-210215.

Dollo, M. (2007). Traditional farmers' groups supporting sustainable farming. *LEISA Magazine*, *23*(1), march 2007.

Dollo, M., Samal P. K., Sundriyal, R. C. & Kumar, K. (2009). Environmentally sustainable traditional natural resource management and conservation in Ziro valley, Arunachal Himalaya, India. *Journal of American Science*, *5*(5), 41-52.

Dutta, J., Muang, O., Balasubramanian, D., Wann, F. M., Pangging, G. & Khan, M. L. (2012). The 'Tani' Mega-cultural Landscape. In Ramakrishnan, P.S., Saxena, K.G., Rao, K.S., Sharma, G. (Eds). *Cultural Landscapes The basis for linking biodiversity conservation with sustainable development*. National Institute of Ecology, New Delhi 110016.

Dwijendra, A., Putra, D., Wirawibawa. & Darma, A. S. (2017). Transformation of Traditional Housing in Traditional Village of Gunung Sari, Jatiluwih, Tabanan Regency. *Journal of Civil Engineering, Architecture and Built Environment*, *1*(1), 6-11.

Farooqy, P. (2017). Symbiosis Between Nature & Culture – A Case Study of the Apatani Cultural Landscape, India. *Journal of World Heritage Studies, Special Issue.* 51-55.

Finlayson, R. & Paramita, E. (2013). *Subak, Bali Irrigation System, Being on the Verge of Destruction.* Accesed August 30, 2013. http://www.worldagroforestry.org/news/subak-sistem-irigasi-khas-bali-berada-di-ambang-kehancuran. (in Indonesian).

Garaway, C. J., Photitay, C., Roger, K., Khamsivilay, L. & Halwart, M. (2013). Biodiversity and nutrition in rice-based ecosystems; the case of Lao PDR. *Human Ecology*, *41*(4), 547-562. http://dx.doi.org/10.1007/s10745-013-9602-z.

Gu, H., Jiao, Y. & Liang, L. (2012). Strengthening the socioecological resilience of forest-dependent communities: the case of the Hani Rice Terraces in Yunnan, China. *Forest Policy and Economics*, *22*, 53-59. http://dx.doi.org/10.1016/j.forpol.2012.04.004.

Hayama, A. (2003). Local forest management in the rice terrace area of Banaue, the Philippines. In Inoue, M. & Isozaki, H. (Eds.), *People and forest—policy and local reality in Southeast Asia, the Russian Far East, and Japan*. 275-286. Institute for Global Environmental Strategies 3. Dordrecht, Netherlands, Springer. http://dx.doi.org/10.1007/978-94-017-2554-5_18.

Hobart, M. (1990). The patience of plants. A note on agency in Bali. Review of Indonesian and Malaysian Studies. Special Bali Edition. *WINTER*, *24*(2), 90-135. Accessed 08/12/2020.https://www.researchgate.net/publication/ 284588005_ The_Patience _of_ Plants_A_Note_on_Agency_in_Bali.

Jha, K. K. (2015). Non-timber Forest Products, Their Vulnerability and Conservation in a Designated UNESCO Heritage Site of Arunanchal Pradesh, India. *Notulae Scientia Biologicae*, *7*, 444-455.

Jha, K. K. (2016). Some Marketing Aspects of Important Non-Timber Forest Products in a Proposed UNESCO Heritage Site of Arunanchal Pradesh, India. *Journal of Plant Chemistry and Ecophysiology*, *1*(2), 1007.

Kacha, D. (2018). The Appertains Cultural Landscape of Zero Valley in Arunachal Pradesh. *IOSR Journal of Humanities and Social Science*, *23*(5), 44-49.

Kala, C. P. (2005). Ethnomedicinal botany of the Apatani in the eastern Himalayan region of India. *Journal of Ethnobiology and Ethnomedicine*, *1*, 11. https://doi.org/10.1186/1746-4269-1-11.

Kani, T. (2012). *Cultural heritage of Hong village of the Apatani valley.* Itanagar: Hong Welfare Association Capital Complex.

Khongsai, M., Saikia, S. P. & Kayung, H. (2011). Ethnomedicinal plants used by different tribes of Arunachal Pradesh. *Indian Journal of Traditional Knowledge*, *10*(3), 541-546.

Koj, D. (2018). Women in Traditional Polity: Buliang of Apatani. *International Journal of Research in Humanities, Arts and Literature*, *6*(3), 31-38.

Kumar, A. & Ramakrishnan, P. S. (1990). Energy Flow through an Apatani Village Ecosystem of Arunachal Pradesh in Northeast India. *Human Ecology*, *18*(3), 315-336.

Lansing, J. S. (1987). Balinese "water temples" and the management of irrigation. *American Anthropologist*, *89*(2), 326-341. http://dx.doi.org/10.1525/aa.1987.89.2.02a00030.

Lansing, J. S. & Kremer, J. N. (1993). Emergent properties of Balinese water temple networks: coadaptation on a rugged fitness landscape. *American Anthropologist*, *95*(1), 97-114. http://dx.doi.org/10.1525/aa.1993.95.1.02a00050.

Lestari, N. P. N. E., Sunia, W., Wijayanthi, N. W. & Sinarti, N. N. (2017). Empowerment of "Subak" in Development of Sustainable Eco-Tourism Village in Bali Indonesia. *International Journal of Economics, Commerce and Management*, *5*(6), 796-810.

Lonely Planet (Accessed 24/10/2019). *Local life and Religion.* Retrieved from https://www.lonelyplanet.com/indonesia/bali/background/other-features/a/nar/daa4612f-26ba-4871-a7e4-cc2f89aa171f/1323403.

Lorenzen, R. P. & Lorenzen, S. (2011). Changing realities-perspectives on Balinese rice cultivation. *Human Ecology*, *39*(1), 29-42. http://dx.doi.org/10.1007/s10745-010-9345-z.

Lu, J. & Li, X. (2006). Review of rice–fish-farming systems in China —one of the Globally Important Ingenious Agricultural Heritage Systems (GIAHS). *Aquaculture*, *260*(1-4), 106-113. http://dx.doi.org/10.1016/ j.aquaculture. 2006.05.059.

Luchman, H., Kim, J. E. & Hong, S. K. (2009). Cultural landscape and ecotourism in Bali Island, Indonesia. *Journal of Ecology & Field Biology*, *32*(1), 1-8.

McDonald, C. (2017). *Illich in the Valley: Applying conviviality as lens of development in tribal India*. Master thesis, Lund University, Lund, Sweden.

McKay, D. (2003). Cultivating new local futures: remittance economies and land-use patterns in Ifugao, Philippines. *Journal of Southeast Asian Studies*, *34*(2), 285-306. http://dx.doi.org/10.1017/s0022463403000262.

MHDCRI. (2015). *Globally Important Agriculture Heritage System (GIAHS) Initiative. Integrated Agricultural System of Tri Hita Karana – Tri Mandala in Bali, Indonesia.* Coordinating Ministry of Human Development and Culture Republic of Indonesia. Accessed on 01 December 2019. http://www.fao.org/3/a-bp891e.pdf.

Mize, T., Kanwal, K. S., Rangini, N., Yama, L., Patuk, O. & Lodhi, M. S. (2016). The Current Development of Ecotourism in Ziro Valley and its significance in Arunachal Pradesh, India. *IJARIIE*, *2*(6), 1735-1743.

Miura, K. & Sarjana, I. M. (2016). The World Heritage Nomination of Balinese Cultural Landscapes, Local Struggles and Expectations. In King, V.T. (Ed.), *UNESCO in Southeast Asia: World heritage sites in comparative perspective*, chapter 12, 274-290. Copenhagen. Nordic Institute of Asian Study Press.

Nimachow, G., Rawat, J. S., Dai, O. & Loder, T. (2010). A sustainable mountain paddy-fish farming of the Apatani tribes of Arunachal Pradesh, India. *Aquaculture Asia Magazine*, *15*(2), 25-28.

Norken, I. N., Suputra, I. K. & Arsana, I. G. N. K. (2015). *Aspects of Traditional Religious Activities on Subak Irrigation: Case Study on Subak Piling, Penebel District, Tabanan Regency.* Civil Engineering Program, University of Udayana, Denpasar. (in Indonesian).

Norken, I. N., Suputra, I. K. & Arsana, I. G. N. K. (2016). Challenges to the Conservation of Subak System as World Cultural Heritage in Bali. *Journal of*

Agricultural Science and Technology, *B*, *6*, 215-225. https://doi.org/10.17265/2161-6264/2016.04.001.

Norken, I. N., Suputra, I. K. & Arsana, I. G. N. K. (2017). Institutional and Regulatory Roles in Maintaining Sustainability of Subak as a World Cultural Heritage in Bali. *Asian Agri-History*, *21*(4), 245-254.

Nozawa, C., Malingan, M., Plantilla, A. & Ong, J. E. (2008). Evolving culture, evolving landscapes: the Philippine rice terraces. In Amend, T., Brown, J., Kothari, A., Phillips, A. & Stolton, S. (Eds). *Protected landscapes and agrobiodiversity values. Volume 1. Protected landscapes and seascapes.* 71-93. International Union for Conservation of Nature (IUCN) and Deutsche Gesellschaft für Technische Zusammenarbeit (GTZ), Heidelberg, Germany.

Pant, R. (2000). *Apatani valley, Lower Subansiri.* CCA/Arunanchal/CS1/Lower Subansiri/apatani/ sustainable forest and agriculture management. Retrieved from http://www.kalpavriksh.org/images/CCA/Directory/ Arunachal_CaseStudy _Apatanivly Lower Subansiri.pdf.

Pastika, M. M. (2014). *Decision of the Governor of Bali No. 11/03-H/HK/2014.* Coordination Forum of World Cultural Heritage Management, Cultural Landscape of Bali Province, Governor of Bali, Denpasar. Retrieved from http://whc.unesco.org/document/134917. (in Indonesian).

Perme, N., Choudhury, S. N., Choudhury, R., Natung, T. & De, B. (2015). Medicinal Plants in Traditional Use at Arunachal Pradesh, India. *International Journal of Phytopharmacy*, *5* (5), 86-98.

Pitana, I. G. (1993). Subak, Traditional Irrigation System in Bali: A General Description. In Pitana, I. G. (Ed.), *Subak, Traditional Irrigation System in Bali*, Denpasar: Upada Sastra Publisher. (in Indonesian).

Pringle, R. (2004). *A short history of Bali.* Allen & Unwin, 83 Alexander Street Crows Nest NSW 2065 Australia.

Rasna, I. W. & Tantra, D. K. (2017). Medical Plants in Usadha: Loloh as Balinese Medicine and Traditional Herbal Product in Educational Perspective. 2nd International Conference on Innovative Research Across Disciplines (ICIRAD 2017). *Advances in Social Science*, *Education and Humanities Research*, *134*, 189-194.

Rechlin, M. A. & Varuni, V. (2006). A passion for pine: Forest conservation practices of the Apatani people of Arunachal Pradesh. *Himalaya*, *the Journal of the Association for Nepal and Himalayan Studies*, *26*(1), 19-24.

Regan, R. G. (2013). *The Curious Case of Bali's Subak System World Heritage Site: A Meditation on Heritage, Agency, & Development in the Light of Complexity.* Internet document accessed on 06/12/2019 file:///C:/Users/ K%20K%20Jha/Downloads/REGAN_HDCA_PAPER%20(1).pdf.

Salamanca, A. M., Nugroho, A., Osbeck, M., Bharwani, S. & Dwisasanti, N. (2015). *Managing a living cultural landscape: Bali's subaks and the UNESCO World Heritage Site.* Stockholm Environment Institute – Asia, Bangkok.

Sepe, J. (2000). *The Impact of the Green Revolution and Capitalized Farming on the Balinese Water Temple System.* Retrieved from http://ecletic.ss.uci.edu/-drwhite/ Anthro129/ Balinesewater templesJonathanSepe.htm 07/12/2019.

Sohkhlet, B. & Lalayang, D. (2017). Significance of the Buliang and need for its revitalization in sociocultural milleu of the Apatanis. *International Journal of Humanities and social Science Invention*, *2*(7), 51-56.

Sordalen, T. K. (2011). *A Traditional Irrigation System in Bali; a Study of its Transferability to Flores.* Master Thesis. University of Agder.

Srivastava, R. C., Singh, R. K. & Apatani Community & Mukherjee, R. K. (2010). Indigenous biodiversity of Apatani plateau: Learning on biocultural knowledge of Apatani tribe of Arunanchal Pradesh for sustainable livelihoods. *Indian Journal of Traditional Knowledge*, *9*(3), 432-442.

Suartika, G. A. M. (2018). Global and Domestic Spheres: Impact on The Traditional Settlement of Penglipuran in Bali. *2nd Geoplanning-International Conference on Geomatics and Planning. IOP Conferance Series: Earth Environment Science* 123 012014. https://doi.org/10.1088/1755-1315/123/1/012014.

Sundriyal, R. C., Upreti, T. C. & Varuni, R. (2002). Bamboo and cane resource utilization and conservation in the Apatani plateau, Arunachal Pradesh, India: implications for management. *Journal of Bamboo and Rattan*, *1*(3), 205–246.

Sujarwo, W., Arinasa, I. B. K., Caneva, G. & Guarrera, P. M. (2014). Traditional knowledge of wild and semi-wild edible plants used in Bali (Indonesia) to maintain biological and cultural diversity. *Plant Biosystems*, *150*(5), 971-976. https://doi.org/10.1080/11263504.2014.994577.

Sujarwo, W. & Caneva, G. (2015). Ethnobotanical Study of Cultivated Plants in Home Gardens of Traditional Villages in Bali (Indonesia). *Human Ecology*, *43* (5), 769-778.

Sushila, J. (1987). *Typical Characteristics of Subak Irrigation System in Bali.* Sub-department of Irrigation, Public Works Department of Bali Province, Denpasar. (in Indonesian).

Sutawan, N. (2008). *Organization and Management of Subak in Bali.* Denpasar, Indonesia. Pustaka Bali Post Publisher.

Sutama, I. N. (2016). *Awig-awig of Subak Jatiluwih, Jatiluwih Village, Tabanan Regency, Bali, Indonesia.*

Sutomo. (2007). Exploration and inventory of the "usada bali" medicinal plants in some areas of buleleng district, Bali. In Siregar, M., Siregar, H.M., Arinasa, I. B. K. & Lestari, W. S. (Eds.), *Seminar Konservasi Tumbuhan Usada Bali dan Peranannya dalam Mendukung Ekowisata 2007*; 92-7. UPT Balai Konservasi Tumbuhan Kebun Raya"Eka Karya" Bali Lembaga Ilmu Pengetahuan Indonesia, Kebun Raya Bali. Accessed 08/01/20120. https://www.research gate.net/publication/263285453_Ethnobotanical_Study_of_Usada_Medicinal _Plants_in_Buleleng_District_Bali.

Syuaib, M. F. (2009). Perspective of sustainable agriculture in Indonesia: Keep growing in harmony with environment. In: *Proceedings of Postgraduate GP Education Workshop on From Environmental to Sustainable Science: Thinking the Shift and the Role of Asian Agricultural Science*, 12–13 January 2009; 93–99, Ibaraki, Japan.

Tayo, T., Safi, V., Tadang, M., Heli, T., Tabyo, T. & Longjam, N. (2017). Integrated Paddy, Fish and Finger Millets Cultivation by Apatani Tribes in the Eastern Himalayan Region- Arunachal Pradesh. *Indian Journal of Hill Farming*, *30*(1), 63-69.

Thakuria, V. (2017). Stakeholders Sustenance in Tourism Progression in Ziro Valley of Arunachal Pradesh, India. *International Journal of Scientific Research and Management*, *5*(10), 7324-7328.

Tirtawati, N. M., Dianasari, D. A. M. L. & Saputra, I. G. G. (2019). The Challenge of Young Generation Involvement in Sustainable Tourism Development: Case Study of Jatiluwih Tourism Village Bali-Indonesia. *Journal of Advanced Management Science*, *7*(1), 25-32.

Truong, V. D., Hall, C. M. & Garry, T. (2014). Tourism and poverty alleviation: perceptions and experiences of poor people in Sapa, Vietnam. *Journal of Sustainable Tourism*, *22(7*), 1071-1089. http://dx.doi.org/10.1080/09669582. 2013.871019.

Ukirsari, M. (2012). UNESCO Plaque, Ecognition Subak as a World Heritage 2012. Accessed September 25, 2012. http://nationalgeographic.co.id/berita/2012/09/ plakat-unesco-pengakuan-subak-sebagai-warisan-dunia-2012. (in Indonesian).

Wiguna, A. A., Lorenzen, R. P. & Lorenzen, S. (2005). Past, Present and Future – Perspectives of Balinese Rice Farming. *International Rice Conference*, 12-14 September 2005, New Delhi.

Willer, H. & Kilcher, L. (2011). *The World of Organic Agriculture. Statistics and Emerging Trends.* International Federation of Organic Agriculture Movements (IFOAM): Bonn, Germany; Research Institute of Organic Agriculture (FiBL): Frick, Switzerland, 2011. http://orgprints.org/19310/.

Windari, R. A. & Hadi, I. G. A. A. (2017). Regulations Availability in the Management of Community Based Tourism (CBT) Villages in the Village of Jatiluwih. *4th Int'l Conference on Studies in Law, Education, Business & Corporate Social Responsibilities (LEBCSR-17)*, Aug. 8-9, 2017 Singapore.

Yakang, B., Gajurel, P.R., Potsangbam, S. & Bhuyan, L. R. (2013). Account of common and traditional non-timber forest products used by Apatani tribe of Arunachal Pradesh, India. *Pleione*, *7*(2), 514-529.

Yekti, M. I. (2017). *Role of reservoir operation in sustainable water supply to subak irrigation schemes in yeh ho river basin.* CRC Press/Balkema, Leiden, The Netherlands. http://dx.doi.org/10.18174/404538.

Yekti, M. I., Schultz, B. & Hayde, L. (2013). Development of a conceptual approach to manage flow of Subak irrigation schemes in Bali, Indonesia. *Proceeding 1st World Irrigation Forum*, 29 September - 3 October 2013, Mardin, Turkey.

Yudantini, N. M. & Jones, D. (2015a). The Conservation of Balinese Traditional Architecture: The Integration of Village Pattern and Housing Pattern in Indigenous Villages. *Applied Mechanics and Materials*, *747*, 84-87.

Yudantini, N. M. & Jones, D. (2015b). The role of traditional regulation in the indigenous villages for Conservation of village pattern. *21st International Sustainable Development Research Society Conference Tipping Point: Vulnerability and Adaptive Capacity*, ISDRS Society 2015, Deakin University, Geelong, VIC-Australia, 10-12 July 2015. http://programme.exordo.com/isdrs 2015/delegates/presentation/279/.

Yudantini, N. M. (2018). The spatial and settlement pattern in Mabi Hamlet, Belantih Village, Kintamani: conservation of tangible and intangible of Balinese traditional architecture in Bali Aga villages. *IOP Conf. Series: Earth and Environmental Science*, *213*, (2018), 012025 IOP Publishing. https://doi.org/doi:10.1088/1755-1315/213/1/012025.

In: Critical Research Techniques in Animal … ISBN: 978-1-53619-720-4
Editors: Kaushalendra Kumar Jha et al.

Chapter 8

THERMAL DRONES AND LARGE MAMMALIAN CARNIVORES

Michael O'Neal Campbell*
Simon Fraser University, British Columbia, Canada

ABSTRACT

Thermal infrared cameras mounted on drones have emerged as powerful tools for the detection, recording and later information syntheses of animal species. Large mammalian carnivores (LMCs) are a key foraging guild that may be studied by these methods, being generally elusive and evasive, difficult to approach and sensitive to human presence, arguably rivals for human life spaces and safety, and dominant actors in current conservation policy and social media. This chapter looks at the current status and capability of thermal drone technology for studying LMC ecologies and the problems, advantages and prospects of this application, using some current literature and social media sources. International examples are cited, and supportive cases of applications to smaller mammals are presented as test cases. Evidence derived from these sources indicates thermal infrared drones are poised to make incisive contributions to the knowledge of the ecology of LMCs, but some serious albeit surmountable limitations exist. Key issues concern the remoteness and irregularity of LMC movements and habitats, the ability of the drone to penetrate the highly diverse and sometimes covered vegetation of carnivore habitats, the legal restrictions on drone operation in urban areas and the possible dangers to drone operators in LMC proximity. Current evidence indicates that these restrictions may be managed with political will, and the technological and logistical issues surrounding drone use are being ameliorated. More knowledge may be required of the impacts of such technologies on LMCs and their ecology. Such technologies may also be tested on smaller animals for preliminary assessments. These findings make an important contribution to wildlife ecology and conservation.

---

* Corresponding Author's E-mail: ecol55@hotmail.com.

Introduction

Drones, unmanned aerial vehicles or uncrewed aerial vehicles (UAV) have gradually emerged as important tools for ecological research (Anderson & Gaston, 2013; Bushaw et al. 2019). The sophistication of such "eco-drones" or "conservation drones" (Ivošević et al. 2015, 113) has increased with the increasingly complex nature of the ecological applications. New technological advances include autopilot and advanced computer software systems, higher resolution Global Positioning System (GPS) receivers, compasses, video cameras and photographic technology, and crucially, infrared and/or ultraviolet optical sensors, physical sensors for measuring temperature and humidity and sound sensors (Ivošević et al. 2015; Ditmer et al. 2015a; Bushaw et al. 2019). The advantages of drones over terrestrial survey techniques create "the potential to revolutionize the way research is conducted in many scientific fields" as drones "can access remote or difficult terrain, collect large amounts of data for lower cost than traditional aerial methods, and facilitate observations of species that are wary of human presence" (Ditmer et al. (2015a, 2278; see also Anderson & Gaston, 2013; Marris, 2013; Vermeulen et al. 2013; Bennitt et al. 2019).

Some drone techology can calibrate for extreme detail. For example, cameras installed on drones have been developed that can take photographs of animal tracks and use algorithms to identify the relevant animals. ConservationFIT developed image analyzing techology that can "read" digital images of animal footprints captured from smartphones, cameras or drones and accurately identify the species, sex and age of the animal that made the tracks, and even match tracks to individual animals" (Lucas, 2017). This software is generic enough to identify the footprints of any species, using preloaded images to create progressively accurate algorithms. This is termed the footprint identification technique (FIT) algorithm. Field-validation on numerous species was tested on large mammalian carnivores (LMCs) such as African lions, cougars, cheetahs, snow leopards, tigers, giant pandas (*Ailuropoda melanoleuca*, David, 1869), tapirs (*Tapirus*, Brünnich, 1772), red wolves (*Canis lupus rufus*, Audubon & Bachman, 1851) and smaller animals such as coyotes (*Canis latrans*, Say, 1823). This sophsticated technology, using images taken from handheld instruments or drones allows the tracking of elusive and rare animals (Lucas, 2017).

Thermal infrared imaging or infrared thermography cameras on drones goes even further, as it is based on the mounting special cameras that use the part of the electromagnetic spectrum outside the visible light bands. Thermal drones have emerged as a key research tool, prinarily due to their ability to monitor animal movements in darkeness. Thermal technology was recently added to drones (unmanned aerial vehicles or uncrewed aerial vehicles, UAV or remotely piloted aircraft, RPA, which are defined as remotely controlled aircraft without onboard humans), which also include global positioning systems (GPS) and ordinary field cameras (Witczuk, 2018). This thermal technology allows the image and video capture of scenes and animals in darkness, relying on the infrared bands rather than the light bands of the electromagnetic spectrum to detect minute temperature differences. Warm blooded animals can be differentiated from their cooler environment, even to near total darkness (Spaan et al. 2019). Crucially, many large carnivores are totally or partly nocturnal, increasing the utility of such methods (Campbell, 2017). Such methods work during both night and day. As long as temperature differences between animals and the environment can be discerned, "the detection of heat radiation and the high contrast of thermal imaging cameras allow users to see the outlines of (warm-blooded) animals that camouflage well in their environments. It can also help distinguish animals (or people) in foggy or smoky conditions" (Fernando (2017).

All cameras are sensors that emit photons that travel from the emitting sensor to the target object, using part of the electromagnet spectrum. "Visible light falls between infrared and ultra violet and constitutes only a tiny fraction of energy. Radio waves, microwaves, infrared waves, ultraviolet waves X-rays and gamma rays make up the rest of the electromagnetic spectrum" (Livingston, 2015, 12). Table 1 below shows the position of light and infrared bands in the spectrum as defined by a source from the National Aeronautics and Space Administration NASA (2018). The optical range includes the visible bands of the spectrum, from red to violet in rainbow colors. As the thermal infrared (TIR) imaging allow the measurement of temperature difference and hence can detect mammals in the cooler temperatires of their environment, hence can also detect such animals in the night or total darkness (Bushaw et al. 2019). This is crucial for the study of noturnal animals, such as large cats (Campbell, 2017). However, Bushaw et al. (2019) caution on the limitations of thermal imaging:

> "The amount of radiation emitted is dependent on both the temperature of the object and emissivity, which is a measure of the reflectivity of an object. The image generated by the thermal camera is therefore not a depiction of the absolute temperature or infrared radiation of an organism, but a combination of the targets' radiation, emissivity and environmental factors such as humidity and cloud cover, and (critically) the temperature of the background against which the animal is observed."

Table 1. The electromagnetic spectrum

Name	**Wavelength (m)**
Radio	$> 1 \times 10^{-1}$
Microwave	1×10^{-3} - 1 x $^{10-1}$
Infrared	7 x 10-7 - 1 x 10-3
Optical	4 x 10-7 - 7 x 10-7
Ultraviolet UV	1 x 10-8 - 4 x 10-7
X-ray	1 x 10-11 - 1 x 10-8
Gamma-ray	< 1 x 10-11

Source: Adapted from NASA (2018).

Source: Reproduced with permission of DJI (2019).

Figure 1. Phantom 4 Pro and Infrared. This drone has a forward vision, backward vision downward vision system, and an infrared sensing system.

Figure 1 shows an example of a drone that mounts an infrared (IR) camera. The basis of the infrared recording of animals is the creation of algorithms for animal or speccies identification. Quick (2018) describes technologies developed by astrophysicists and ecologists at Liverpool John Moores University. This involved firstly the capture of the "unique thermal profiles of various animals, including rhinos and baboons, and build up a library of different animal thermal signatures" and then "field tests to detect endangered animals in their native habitats" (Quick, 2018). The field tests focussed on the very small, endangered Riverine rabbits, which were detected when the drone flew at an elevation of 20 meters (66 ft). Problems included vegetation obstructrion, requiring software to detect animals behind vegetation. Temperature similarities between animals and the environment required compensatory software. Studies examining drone-based infrared imaging for ecological studies, are summarized below and listed in Table 2. Bushaw et al. (2019: 28) argue "existing field survey techniques such as track plates and motion-sensitive camera traps are time-consuming and expensive, and yet still yield data prone to systematic errors." By contrast, unmanned aerial vehicles (UAVs) are "eclipsing the efficiency and even accuracy of traditional methods" even in the assessment of nocturnal, elusive animals like coyotes (Bushaw et al. 2019: 28).

Table 2. Examples of drones used for infrared-based studies

No.	Animal	Drone	Camera
1.	Koala bear	S800 EVO Hexacopter	Forward-looking infrared (FLIR) camera, Tau 2-640 model
2.	Red deer, Roe deer	AVI-1 aircraft, Skywalker X8 Flying Wing	Thermal infrared (TIR) camera, IRMOD v640 model
3.	Monkeys, kinkajou	DJI Matrice 600Pro	DJI Zenmuse XT® V2.0 FLIR, TIR radiometric sensor, Canon EOS 5DS® RGB
4.	Bornean Orangutan	Tarot X4	Dual thermal-visible 33 spectrum camera. Thermal Capture Fusion Zoom
5.	Meso-carnivore	DJI Inspire 1 quadcopter UAV	DJI Zenmuse XT2 R thermal imaging camera
6.	Spider monkey	Custom-built	TeAx Fusion Zoom dual vision camera, FLIR Tau2 640 core TIR, TAMRON RGB

Note: The numbered points in column 1 correspond to the following numbered references. 1. Gonzalez et al. (2016); 2. Witczuk et al. (2018); 3. Kays et al. (2019); 4. Burke et al. (2019); 5. Bushaw et al. (2019); 6. Spaan et al. (2019). The full scientific name for the koala is Phascolarctos cinereus (Goldfuss, 1817), red deer is Cervus elaphus L.1758, the roe deer is Capreolus capreolus (Linnaeus, 1758). The monkeys were Mantled howler monkey (Alouatta palliata, Gray, 1849), Geoffroy's spider monkey or Black-handed spider monkey (Ateles geoffroyi, Kuhl, 1820), and the kinkajous were Potos flavus (Schreber, 1774). Bornean orangutan is Pongo pygmaeus (Linnaeus, 1760). Spider monkey is Ateles spp. Mesocarnovore refers to medium sized carnivores such as the coyote (Canis latrans, Say, 1823), red fox (Vulpes vulpes, Linnaeus, 1758), striped skunk (Mephitis mephitis. Schreber, 1776) or raccoon (Procyon lotor, Linnaeus, 1758).

Bushaw et al. (2019) list some of the advantages of using thermal imaging technology as such tools have been used for ecological research for years, but an important development was the "the miniaturization of thermal imaging cameras" which "allowed them to be attached to an UAV and used for everything from monitoring wildfires, flying search and rescue missions, searching farm land for animals before mowing operations, and conducting wildlife surveys" (see also Haschberger et al. 1996; Ambrosia et al. 2003; Kastberger et al. 2003; Collier et al. 2007; Rudol & Doherty, 2008; . Israel, 2011).

Witczuk (2018, 5504) used thermal infrared imaging tools on drone platforms to study large ungulates and concluded that such technology "may have great potential as a tool for ungulate surveys" enabling "safe operations at low flying altitudes, and at night – a time that often offers the optimal conditions for wildlife monitoring" and were able to demonstrate that "ungulate thermal signatures are visible both in leafless deciduous and in pine-dominated coniferous forests" especially at dawn, dusk and night. The resolution of approximately 10 cm enabled the identification of large mammals, but there were "difficulties in species identification due to relatively low resolution of TIR cameras, regulations limiting drone operations to visual line of sight, and high dependence on weather."

Spaan et al. (2019) also contrast the use of a thermal infrared (TIR) camera on a drone with a ground survey, arguing that the latter is hampered by high costs and logistical difficulties. For a case study of spider monkeys (*Ateles geoffroyi*, Kuhl, 1820) the technology could determine the number of monkeys in forested vegetation. Drone counts were approximately like ground counts, but larger numbers of monkeys were counted with the drone counts in 83% of the surveys, mostly due to the large ground coverage. The impact of this study concerned the focus on arboreal mammals, differing from other studies that surveyed ground dwelling small mammals and larger ungulates (citing Witczuk et al. 2018; Burke et al. 2019) and linking with others that focused on arboreal species such as koalas, howler monkeys, spider monkeys, kinkajous, orangutans (Gonzalez et al. 2016; Burke et al. 2019; Kays et al. 2019). The authors concluded that "the use of a TIR camera is a feasible and valid method of obtaining data on the presence and number of spider monkeys at their sleeping sites in a closed-canopy forest" (Spaan et al. 2019).

These studies exemplify and describe the utility of thermal equiped drones for the study of wildlife. However, few studies have documented the application of such technology for the study of LMCs. The application to smaller mammals is useful, as such animals are more difficult to detect (due to their size and their ability to hide in vegetation) and require extremely sharp resolutions, especially when species identification is required. However, considering the serious problems faciing large carnviores (human-wildlife conflicts, local and possible global extinction, and local rarity) studies of such species must attain prominence (Campbell & Lancaster, 2010; Campbell & Torres Alvarado, 2011; Campbell, 2012, 2014, 2015; Kuijper et al. 2016). For example, Inskip and Zimmerman (2009: 18) "found evidence of conflict affecting over 75% of the world's felid species. The severity of conflict increases with felid body mass and is of greatest conservation significance to nine species: caracal, cheetah, Eurasian lynx, jaguar, leopard, lion, puma, snow leopard and tiger.' Suryawanshi et al. (2014: 1666) add that "The threat posed by large carnivores to livestock and humans makes peaceful coexistence between them difficult."

Tracking such animals at the individual, micro-level is crucial as with human population expansion into LMC habitats, conflicts may emerge that require assessments of particular individual animals, requiring a degree of identification above that of species identification or general size assessments. The justifications would be the emerging confliucts between people and LMCs, (Woodroffe, 2005; Campbell & Torres Alvarado, 2011; Kansky et al. 2014) and the need to secure public support, as people's reactions to individual LMC amy influence attitudes ranging from conserrvation support to direct hostility (Grumbine, 1990; Karanth & Stith, 1999; Robinson & Bennett, 2000; Balme et al., 2010). However, few such studies have been undertaken. The rest of this chapter will look at the parameters of large carnovore ecology that may be measured by drones.

LARGE CARNIVORE SURVEYS

The LMCs considered in this chapter occur on all the large continents, North and South America and Eurasia, but are mostly rare or even threatened (Campbell, 2017). These are the Ursines, the large Felines Canids and hyenas, listed in Table 2 below. The classification of "large mammalian carnivore" is fairly arbitrary, but some writers have attempted classifications (Campbell, 2017). For example, Bushaw et al. (2019) contrast mesocarnivores with apex predators and define the former as "carnivores with a body weight <15 kg) such as coyotes," hinting that the apex predators are those with a body weight >15kg.

Table 2. Large Terrestrial Mammal Carnivores of the World

Species	Scientific name	Range	Weight (kg)
Polar bear	Ursus maritimus, Phipps, 1774		150 - 700
Brown bear	Ursus arctos, L. 1758	Eurasia, N.America	152 - 217
Black bear	Ursus americanus, Pallas, 1780	N. America	41 - 250
Spectacled bear	Tremarctos ornatus, Cuvier, 1825	S. America	35 - 200
Asiatic black bear	Ursus thibetanus, G. Cuvier, 1823	Asia	40 - 200
Sloth bear	Melursus ursinus, Shaw, 1791	Asia	55 - 145
Lion	Panthera leo, Linnaeus, 1758	Africa, Asia	90 - 200
Tiger	Panthera tigris, Linnaeus, 1758	Asia	65 - 306
Leopard	Panthera pardus, Linnaeus, 1758	Africa, Asia	28 - 90
Jaguar	Panthera onca, Linnaeus, 1758	N. & S. America	36 - 158
Snow leopard	Panthera uncia, Schreber, 1775	Asia	22 - 75
Cougar	Puma concolor, Linnaeus, 1771	N. & S. America	29 - 100
Cheetah	Acinonyx jubatus, Schreber, 1775	Africa, Asia	21 - 72
Clouded leopard	Neofelis nebulosi, Griffith, 1821	Asia	12 - 23
Wolf	Canis lupus, Linnaeus, 1758	Eurasia, N.America	12 - 79
African wild dog	Lycaon pictus, Temminck, 1820	Africa	18 - 36
Spotted hyena	Crocuta crocuta, Erxleben 1777	Africa	41 - 64
Striped hyena	Hyaena hyaena, Linnaeus, 1758	Africa, Asia	22 - 55
Brown hyena	Hyaena brunnea, Thunberg, 1820	Africa	38 - 44

Sources: Mazák, V. 1981; Nowell & Jackson, 1996; Heptner & Naumov, 1998. Mills & Hofer, 1998; Macdonald, 2001; Macdonald & Norris, 2001; Derocher & Wiig, 2002; Sunquist & Sunquist, 2002a,b; Feldhamer et al. 2003; Estes, 2004; Nowak, 2005; Swenson et al. 2007; Campbell, 2012, 2017; Campbell & Torres Alvarado, 2011; Hunter, 2011; Johnsingh & Manjrekar, 2013; West & Packer, 2013; Castelló, 2018).

As shown in Table 2, by this clasification, all the mammals selected for the current study cluster in weight well above the demarcated limit of Bushaw et al. (2019). Several species have a pronounced size range, largely due to sexual dimorphism, with males generally much larger than females and subspecies size ranges in some cases quite pronounced. Some of the subspecies size ranges may be due to climate in accordance with Bergmann's rule (Meiri & Dayan, 2003).

Two key areas for LMC research are animal recording in remote or complex landcover, where orthodox research techniques are problematic, and animal penetration into human dominated (especially urbanized) landcover, where elusive animals are difficult to detect (Campbell, 2017). Drones are applicable to such research because their remote, aerial location may be undetected by the LMC target, which not therefore take the evasive action common against more visible human or large aircraft trackers. Thermal drones have the added advantage in that they may operate during night and total darkness, which coincides with foraging periods for most LMCs (especially felines) and the lowest visbility of the drone. Additionally, declines in LMC populations contributes to the greater difficulty in tracking these animals in their native habitat, their tendency to move into more protective, remote locations and their lower tolerence of human presence (Inskip & Zimmermann, 2009; Hunter, 2011; Campbell, 2017). Advanced drone tracking technologies are also needed, as increased urbanization and the adaptatblility of some LMCs to human presence and life spaces leads to LMc penetration of urban substrates, with resultant human/animal conflicts (Campbell, 2014). Selected, illustrative examples are cited below. These are not exhaustive, but reflect some of the key issues that have emerged in recent drone-based studies of LMCs.

Tracking of Elusive LMCs in Remote Habitat

The drone-measurable parameters for LMCs, with primary utlility for thermal infrared techology would principally concern the tracking of animals during night, periods of poor visibility or in areas of limited access for foot observation. Issues may concern migration movements, urban scavenging, interspecific and intraspecific conflicts, predation, local extinction and possible reintroduction, and implications for human quality of life. For example, GeoNet (2019) describes a project using a UAV (DJI - Matrice 210 drone, equiped with a radiometric infrared sensor (Zenmuse XT2) and the 30X zoom Zenmuse camera to study snow leopards in Mongolia. This project is led by top-class researchers Jackson, Hunter and Munkhtsog (GeoNet, 2019). The website notes that "both sensors offer capabilities to improve upon current censusing methods that rely on visual observation along a 2km wide by 4-5 km long transect. The infrared sensor detects warm-blooded animals especially well early morning late evening. The zoom camera can clearly magnify ungulate-sized animals from 0.75 km distance." The use of drones is justified through the difficulties encountered for other research methods. The website notes that "this iconic feline is incredibly elusive, rarely sighted and known for occupying home ranges in imposing habitat as large as 1,000 sq km.

Snow leopards are sparsely distributed at elevations of 2,000 – 5,400 m in the mountains of 12 Asian countries: Afghanistan, Bhutan, China, India, Kazakhstan,

Kyrgyz Republic, Mongolia, Nepal, Pakistan, Russia, Tajikistan, and Uzbekistan." Furthermore, the primary goal of this study is to validate drone technology as an effective tool to advance snow leopard conservation by improving the censusing of its large prey species like ibex, argali (photo, right) and blue sheep." To this end there are four objectives: to "successfully deploy and test a drone in snow leopard prey habitat"; to "test thermography for detecting wild argali and ibex (early morning, late evening)"; to "test a drone-mounted zoom lens for detecting wild argali and ibex (daylight); and to "access ungulate survey accuracy with and without drone support" (GeoNet, 2019).

Another study tracked polar bears in the Arctic ice using their body temperature contrast with their cold environment. Intel (2017) argues for the use of the Intel Falcon 8+ thermal drones for such studies, in that

> Traditional methods involving helicopters for exploration are invasive and costly. Paired with the treacherous arctic conditions, including freezing temperatures and heavy winds, exploration from a boat in these remote areas is challenging even for the most experienced researchers. In addition, the steel found in most boats can cause magnetic fields that challenge drone compasses making it extremely difficult to take off and land on a moving boat. Deploying drones is an alternative that provides greater access to researchers and wildlife experts.

In this study the effect of drones on bears was assessed and found to be insignificant (Intel, 2017). The thermal photography was able to detect bears in the ice, even when they could not be detected with the naked eye. The data gathered shed light on the bears' ecological behavior, breeding, feeding and migration habits and crucially may be useful in the determination of the impacts of climate change (Sigfusson, 2017).

Another planned study envisages using thermal cameras on drones to spot hibernating bears (SPI Infrared, 2014). Hibernating bears burn calories during sleep, allowing thermal equipment on passing drones to detect their heat signatures. The thermal camera manufactured by SPI Corp, the M1D Thermal Camera for drones is ideal for this purpose. The object in detecting hibernating bears is to monitor their locations, to provide this information to the oil and gas companies that work in these environments. These companies are legally required by federal law "to maintain a distance of one mile away from polar bear dens and one half mile away from grizzly dens between the months of November to April so as not to disturb their hibernation with noise or vibrations from drilling operations or vehicle movements" (SPI Infrared, 2014).

Another study used a thermal drone to record a lion pride's night hunting in Kruger Nationl Park, South Africa (Kesteloo, 2018). Being undetected by the lions, the drone arguably captured scenes undeliverable by other research methods. Kesteloo (2018) argues that it "provides a totally new perspective on the pride's movement through the South African bush. The lions are playful at times but seem hungry and are on the hunt looking for something to eat." This was a pride that had been monitored using older assessment methods, but the drone-based methods allowed new insights due to the lions unawareness of close human observation. Observation of lion behvior during drone approach and surveys alos revealed that lions were not affected by drone noise, despite their night vision (Kesteloo, 2018).

Thermal drones have also become a key mainstay of the Indian anti-poaching methods, especially for night observations when most poaching of large mammals such as nocturnal tigers occurs (Jyoti, 2019). Jyoti (2019) justifies the use of thermal drones: "systems combining artificial intelligence, drones and infra-red imaging have been designed to monitor animals at night in order to prevent poaching. The drones can survey large areas of difficult terrain from above, allowing ecologists to access hard to reach areas and monitor wildlife without disturbing the animals." However, recent legislation in India, termed 'The Drone Regulations 1.0' regulated by the Directorate General of Civil Aviation (DGCA) was implemented in December 2018. It is arguably negative for the work of certain wildlife agencies, such as the National Tiger Conservation Authority (NTCA) and Wildlife Institute of India (WII). These agencies, under the Union Environment Ministry, require the right to use drones nocturnally, both in protected areas and along international borders that share the Indian tiger reserve boundaries. They may also be hampered in their work by the requirement that drone operators keep a constant "full visual line of sight" with the drone. The policy restrictions were also supported by a No Permission- No Takeoff' (NPNT) clause, for drone operation, in addition to the only flight during daylight restriction. Cited reserves most affected include those of Dudhwa, Pilibhit, Valmiki, Buxar and Sundarbans, (Jyoti, 2019).

The theme of monitoring not just the LMCs and other animals, but the poachers as well is developing internationally. The World Wildlife Fund WWF (2016) states "WWF's thermal infrared camera imaging and human detection software stops poachers in their tracks." The citation is that the WWF "installed a new thermal infrared camera that can identify poachers from afar by their body heat—even in the dead of night—and it has since transformed the way rangers track down and apprehend criminals..." The thermal streaming video is powerful enough to scan warm-blooded animals and people for over one kilometer, even in near total darkness. The report concludes that the WWF "is working with FLIR Systems Inc., as part of a new collaboration to broaden the use of thermal imaging, and African Parks, UDS and Lindbergh Foundation's Air Shepherd to install similar thermal imaging technology in drones. Anti-poaching drone test flights began in Zimbabwe and Malawi in October of this year" (2016) (WWF (2016).

Another developing theme is the use of thermal infrared imaging to monitor LMC health. Gregory (2018) cites the example of wolves in Yellowstone National Park, which are infected with sarcoptic mange, a contagious disease caused by the mite *Sarcoptes scabiei* (Linnaeus, 1758). This infection causes extreme itching, Gregory (2018) notes that this is a "form of scabies that causes canines to lose their hair, which translates into lost calories as the body tries to keep itself warm with holes in its shaggy coat; infrared camera can monitor this "because heat is escaping through bald patches, these spots show up on thermal images of wolves wandering the wilderness" and as " using infrared is less invasive than traps, radio collars, or injections" scientists can "collect body surface temperatures for each sick wolf, which helps them determine how many calories the mangy canines are losing relative to healthy wolves and therefore how long they have to live" (Gregrory, 2018).

Drone Surveys and Urban Wildlife

Urban ecology is emerging as a vital research topic, fueled by the rapid global urbanization, and the consequent intrusion of large animals into urban areas and urban settlements into animal habitats. These developments necessitate the utilization of more sophisticated, flexible and technical research and survey technqiues for the assessment of ecological change (Campbell, 2017, 2018). LMC conflicts with people in dense urban, suburban and periurban areas has increased human/wildlife conflict, which be be investigated by new research methods (Campbell, 2014). Such conflicts may impact on the disputed concept termed human quality of life (QOL) (Campbell, 2017).

Human quality of life is a wide and even polarized concept, but it generally comprises: (a) emotional well-being, (b) interpersonal relationships, (c) material well-being, (d) personal development, (e) physical well-being, (f) self-determination, (g) social inclusion, and (h) human rights. These may vary in different cultural settings, but a global concesus may cluster around these points (Felce & Perry,1995, 1996; Hughes & Hwang, 1996; Cummins, 1997a 1997b; Schalock et al. 2002). Animals that intrude into human life spaces may enhance, complement, degrade or even destroy human quality of life (Campbell & Lancaster, 2010; Schaltegger & Beständig, 2012; Campbell, 2012, 2017).

LMCs create a new and unique dimension, competing with people for landcover intrinsic for QOL, such as urban parks, greenbelts, suburban forest reserves, forests marked for hiking and riverbanks, and even killing companion animals and livestock (Herrero & Higgins, 1999; Whittaker & Burns, 2001; Treves & Karanth, 2003; Kleiven et al., 2004; Loe & Roskaft, 2004; Gore et al., 2006; Spencer et al., 2007; Kuijper et al. 2016; Campbell, 2017). Hence, the presence of LMCs may even remove some of the mentioned components of QOL, either physically or indirectly through human percepations, which may be real or imagined (Elorriaga et al., 2000; Linnell, 2001; Williamson, 2002; Bowman et al., 2001, 2004; Gore et al. 2006; Karanth & Chellam, 2009; Fonturbel & Simonetti, 2011).

Cougar presence in urban areas is one of the more important issues concerning urban wildlife in North America (Campbell, 2014). Measurable parameters for drone-based studies are complicated, due to the elusive and evasive nature of this species. A major dificulty is that cougar presence in urban areas is conditioned by their individualist, non-generalist behavior, this presence prediction difficult (Kertson et al. 2013). Cougars are much less dangerous to people than other larger cats such as the tigers, lions and leopards in the Old World (Campbell & Torres Alvarado, 2011). Compared to North American brown bears, Compared with brown bears, the cougar is generally less feared, despite its strongly predatory assessment (Kellert et al., 1996; Campbell 2012). However, compared with black bears, cougars may be perceived as more dangerous to humans especially children (Beier, 1991; Campbell & Lancaster, 2010; Campbell, 2012).

Cougars may be present in urban green or mixed green and built areas, as transients, occasional inhabitants or permanent residents, all of which may involve hunting, migratory transit and territorial marking (Campbell 2014). Cougars may frequent areas outside urban green spaces, such as mixed green and built areas. For example, Shuey (2008) reports no linear relation between cougar presence and green areas, with more cougar-human encounters in grassland and suburban land-cover types than in evergreen tree growth, with greater cougar presence in built areas than common assessments. Some cougars used residential matrix as corridors between green spaces (Kertson et al345|"., 2011).

Few studies report on the use of thermal drones for the study of such urban cougar dynamics. However, Reyes-Velarde (2018) reports an incident where a young cougar was sighted near office complex in an East Bay suburb, San Francisco, California was tranquilized and relocated to o the Sunol Regional Wilderness in Alameda County, California. Reyes-Velarde (2018) noted that "a drone with a thermal imaging camera from the Alameda County Sheriff's Office was on standby to track the young cat if it didn't fully succumb to the tranquilizer darts and ran off." The Associated Press (2018) noted that "Police in Pleasanton, California, and California Department of Fish and Wildlife officers used a drone with thermal imaging cameras to track the puma, which appears to be about a year old and healthy at about 80 pounds."

Brookes (2018) gives the example of TIR drones being used to answer questions concerning large cats, possibly including cougars being sighted in Australia. Brookes (2018) argues that "deeply ingrained in hills folklore are tales of big cats roaming the bush... Still to this day no one has successfully documented a panther in the wild, despite thousands of sightings... thermal imaging drones and motion activated camera traps were being used by filmmakers to finally capture proof of the legendary big cats once and for all." The habitat under study was ideal for feral domestic cats (*Felis catus*, Linnaeus, 1758) and larger cats such as the cougar. This is recorded as the first time TIR drones were used for such a study in Australia, the justification being that "documented sightings pointed to a healthy, established population of cougars and panthers, and possibly black leopards, surviving and thriving in outer lying areas of Perth, including Yanchep" in Australia (Brookes, 2018).

Limitations on the Use of Thermal Infrared Drones

There are significant legal limitations on the use of drones that may hamper some studies of LMCs. As noted above, severe restrictions have been placed on thermal drone operation over India tiger reserves, despite the obvious utility of these methods (Jyoti (2019). In another example, in the United States, drone use may be restricted near airports. Shepardson (2019) argues "there has been rising concern about drones near U.S. airports." The Federal Aviation Administration (FAA) has informed U.S. airports that only authorized federal agencies could install systems for the detection and destruction of drones and airport authorities require federal consent to install such systems, especially as "they could pose an aviation safety risk by interfering with aircraft navigation and air navigation services" (Shepardson, 2019). In 2018, the US Congress granted the Justice and Homeland Security departments authority to disable or destroy drones (Shepardson, 2019).

The National Park Service banned drones from National Parks (Policy Memorandum 14–05 issued by the NPS on June 19, 2014), without a Special Use Permit, issued only for applications such as search and rescue, research and fire safety (Drone U, 2018). In the United States recreational and commercial uses are distinguished. Recreational rules include flight within sight, drones weighing less than 25 kg, stay 8 km from airports (without notification or permission) and avoid aircraft and emergency response efforts. Commercial rules include flight below 122 meters, drone weights less than 25 kg, flight during daylight to twilight, less than about 162 km, avoid flight over people and operation of drones from a moving vehicle (unless in areas with low human presence) and yield to manned aircraft (UAV Coach, 2019).

In Canada, drones may not be used within 5.6 kilometers from any airport listed as Certified ("Cert") in the Canada Flight Supplement and 1.9 kilometers heliports or aerodromes used by helicopters (Government of Canada, 2019a). Crucially, drones may not be used in Canadian National Parks under Parks Canada supervision. This includes recreational flying of drones, while commercial drone use may be allowed with a permit. Such Restricted Activity Permits may be issued for the use of a drone for "natural or cultural resource management and protection, public safety, law enforcement, or park/site management purposes directly relating to park administrative purposes" (Government of Canada, 2019b).

New regulations were implemented in Canada in 2019, applied to drones weighing "250 grams (g) up to and including 25 kilograms (kg) and are operated within the drone pilot's visual-line-of-sight" (Lee, 2019). The new legislation also gave equal treatment to recreational and commercial drone usage. Under this legislation, drone operators must register and acquire required certificates, and identify their usage as basic, which means the drone is flown in uncontrolled airspace, more than 30 meters horizontally from people and not over people.

Deviating from these rules implies that the operator is conducting advanced operations in controlled airspace, within 30 meters of people and/or over people. Drones must respect human privacy and be flown within sight, below 122 meters elevation, fly away from people, avoid forest fires and social events, stay 5.6 kilometers from airports and 1.9 kilometers from heliports and avoid other aircraft. These regulations may restrict the use of thermal imaging drones for tracking LMCs across landcover, including airports, suburban farms, downtown areas and other places with dense urban substrate and private gardens, and in flight above about 130 meters (Lee, 2019).

Conclusion

Thermal infrared-equiped drones clearly represent *advant-garde* tools for research on LMCs, their behavioral adaptations, landcover occupance and relations with people. This chapter has examined two key research areas where drone-mounted thermal infrared camera may make the biggest contrution: the recording of rare, elusive LMCs in natural habitat, the better to learn more about their threatened and unknown statuses; and the tracking of common LMCs in conflict zones such urban substrates, contributing information to prevailing debates on LMC adaptability, reactions to human presence and consequent human attitudes to conservation and removal of such animals. Evidence indicates that the thermal camera-equiped drone may enable a new look at these issues and provide groundwork for more advanced ecological studies.

Current uses of these advanced drones are in their infancy, hence conclusive assessments are premature. The limitations of this study include the small number of case studies, the broad, inter-continental sources and the greater number of applications to non-LMc species. Much of the knowledge of the effectiveness of such studies may be based on the assessment of studies of non-LMC species. While such an approach may offer some preliminary information, this may be limted as studies of LMCs are unique; LMCs are large and potentially dangerous to people, are generally much rarer than smaller species, may be more elusive and evasive and have large hunting and breeding ranges. Limitations on the use of the drones however center not so much on the LMCs and their habitats, but on the legal restictions of drone use. Perhaps the increased interactions between people and LMCs in urban areas and the requirements of global conservation politics, some of the restrctions may be relaxed to allow more detailed studies of LMCs in different contexts.

Further developments for drones may involve the contruction of ever smaller, quieter, more mobile machines and sharper cameras, possibly equiped with even sharper soun recording equipment, more accurate global positioning systems, and more support technology such as fixed field or motion cameras, satellite images and aerial photographs and more detailed ecological studies of the studied contexts. Comparative studies of single species across multiple landcover types and international cases and of different species in similar habitat types may also make a contribution. Further studies may assess the impact of such developments through the near future.

References

Ambrosia, V. G., Wegener, S. S., Sullivan, D. V., Buechel, S. W., Dunagan, S. E., Brass, J. A., Stoneburner, J. & Schoenung, S. M. 2003. Demonstrating UAV-acquired real-time thermal data over fires. *Photogrammetric Engineering and*

Remote Sensing (PE&RS), 69, 391–402. https://doi.org/10.14358/PERS.69.4.391.

Anderson, K. & Gaston, K. J. 2013. Lightweight unmanned aerial vehicles will revolutionize spatial ecology. *Frontiers in Ecology and the Environment*, 11(3): 138-146. https://doi.org/10.1890/120150.

Associated Press 2018. *California Mountain Lion Caught Using Thermal Imaging Drone.* Retrieved from https://www.apnews.com/0254751f2e4c4027a1e7c94a24763b98.

Balme, G. A., Slotow, R. & Hunter, L. T. B. 2010. Edge effects and the impact of nonprotected areas in carnivore conservation: leopards in the Phinda–Mkhuze Complex, South Africa. *Animal Conservation,* 13: 315-323. https://doi.org/10.1111/j.1469-1795.2009.00342.x.

Beier, P. 1991. Cougar attacks on humans in the United States and Canada. *Wildlife Society Bulletin* 19, 403-412. https://www.jstor.org/stable/3782149.

Bennitt, E., Bartlam-Brooks, H. L. A., Hubel, T. Y. & Wilson, A. M. 2019. Terrestrial mammalian wildlife responses to Unmanned Aerial Systems approaches. *Scientific Reports*, 9(1):2142. http://www.doi:10.1038/s41598-019-38610-x.

Boitani, L. 1984. *Simon & Schuster's Guide to Mammals*. New York, NY: Simon & Schuster/Touchstone Books.

Bowman, J. L., Leopold, B. D., Vilella, F. J., Gill, D. A. & Jacobson, H. A. 2001. Attitudes of landowners toward American black bear compared between areas of high and low bear populations. *Ursus*, 12, 153-160. https://www.jstor.org/stable/3873243.

Bowman, J. L., Leopold, B. D., Vilella, F. J. & Gill, D. A. 2004. A spatially explicit model, derived from demographic variables, to predict attitudes toward black bear restoration. *Journal of Wildlife Management*, 68(2): 223-232. https://doi.org/10.2193/0022-541X(2004)068[0223:ASEMDF]2.0. CO;2.

Burke, C., Rashman, M. F., Longmore, S. N., McAree, O., Glover-Kapfer, P., Ancrenaz, M. & Wich, S. A. 2019. Successful observation of orangutans in the wild with thermal-equipped drones. *Journal of Unmanned Vehicle Systems*, 1-23. https://doi.org/10.1139/juvs-2018-0035.

Bushaw, J. D., Ringelman, K. M., & Rohwer, F. C. 2019. Applications of Unmanned Aerial Vehicles to Survey Mesocarnivores. *Drones*, 2019, 3(1), 28. https://doi.org/10.3390/drones3010028.

Campbell, M. & Lancaster, B. 2010. Public attitudes towards black bears Ursus americanus and cougars *Puma concolor* on Vancouver Island. *Journal of Society and Animals* 10(1): 47-51. https://doi.org/10.1163/106311110X12586086158448.

Campbell, M. & Torres Alvarado, M. E. 2011. Public perceptions of jaguars Panthera onca, pumas Puma concolor and coyotes Canis latrans in El Salvador. *Area* 43(3): 250 – 256. https://doi.org/10.1111/ j.1475-4762.2011.00996.x.

Campbell, M. 2012. The Relevance of age and gender for public attitudes to brown bears (*Ursus arctos*), black bears (*Ursus americanus*), and cougars (*Puma concolor*) in Kamloops, British Columbia. *Society and Animals* 21(4): 341 – 359. https://doi.org/10.1163/15685306-12341260.

Campbell, M. 2014. Urban planning for cougar presence in North America: practices, challenges and benefits. In Amado, M. P. (Ed.) *Urban planning: Practices, challenges and benefits*, pp. 19-32. New York, NY: Nova Science Publishers.

Campbell, M. 2015. The factors for the extinction of jaguars and cougars in El Salvador. Journal of Biodiversity, *Bioprospecting & Development*, 3; 154. https://doi.org/10.4172/2376-0214.1000154.

Campbell, M. 2017. *Biological Conservation for the 21st Century: A Conservation Biology of Large Wildlife*. New York, NY: Nova Science Publishers.

Campbell, M. 2018. *Geomatics and Conservation Biology*. Nova Science Publishers, New York.

Castelló, J. R. 2018. *Canids of the World: Wolves, Wild Dogs, Foxes, Jackals, Coyotes, and Their Relatives*. Princeton: Princeton University Press.

Christiansen, P., Steen, K. A., Jørgensen, R. N. & Karstoft, H. 2014. Automated Detection and Recognition of Wildlife Using Thermal Cameras. *Sensors*, 14, 13778-13793. https://www.doi:10.3390/s140813778.

Collier, B. A., Ditchkoff, S. S., Raglin, J. B. & Smith, J. M. 2007. Detection probability and sources of variation in white-tailed deer spotlight surveys. *Journal Wildlife Management*, 71, 277–281. Retrieved from http://www.jstor.org/stable/4495174.

Cummins, R. A. 1997a. Assessing quality of life. In Brown, R. J. (Ed.) *Assessing Quality of Life for People with Psychiatric Disabilities*, 116-150. Cheltenham: Stanley Thornes (Publishers).

Cummins, R. A. 1997b. *Comprehensive Quality of Life Scale – Adult. School of Psychology*. Melbourne: Deakin University.

Derocher, A. E. & Wiig, Ø. 2002. Postnatal growth in body length and mass of polar bears (Ursus maritimus) at Svalbard. *Journal of Zoology*, 256 (3): 343–349. https://doi.org/10.1017/ S0952836902000377.

Drone U. 2018. *What Are the Rules for Flying Your Drone in National Parks and National Forests?* Retrieved from https://medium.com/@askdroneu/what-are-the-rules-for-flying-your-drone-in-national-parks-and-national-forests-1d27554984ad.

Elorriaga, J., Garcia, L., Martinez, J. & Unamunzaga, E. 2000. Quality of life of persons with mental retardation in Spain. In Keith, K. D. & Schlock, R. L. (Eds.) *Cross-cultural perspectives on quality of life*. Washington, DC: American Association on Mental Retardation.

Estes, R. D. 2004. Cheetah Acinonyx jubatus. In *The Behavior Guide to African Mammals: Including Hoofed Mammals, Carnivores, Primates* (4th ed.), 377–383. Berkeley, CA: University of California Press.

Felce, D. & Perry, J. 1995. Quality of life: its definition and measurement. *Research in Developmental Disabilities*, 16(1), 51-74. https://doi.org/10.1016/0891-4222(94)00028-8.

Felce, D. & Perry, J. 1996. Assessment of quality of life. In Schalock, R. L. & Siperstein, G. N. (Eds.), *Quality of Life Volume I: Conceptualization and Measurement,* 63-70. Washington DC: American Association on Mental Retardation.

Feldhamer, G. A., Thompson, B. C. & Chapman, J. A. 2003. *Wild Mammals of North America: Biology, Management, And Conservation*. Baltimore. MD: Johns Hopkins University Press.

Fernando, S. 2017. *Seeing in the Dark and More: Facts and FAQs About Thermal Imaging.* Retrieved https://news.mongabay.com/2017/03/facts-and-faqs-about-thermal-imaging/.

Fonturbel, F. E. & Simonetti, J. A. 2011. Translocations and human-carnivore conflicts: problem solving or problem creating. *Wildlife Biology*, 17: 217-224.

GeoNet 2019. *Drones for Snow Leopard Conservation: Detecting and Counting Prey for an Endangered Species in Mongolia*. Retreived from https://community.esri.com/docs/DOC-13613-drones-for-snow-leopard-conservation.

Gore, M. L., Knuth, B. A., Curtis, P. D. & Shanahan, J. E. 2006. Education programs for reducing black bear-human conflicts: Indicators of success? *Ursus*, 17(1): 75-80.

Gonzalez, L. F., Montes, G. A, Puig, E., Johnson, S., Mengersen, K., Gaston, K. J. 2016. Unmanned Aerial Vehicles (UAVs) and artificial intelligence revolutionizing wildlife monitoring and conservation. *Sensors,* 16: 97. https://doi.org/10.3390/s16010097.

Government of Canada 2019a. *Where to Fly Your Drone*. Retrieved from https://www.tc.gc.ca/ en/services/aviation/drone-safety/where-fly-drone.html.

Government of Canada 2019b. *Use of Unmanned Aerial Vehicles (UAV) or drones in Canada's national parks.* Retrieved from https://www.pc.gc.ca/en/pn-np/ab/banff/info/permis-permit/drone.

Gregory, V. 2018. *Infrared Imaging Helps Save Yellowstone's Wolves*. Retrieved from https://www.wired.com/2012/05/st-photo-wolves/

Grumbine, R. E. 1990. Viable populations, reserve size, and federal lands management: a critique. *Conservation Biology*, 4: 127-134. https://doi.org/10.1111/j.1523-1739.1990.tb00101.x.

Haschberger, P.; Bundschuh, M.; Tank, V. 1996. Infrared sensor for the detection and protection of wildlife. *Optical Engineering*, 35, 882–889. https://doi.org/10.1117/1.601005.

Heptner, V. G. & Naumov, N. P. 1998. *Mammals of the Soviet Union Vol. II Part 1a, Sirenia and Carnivora (Sea cows; Wolves and Bears)*. Boca Raton: Science Publishers, Inc.

Herrero, S. & Higgins, A. 1999. Human injuries inflicted by bears in British Columbia: 1960 - 1997. *Ursus*, 11: 209-218. https://www.jstor.org/stable/3873003.

Hughes, C. & Hwang, B. 1996. Attempts to conceptualize and measure quality of life. In Schalock, R. L. (Ed.), *Quality of Life. Vol. I: Conceptualization and Measurement*, 51-61. Washington, DC: American Association on Mental Retardation.

Hunter, L. 2011. *Carnivores of the World*, Princeton: Princeton University Press.

Inskip, C. & Zimmermann, A. 2009. Human–felid conflict: a review of patterns and priorities worldwide. *Oryx* 43, 18–34. https://www.doi:10.1017/S003060530899030X

Intel 2017. *From Polar Bears to Whales, Intel Pushes the Boundaries of Wildlife Research with Drone and Artificial Intelligence*. Retrieved from https://newsroom.intel.com/news-releases/from-polar-bears-to-whales-intel-pushes-the-boundaries-of-wildlife-research-with-drone-and-artificial-intelligence/#gs.x9oo3y

Israel, M. A. 2011. UAV-based roe deer fawn detection system. The International Archives of the Photogrammetry. *Remote Sensing and Spatial Information Sciences*, 38: 1–5.

Johnsingh, A. J. T. & Manjrekar, N. (Eds.). 2013. *Mammals of South Asia*. Himayatnagar: Universities Press.

Jyoti, A. 2019. Anti-poaching drones face shooting down by DGCA. *The Pioneer*. Retrieved from https://www.dailypioneer.com/2019/pioneer-exclusive/anti-poaching-drones-face-shooting-down-by-dgca.html.

Kansky, R., Kidd, M. & Knight, A. T. 2014. Meta-analysis of attitudes toward damage causing mammalian wildlife. *Conservation Biology*, 28: 924–938. http://www.doi:10.1111/cobi.12275 PMID: 24661270.

Karanth, K. U. & Chellam, R. 2009. Carnivore conservation at the crossroads. *Oryx*, 43: 1-2. https://doi.org/10.1017/S003060530843106X.

Karanth, K. U. & Stith, B. M. 1999. Prey depletion as a critical determinant of tiger population viability. In Seidensticker, J., Christie, S. & Jackson, P. (Eds.), *Riding the Tiger: Tiger conservation in Human-Dominated Landscapes*, 100-113. Cambridge, UK: Cambridge University Press.

Kastberger, G. & Stachl, R. 2003. Infrared imaging technology and biological applications. *Behavior Research Methods, Instruments & Computers*, 35: 429–439. http://www.doi:10.3758/BF03195520.

Kays, R., Sheppard, J., Mclean, K., Welch, C., Paunescu, C., Wang, V., Kravit, G. & Crofoot, M. 2019. Hot monkey, cold reality: Surveying rainforest canopy mammals using drone-mounted thermal infrared sensors. *International Journal of Remote Sensing*, 40: 407–419. https://doi.org/10.1080/ 01431161.2018.1523580.

Kellert, S. R., Black, M., Rush, C. R. & Bath, A. J. 1996. Human culture and large carnivore conservation in North America. *Conservation Biology*, 10(4), 977-990. https://doi.org/10.1046/j.1523-1739.1996.10040977.x.

Kertson, B.N., Spencer, R.D., Marzluff, J.M., Hepinstall-Cymerman, J. & Grue, C. E. 2011. Cougar space use and movements in the wildland–urban landscape of western Washington. *Ecological Applications*, 21: 2866–2881. https://doi.org/10.1890/11-0947.1.

Kertson, B. N., Spencer, R. D. & Grue. C. E. 2013. Demographic influences on cougar residential use and interactions with people in western Washington. *Journal of Mammalogy*, 94(2): 269–281.

Kesteloo, H. 2018. *A thermal drone shows the Nkuhuma Pride of lions hunt at night in Kruger National Park*. Retrieved from https://dronedj.com/2018/07/23/a-thermal-drone-shows-the-nkuhuma-pride-of-lions-hunt-at-night-in-kruger-national-park/.

Kleiven, J., Bjerke, T. & Kaltenborn, B. P. 2004. Factors influencing the social acceptability of large carnivore behaviours. *Biodiversity and Conservation*, 13: 1647- 1658. https://doi.org/10.1023/B:BIOC.0000029328.81255.38.

Kruuk, H. 1972. *The Spotted Hyena: A Study of Predation and Social Behaviour*. Berkeley, CA: University of California Press.

Kuijper, D. P. J., Sahlén, E., Elmhagen, B., Chamaille´-Jammes, S., Sand, H., Lone, K. & Cromsigt, J. P. G. M. 2016. Paws without claws? Ecological effects of large carnivores in anthropogenic landscapes. *Proceedings of the Royal Society of Biology*, 283. http://dx.doi.org/10.1098/rspb.2016.1625.

Lee, I. 2019. *Canada Announces New Drone Laws That Allow Recreational and Commercial Pilots to Operate Under Same Rules* | Plans to Implement June 2019. Retrieved from https://uavcoach.com/canada-new-drone-laws/.

Linnell, J. D. C., Swenson, J. E. & Anderson, R. 2001. Predators and people: conservation of large carnivores is possible at high human densities if management policy is favorable. *Animal Conservation*, 4: 345-349.

Livingston, S. 2015. Commercial remote sensing satellites and the regulation of violence in areas of limited statehood. *CGCS Occasional Paper Series on ICTs,Statebuilding, and Peacebuilding in Africa* No. 5. Philadelphia, PA: University of Pennsylvania.

Loe, J. & Roskaft, E. 2004. Large carnivores and human safety: A review. *Ambio*, 33(6): 283-287.

Lucas, T 2017. New Project Uses Phones and Drones to Monitor Endangered Species. Retreived from https://nicholas.duke.edu/about/news/new-project-uses-phones-and-drones-monitor-endangered-species.

Macdonald, D. 2001. *The New Encyclopedia of Mammals*. Oxford: Oxford University Press.

Macdonald, D. W. & Norris, S. 2001. *Encyclopedia of Mammals*. Oxford: Oxford University Press.

Marris, E. 2013. Fly, and bring me data – Unmanned aerial vehicles are poised to take off as popular tools for scientific research. *Nature*: 498: 156–158.

Mazák, V. 1981. Panthera tigris. *Mammalian Species*. 152 (152): 1–8. https://www.doi:10.2307/ 3504004.

Meiri, S. & Dayan, T. 2003. On the validity of Bergmann's rule. *Journal of Biogeography,* 30 (3): 331–351. https://www.doi:10.1046/j.1365-2699. 2003.00837.x.

Mills, G. & Hofer, H. 1998. *Hyaenas*. Gland: IUCN/SSC Hyaena Specialist Group

National Aeronautics and Space Administration NASA (2018). *Regions of the Electromagnetic Spectrum.* Retrieved from https://imagine.gsfc.nasa. gov/science/toolbox/spectrum_chart.html

Nowak, R.M. 2005. *Walker's Carnivores of The World*. Baltimore, MD: Johns Hopkins University Press.

Nowell, K. & Jackson, P. 1996. *Wild Cats*. Gland, Switzerland: International Union for Conservation of Nature and Natural Resources.

Quick, D. 2018. *Astronomy and Ecology Come Together in Drone Designed to Tackle Poaching Problem.* Retrieved from https://newatlas.com/animal-poaching-astro-ecology-drone/54066/

Reyes-Velarde, A. 2018. *Young Mountain Lion Found Lounging In Pleasanton is Released to the Wild.* Retreived from https://www.latimes.com/local/lanow/la-me-ln-mountain-lion-pleasanton-20181030-story.html

Robinson, J. G. & Bennett, E. L. 2000. *Hunting For Sustainability in Tropical Forests.* Irvinton: Columbia University Press.

Rudol, P. & Doherty, P. 2008. Human body detection and geolocalization for UAV search and rescue missions using color and thermal imagery. In *Proceedings of the 2008 IEEE Aerospace Conference*, 1–8. Big Sky, MT.

Schalock, R. L, Alonso, M. A. V. & Braddock, D. L. 2002. *Handbook on Quality of Life for Human Service Practitioners.* Washington DC: American Association on Mental Retardation.

Schaltegger, S. & Beständig, U. 2012. *Corporate Biodiversity Management Handbook: A Guide for Practical Implementation.* Federal Ministry of the Environment, Berlin.

Shuey, M. L. 2008. The influence of landscape on cougar-human encounter hazard in Boulder and Colorado Springs, Colorado. *ETD Collection for Texas State University.* https://digital.library.txstate. edu/handle/10877/3004.

Sigfusson, L. 2017. *Drones Are Keeping Watch on the Arctic's Polar Bear Population.* Retrieved from http://blogs.discovermagazine.com/drone360/2017/10/04/polar-bear-drone/#.XVo4sOhKi70.

Shepardson, D. 2019. *FAA to U.S. airports: Do Not Install Drone Countermeasures On Your Own.* Retrieved from https://www.reuters.com/article/us-usa-drones-airports/faa-to-u-s-airports-do-not-install-drone-countermeasures-on-your-own-idUSKCN1SE2QZ.

Spaan, D., Burke, C., McAree, O., Aureli, F., Rangel-Rivera, C. E., Hutschenreiter, A., Longmore, S. N., McWhirter, P. R. & Wich, S. A. 2019. Thermal infrared imaging from drones offers a major advance for spider monkey surveys. *Drones*, 32:34. https://doi.org/10.3390/drones3020034.

Spencer, R. D., Beausoleil, R. A. & Martorello, D. A. 2007. How agencies respond to human black bear conflicts: A survey of wildlife agencies in North America. *Ursus*, 18(2): 217-229.

SPI Infrared 2014. Thermal Cameras on Drones to Spot Hibernating Bears in Study. Retrieved from https://www.x20.org/thermal-cameras-on-drones-bears/

Sunquist, M. & Sunquist, F. 2002a. Snow leopard *Uncia uncia* (Schreber, 1775). *Wild Cats of the World*, 377–394. Chicago: University of Chicago Press.

Sunquist, M. & Sunquist, F. 2002b. Clouded leopard *Neofelis nebulosa* (Griffiths, 1821). *Wild Cats of the World*, 278–284. Chicago: University of Chicago.

Suryawanshi KR1, Bhatia S, Bhatnagar YV, Redpath S, Mishra C. 2014. Multiscale factors affecting human attitudes toward snow leopards and wolves. *Conservation Biology*, 28(6):1657-66. https://doi.org/10.1111/cobi.12320.

Swenson, J. E., Adamič, M., Huber, D. & Stokke, S. 2007. Brown bear body mass and growth in northern and southern Europe. *Oecologia*, 153 (1): 37–47. https://www.doi:10.1007/s00442-007-0715-1.

Treves, A. & Karanth, K. U. 2003. Human-carnivore conflict and perspectives on carnivore management worldwide. *Conservation Biology*, 17, 1491-1499. https://doi.org/10.1111/j.1523-1739.2003.00059.x.

UAV Coach 2019. *Drone Laws in the United States of America*. Retrieved from https://uavcoach.com/drone-laws-in-united-states-of-america/.

Vermeulen, C., Lejeune, P., Lisein, J., Sawadogo, P. & Bouché, P. 2013. Unmanned aerial survey of elephants. *PLoS One*, 8(2):e54700. https://doi.org/10.1371/journal.pone.0054700.

West, P. M. & Packer, C. 2013. Panthera leo Lion. In Kingdon, J., Happold, D., Butynski, T., Hoffmann, M., Happold, M. & Kalina, J. (Eds.). *Mammals of Africa*, 150–159. London, UK: Bloomsbury Publishing.

Whittaker, D. G. & Burns, A. G. 2001. Black bear status in western North America: Summary of western state and province bear status report surveys. *Western Black Bear Workshop*, 7: 32-55.

Williamson, D. F. 2002. *Status, Management, and trade of the American Black Bear (Ursus americanus) in North America*. Washington D. C.: World Wildlife Fund.

Witczuk, J., Pagacz, S., Zmarz, A. & Cypel, M. 2018. Exploring the feasibility of unmanned aerial vehicles and thermal imaging for ungulate surveys in forests – preliminary results. *International Journal of Remote Sensing*, 39:15-16, 5504-5521. https://doi.org/10.1080/01431161.2017. 1390621.

Woodroffe, R., Thirgood, S. & Rabinowitz, A. 2005. The impact of human-wildlife conflict on natural systems. In Woodroffe, R., Thirgood, S. & Rabinowitz, A. (Eds.). *People and Wildlife Conflict or Coexistence?* 1 -12. Cambridge, UK: Cambridge University Press.

World Wildlife Fund 2016. *New Anti-Poaching Technology Leads to Dozens of Arrests of Wildlife Criminals in Africa*. Retrieved from https://www.worldwildlife.org/stories/new-anti-poaching-technology-leads-to-dozens-of-arrests-of-wildlife-criminals-in-africa.

In: Critical Research Techniques in Animal ...
Editors: Kaushalendra Kumar Jha et al.
ISBN: 978-1-53619-720-4

Chapter 9

REHABILITATION OF MINE SPOILS: A REVIEW ON PROBLEMS AND STRATEGIES WITH CASES OF INDIAN MINES

***Abhay Kumar Singh*[1,3] *and Kaushalendra Kumar Jha*[2,3*]**
[1]Ministry of Mines, Government of India, New Delhi, India
[2]Indian Institute of Forest Management, Bhopal, India
[3]Department of Forest, Environment and Climate Change, Uttar Pradesh, Lucknow, India

ABSTRACT

Mining has been an inevitable activity for the development of any civilization throughout history. With increasing demands for minerals and metals in the second half of the last century, the pace of excavation for these raw materials increased markedly and the ecological restoration of intensively mined areas fell behind the excavation in maximum parts of the globe. The possible restoration of such ecologically damaged areas is a major concern for ecologists, and a crucial challenge in several countries. Reversal of soil strata, instability of slopes, and poor quality of left-over soils, and other adverse factors have been linked to other environmental, social and political issues, these requiring rapid and efficient solutions. The juxtaposition of mining areas with forest ecosystems and human habitation aggravates the problem. This chapter examines these issues in a case of study of bauxite mining in India. Topics discussed include the problems of mining and strategies to deal with specific targets of rehabilitation, and the evidence of these problems presented in the Indian and international literature. The methodology for this review uses abstract book consultation, and online and hardcopy literature searches. From amongst the numerous searches only 252 documents were incorporated in this synthesis. The review begins by examining the legal provisions for tackling the mining problems in India. This is followed by

* Corresponding Author's E-mail: jhakk1959@gmail.com.

the assessment of rehabilitation problems, such as plantation versus mining spoil, mine-spoil impoverishment, erosion and water conservation, and mine-spoil toxicity. Management interventions include top-soil replacement, soil amendments, seeding and planting, and species selection and revegetation. This discussion is followed by a case study of bauxite mine rehabilitation in Koraput District, Odisha, India. It is argued that a classical restoration model, such as the Bradshaw model would be exceedingly difficult, if not impossible to implement in this area, because the original ecosystem of the area would have to be recreated. An incomplete restoration may be termed as 'rehabilitation'. The chapter notes that the extraction method (and the refilling of the extracted area) for bauxite, using the trench method is less problematic than the opencast mining of iron ore or coal. The gradient of the filled trenches is usually gentle in bauxite mines, making it easier to rehabilitate with plantation development. The chapter concludes with an assessment of the application of the classical Bradshaw model of rehabilitation for the stabilization of disturbed areas, and the social and environmental implications of such a development. This assessment is useful for habitat, plant and animal ecology.

INTRODUCTION

Minerals are vital raw materials for infrastructure, capital goods and basic industries, and hence, are valuable resources for development. The extraction and management of minerals must be integrated with the overall strategy of national economic development. For example, India has substantially exploited mineral resources, and has the geological environment for many others. The role of minerals and metals in the evolution of human civilization in this country can be gauged from the fact that various stages of civilization have been named with the progress in the usages of minerals and metals at various points of time, such as Stone Age, Iron Age and Bronze Age. The presence of old mine workings and slag heaps show that mining industry did flourish in olden times in this country. However, as early as 1993, The National Mineral Policy (NMP) was announced to cope up with the basic structural reforms in fiscal, industrial and trade regimes for globalization of economy, initiated by the Government of India in 1991. The process of liberalization proceeded with increasing allowances in foreign direct investments which resulted into quantum jump in number of applications for reconnaissance permits, prospecting licences and mining leases for various minerals.

Development planning in India has always been driven by an obsession to catch up with the developed nations, making the economic development antagonistic to conservation priorities, and leading to ecosystem degradation and the depletion of biodiversity. Wali (1987) has defined ecosystem degradation as an event or a series of events that results in altering, both spatially and temporally, the relationships of organisms and their habitats from their natural state. Mining, both surface and underground, drastically disturbs ecosystems. Surface mining of minerals creates vast stretches of derelict lands which are technically speaking areas of 'no value', or to be more precise, areas of 'negative value' which means that whereas the area after mining has lost its all ecological and socio-economic yield capability, it may also be a threat to the ecological and socio-economic stability (Soni et al., 1992). Due to surface mining, the geology-soil-plant stability circuit is disrupted and the flora-fauna-hydrological relations and soil biological systems are drastically disturbed. Soil is lost both in pedological and biological senses (Jha and Singh, 1992). It is widely recognized and accepted by industrial and regulatory bodies alike that mining and mineral processing activities create damage to the surrounding land, air, and water that cannot be reversed by nature, but requires the aggressive application of carefully planned restoration programmes (Ripley et al., 1996).

The Stockholm Conference on the Human Environment 1972, considered as the official start of international environment awareness resulted in The Brundtland Commission's 1987 report, *Our Common Future*, which motivated both the governments and non-governmental organizations throughout the globe to think over the concept of sustainable development seriously. Clüsener-Godt (2004) envisaged that sustainable development is concerned both with biodiversity and cultural diversity, in so far as it seeks ecosystem-specific and culture-specific responses to peoples' needs based on environmentally sound resource-use patterns. International debates on sustainable development and conservation of biodiversity gathered speed, focusing on the need to ensure that 'the needs of the present' can be met 'without compromising the ability of future generations to meet their needs' (Treweek, 1996). The idea of sustainable development explicitly recognizes the limited nature of environmental resources. The United Nations Conference on Environment and Development (UNCED), held in 1992 in Rio de Janeiro, which adopted a global plan of action popularly known as Agenda 21, represented a partial 'coming of age' of the international environmental movement. The Johannesburg World Summit on Sustainable Development, in 2002, recognized that poverty eradication, changing consumption and production patterns, and managing the natural resource base for economic and social development are overarching objectives and essential requirements for sustainable development (MoEF, 2010). It is now well recognized that environmental issues are fundamental to both social and ecosystem wellbeing, as well as to sustainable economic development. World Development Report of the World Bank focused in 1992 on the link between development and the environment and highlighted opportunities for 'win-win' policies that are good both for the environment and economic development.

Rehabilitation/reclamation of mines, which may act as a link between development and the environment as far as mining sector is concerned, is now considered an essential element of resource management and is an ongoing activity throughout the life of the operation of mining. Mining is a temporary land use and the mining industry, governmental land use management agencies, and the public have long recognized the need to consider criteria to determine when rehabilitation, reclamation, or restoration is complete, and the mine sites can be released back to public management (Gardner and Bell, 2007). Therefore, the exploitation of minerals must be guided by long term national goals and perspectives, partly to ensure future social and environmental benefits. Mining areas often contain other resources, creating conflicting landuse choices. Some such areas are ecologically fragile and some are biologically rich. Therefore, prevention and mitigation of adverse environmental effects due to mining of minerals and rehabilitation of affected areas should be in accordance with the latest internationally acceptable norms, and modern afforestation practices should form integral part of mine development strategy in every instance. National Mineral Policy, 2008 has also envisaged that no mining lease would be granted to any party, private or public, without a proper mining plan including the environmental management plan (EMP) approved and enforced by statutory authorities. EMP should adequately provide for controlling the environmental damage, restoration of mined areas and afforestation according to the prescribed norms.

Governing Principles and Laws

Though environmental awareness has been an integral part of ancient culture of India, the impetus came after Stockholm declaration in 1972. After this, various Acts and Rules governing protection of environment were promulgated. Directives for the State and its citizens have been laid in Articles 48A and 51A of The Constitution of India for protection and improvement of the environment and safeguarding the forests and wildlife of the country. For regulating the development of mines and minerals in the country, the Mines and Minerals (Development and Regulation) Act (MMDR Act), 1957, the Mineral Concession Rules (MCR), 1960 and the Mineral Conservation and Development Rules (MCDR), 1988 were promulgated. Sections 12 and 18 of the MMDR Act, 1957, Rules 22(5) (iv) & 22 A (1) of the MCR, 1960 and Rules 31 to 41 of the MCDR, 1988 are of interests for the present subject of ecological rehabilitation after mining. Similarly, Sections 3, 5, 9 and 15 of the Environment (protection) Act, 1986 are concerned with the protection and improvement of the environment. A notification (January 1994) under Environment (Protection) Act, 1986 was issued, mandating environmental impact assessments before clearing new developmental projects and expansion of ongoing projects. Rehabilitation of mines drew strength also from Principles 3 and 4 of the Rio Declaration on Environment and Development, of which India is one of the signatories. This envisaged the right of development with equitability in meeting the needs of present and future generations, constituting environmental protection as an integral part of the development processes.

Rehabilitation Related Problems

Where land has been destroyed by mining or a similar activity, what is left is something exceedingly basic - just the raw materials, the skeleton from which a soil can be formed and which with plants and animals, will develop into an ecosystem

(Bradshaw, 1987). During a mine's operational stage, irrespective of whether it is an underground or open pit mine, the main impact upon local land surfaces arises from the disposal of waste materials (Ripley et al., 1996). Bradshaw (1987) is of the view that the basic principles of land and ecosystem restoration are the same as the basic principles of ecological succession, although in many situations primary rather than secondary succession is dealt with. However, in both the circumstances, factors determining the development of ecosystem from very skeletal beginnings and factors restricting the development are of interest.

Ecosystem development can be quantified in two dimensions, structure and function. In natural succession there is an increase in both dimensions. When ecosystems are degraded by mining or other operations, there is reduction in both dimensions, perhaps almost to zero. The first option with such degraded or derelict ecosystem is to do nothing, in which case it may recover slowly by natural process, though it may also degrade further by erosion or landslip. The second option is to try to build back exactly what was there before. If this is successful then what is achieved is “restoration.” If it is not completely successful then what is achieved can be termed “rehabilitation.” The third option is “replacement,” in which an alternative to the original ecosystem is produced. The whole process is depicted in Figure 1.

Surface materials exposed by mining activities frequently appear sterile, lacking in humus and living matter, and sometimes may be extremely toxic and entirely unsuitable for plant life. In this situation two avenues are open. The first is to rely on the natural processes of weathering, leaching and erosion, but the time required for stabilization and revival of the ecosystem by natural processes could span several human lifetimes, possibly disrupting surrounding ecosystems in the meantime. The second option is to reclaim the site through an active programme involving a combination of techniques including revegetation (Ripley et al., 1996). It is, therefore, desirable that on completion of the mining operations, the mined areas should be returned to a form suitable for further use. In the context of sustainable development, this should conform to the surrounding landscape from ecological, economic and social point of view. From the socio-political view point, it is important that mining as an alternative use be accompanied by rehabilitation techniques which can demonstrate that the mined land can be successfully retuned to agriculture, preferably profitable grain production. Mining cannot be accepted as a land use alternative if it is seen as a terminal use; neither can erosion-inducing agricultural methods be tolerated (Rusell and Roberts, 1986).

However, the waste rock dumps generated after surface mining have various problems of physical, chemical and biological properties, limiting the choice of species to be planted, and the success of such planted species for revegetation of mine spoil. Wali and Kollman (1977) have rightly pointed out that the unrehabilitated post mining physical, chemical and biological characteristics of the mine spoils create a specific ‘mining ecology’. The goal of restoration is usually to develop a long-term sustainable ecosystem native to the area where mining occurred and to accelerate natural successional processes so as to increase biological productivity, reduce rates of soil erosion, increase soil fertility and increase biotic control over biogeochemical fluxes within the recovering ecosystems (Young, 2000; Singh et al., 2002; Maiti and Das, 2004; Maiti and Ghose, 2005; Maiti, 2006; Zedler, 2005; Grant, 2006).

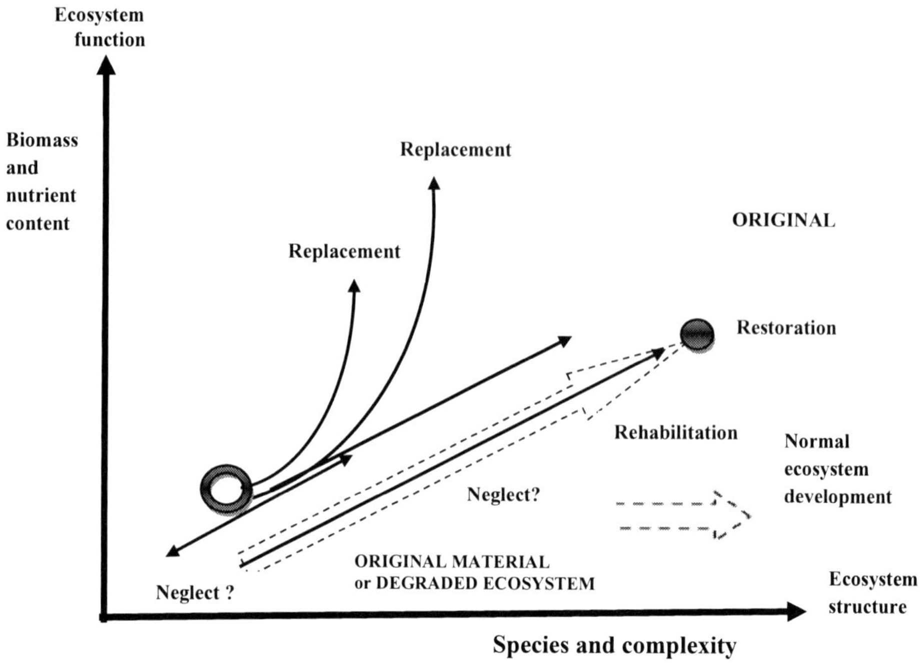

Figure 1. The diagrammatic representation of the ecosystem structure and functioning with respect to mining activities. Adopted from Bradshaw, 1987.

The large-scale land disturbances associated with mining operations and related concerns about the environmental effects have triggered an increasing number of rehabilitation programmes which aim for the restoration of natural ecosystems disturbed by mining. Restoration of mine sites often entails amelioration of physical and chemical characteristics of substrate, ensuring the return of vegetation cover (Bradshaw, 1987). If specific problems hindering ecosystem redevelopment can be identified, a cure can be designed using or mimicking natural processes. According to Dobson et al. (1997), this process of identification and intervention is the essence of ecological restoration.

The most common response to land degradation has been abandonment or reliance on natural succession to restore lost soil fertility, species richness and biomass productivity (Parrotta et al., 1997 a, b). However, the process of natural succession on surface-mined soil is slow due to the removal of topsoil, resulting in elimination of soil seed bank and root stocks and due to soil profile disturbances (Parrotta et al., 1997 b). It has been argued that if environmental change produced by disturbance is large, it may become lethal to greater number of established species than are, or can be, immediately replaced by immigrants (Sheil, 1999). Disturbances such as logging usually cause immediate declines in biodiversity followed by a recovery, although not necessarily of the same species (Noble and Dirzo, 1997). Species richness of a site experiencing disturbance, therefore, will be a cumulative outcome of differential responses of species to disturbance. Some species may tolerate the disturbance and the other may disappear (Sagar et al., 2003) from the ecosystem. Ecosystem, in general, refers to the universal assemblage of soil, air, water, plants and animals along with many other components which are both biotic and abiotic in nature. It denotes the interactions of the biotic community with the abiotic environment. Bradshaw (1984) has stated that the characteristics of ecosystems can be decided by their two main parameters, namely the structure, i.e., species number and organizational complexity, and function, i.e., metabolism, interactions and resource recycling.

Open cast mining results into physically, nutritionally and biologically poor surface materials. The establishment of vegetation cover on coal mine spoil is challenging task due to problems such as compaction, poor water-holding capacity, infertility, high acidity or salinity of soil and extreme temperatures (Joshi et al., 2006). Bradshaw (1983) has also opined that the devastated ecosystems in the aftermath of mining, if left to itself, could have recuperated along the normal natural trajectory of development, but due to extreme poverty of starting materials, the progress will be extremely slow. Soil or the substrate into which plants establish and root is the starting point which may be varied, and the problems they pose can be dependent on the specific site conditions. This starting material is usually extremely skeletal on derelict sites, and its properties and situation determine in a very crucial manner the degree to which an ecosystem can develop naturally on the site, how far this development will progress, and what treatments are necessary to assist its development. However, the very simple needs of plants are: (i) a medium into which they are physically able to root, (ii) an adequate water supply, (iii) an adequate nutrient supply, and (iv) lack of toxicity (Bradshaw, 1987).

The main aim of any restoration process is to create sustainable plant communities representative of the composition and diversity of the surrounding natural plant communities (Jefferson, 2004). Restoration of the original natural or semi-natural vegetation, though often technically very difficult or impossible to achieve, is frequently the ultimate aim of rehabilitation programmes (Dickie et al., 1988.). Bending and Moffat (1999) have indicated that establishing vegetation on spoils often present challenging problems. These include compaction, salinity, acidity, poor water-holding capacity, hostile temperature regime and inadequate supplies of plant nutrients (Doubleday and Jones, 1977). An important goal of ecological rehabilitation is to accelerate natural successional processes so as to increase biological productivity, reduce rates of soil erosion, and increase soil fertility and biotic control over biogeochemical fluxes within the recovering ecosystems (Parrotta, 1992).

Other researchers have also emphasized that establishment of vegetation on abandoned mined lands is hindered by physical factors such as high temperature, low availability of soil moisture. Particularly, in arid and semi-arid areas, limited rainfall during the growing season and high surface temperatures often limit plant establishment and growth. In certain areas, the main factor in preventing vegetation from becoming establishment is acidity. If specific problems hindering ecosystem redevelopment can be identified, a cure can be designed using or mimicking natural process (Maiti, 2003; Singh et al., 2002).

Doerr et al. (1983) demonstrated the feasibility and potential problems involved in revegetating intensively disturbed sites. According to them, alkalinity and salinity of mine spoil, higher rates of litter accumulation, damages by insect and pathogens, and lower spoil nutrient contents are the usual problems affecting revegetation. Power and Bennett (1977) found nutrient deficiencies (N and P), toxicities (Mg, B and Mo), compaction and steep slopes as the major problems. Bradshaw (1987) concluded that the problems these derelict lands pose can be reduced to: (i) physical- structure, either too compact or too open, instability, and moisture level either too low or too high, (ii) low or negligible availability of macro- or micro-nutrients, (iii) toxicity- either too low or too high pH, too high salinity and too high level of heavy metals, and (iv) absence or slow colonization of wild plants and animals.

Plantation versus Mine-Spoil

In surface mining, when overburden is removed, the soil horizons may be inverted and the unweatherd 'C' horizon exposed. This contains unweathered clay with high cation exchange capacity (CEC) and high level of exchangeable bases. The major deficiency is lack of organic carbon and almost complete absence of nitrogen (Das et al., 1992; Mohanty, 2004). Fail Jr. and Wochok (1977) reported that coal spoils characteristically lack nitrogen primarily because of loss of top-soil with its organic matter, together with the fact that nitrogen availability rapidly drops with decrease in pH. A pre-requisite to any revegetation plan is the knowledge of physico-chemical properties of the mine spoils. According to Power (1978), the physico-chemical characters crucial to prediction of the plant growth potential for overburden include texture, pH, electrical conductivity, soluble Ca, Mg, Na, B, cation exchange capacity, exchangeable cations, gypsum and calcium carbonate equivalents, and carbonate, bicarbonate, sulphate, chloride and nitrate contents of the saturation extract. Water holding capacity and infiltration rate are other key variables.

Many obstacles, like low pH, soil compaction, elevated surface temperature, water stress and lack of nutrients, viz. N and P exist to establishing vegetation on previously mined sites. However, considering the stressful abiotic condition and minimal management efforts, the degree of resilience of these systems is encouraging (Holl and Cairns, 1994). Analyzing spoils of four coal-mine sites of Jharkhand, Prasad and Roy (2001) reported that the pH of coal-mine spoils was acidic (5 to 5.6). Those spoils were significantly poorer in available nitrogen, phosphorus, potassium and organic matter than the plant populated control area.

Coarse and medium textured overburden materials with low water holding capacity when compacted exhibit improvement in water holding capacity. However, when heavy clays are compacted, they exhibit undesirably low water transmissivity characteristics. Shukla and Lal (2005) found that the soil disturbance due to mining operation increased soil bulk density at both depths of 0-15 cm and 15-30 cm. Soil quality improved with increase in reclamation duration as evidenced by decline in bulk density, increase in pH, and total nitrogen concentration of soil. With increase in bulk density, water infiltration and movement decreases. Generally, low bulk densities in soils indicate high organic matter content, good granulation, high infiltration, and good aeration, resulting in a good rooting medium. Conversely, higher bulk density values generally indicate low soil aggregation, inhibited root penetration, and low infiltration and permeability rates.

Healthy plant growth is intimately related to the capacity of soil particles to group together into stable structures, i.e., aggregation property of the soil. A fine-textured, well-aggregated soil is ideal for growth, since it has the desirable properties of both a fine-textured soil, in its capacity for holding water and nutrients, and a coarse-textured soil, in its water permeability, aeration, and resistance to compaction. The binding colloids are derived from organic matter; therefore, a newly created surface generally lacks them. Aggregation will improve with time if suitable soil flora and fauna can be established and maintained (Jha and Singh, 1992).

The most suitable soil texture for any site depends, to a large extent, upon the type of vegetation that is desired. The proportion of particles with diameters less than 2 mm (that is, sands, silts, and clays) is most significant to plant growth because small particles, with their greater surface area, contribute most to the water-and nutrient-holding capacity of the substrate (Jha and Singh, 1992). High temperatures and low moisture of surface mine spoil limit plant growth (Richardson, 1958; Richardson and Greenwood, 1967; Bradshaw et al., 1975; Bell and Ungar, 1981) and also reduce decomposer activity (Wieder et al., 1983). Root development is restricted due to changes in physical and chemical properties during spoil handling. Due to high sodium and low soluble salt content clays become dispersed, water movement is restricted and unfavourable conditions for root growth occur. Root growth is restricted in soils with a bulk density higher than 1.6 (Russell, 1977). Power et al. (1978b) have reported 10 to 30% lower bulk density for spoils compared to original undisturbed soil. The bulk density values of spoil and undisturbed soil typically range from 1.1 to 1.4 and 1.4 to 1.7 g cm^{-3}, respectively. When spoils are dry, the heavy machinery used for mining operations causes severe compaction, warranting ripping or other surface disturbances before seeding. Lyle (1987) and Zeleznik and Skousen (1996) suggested that bulk density higher than 1.5 mg/m^3 can severely restrict root growth of many species. A study by Fierro et al. (1999) found that bulk density of unamended mine soil was more or less equal to 1.7mg m^{-3}.

Electrical conductivity (EC) is a common mine spoil variable influencing plant productivity (Andrews et al., 1998, Torbert et al., 1988, McFee et al., 1981). High levels of soluble salts inhibit water and carbon dioxide uptake and also inactivate enzymes affecting protein synthesis and C metabolism. Regression analysis by Rodrigue and Burger (2004) indicated a decrease in site productivity with an increase in the soluble salt concentration.

Mine Spoil Impoverishment

Soil fertility may be defined in terms of the capacity of the soil to supply essential nutrients and water for the growth of plants. In tropical forest areas with highly weathered soils, organic matter plays an important role in soil functioning and forest sustainability. When forests are clear-cut, the soil begins almost immediately to lose organic matter, triggering a series of soil degradation processes, the extent and intensity of which depends on soil management. Most tropical forests grow on highly weathered soils that contain low activity clays and have low natural fertility and pH. Because of this, these systems depend on efficient nutrient cycling based on litter deposition and decomposition. In addition, in these soils, soil organic matter (SOM) plays an important role in the soil cation exchange capacity (CEC), retention of base ion, soil aggregation and is directly related to nutrient availability, especially soil nitrogen. Studies indicate that nitrogen is probably the most important limiting factor to the reestablishment of forest communities (Macedo et al., 2008). Increased nutrient export from the system (Likens et al., 1970; Stark, 1977; O'Neill et al., 1977) and depletion of soil carbon pool, i.e., "energy currency" from the belowground ecosystem (Parkinson, 1979), cause ecosystem disruption. Mine wastes or sub-soils are unlikely to contain any significant quantities of nitrogen in particular, since in most ecosystems nitrogen accumulates in surface soils as a result of biological activity (Bradshaw, 1987). Mining activity alters the flow of N through a stable-plant-microbial system (Reeder and Sabey, 1987). It causes loss of litter layer, which is an integral storage and exchange site for nutrients. Nutrient-holding capacity of mine spoil is drastically reduced and a major disadvantage of surface-mined sites is thus low inherent fertility of the spoil material (Bradshaw, 1983). Mine spoils are nutritionally and microbiologically impoverished (Visser et al., 1979) and soil-forming processes are confined to the top few centimeters even in older spoils (Smith et al., 1971; Anderson, 1977; Down, 1975; Rimmer, 1982). Schafer et al. (1980) also found that organic matter was present only in the upper few centimeters of a 53-year old spoil. Samples of overburdens from chromite and coalmines in eastern India have showed major deficiencies in nitrogen at all these sites, but not of other nutrients (Das et al., 1992).

Reduced nutrient availability was reported even in reclaimed sites compared to soils of natural sites by Toy and Shay (1987). Investigations suggest that nitrogen is usually the major limiting factor (Crocker and Major, 1955; Roberts et al., 1981). It is the only nutrient that changes significantly with ecosystem development and can be shown to be continually limiting (Marrs et al., 1983). Mine dumps usually lack biologically active nitrogen; however, exchangeable ammonium-N may be present which nitrifies to nitrate-N in a few months when exposed to atmosphere (Power et al., 1974). Other researchers have also found nitrogen accumulation as a major factor limiting the rate of vegetation development on mine wastes (Bradshaw et al., 1975; Dancer et al., 1977; Roberts et al., 1981). Bradshaw (1987) is of the view that the exact amount of soil nitrogen capital will depend on rates of decomposition and the particular type of vegetation considered and it will take some time to accumulate.

Bradshaw (1987), supported by findings of Marrs et al. (1980) and Roberts et al. (1980), in well-managed kaolin mine waste reclamation projects in south-western England, concluded that the problem of nitrogen accumulation is reflected not only in the total amount of nitrogen accumulated, but also in the total amount mineralized compared with normal soils. Lack of mineralizable organic N and lower mineralization rates affect the availability of N to plants in mine spoils (Reeder and Berg, 1977). High levels of organic matter in mine soils improve aggregation and infiltration capacities and increase the availability of nutrients (Toy and Shay, 1987).

In a study of lignite mining areas in south India, Sengupta (1991) found that the huge Bucket Wheel Excavator cannot segregate and excavate the top 30 cm soil which has the plant nutrients, humus and microbes, separately. The excavated mine spoil is practically inert with very low organic matter and very poor in available N, P and K. Potassium may also be deficient, either in terms of total or available amounts (Beaton, 1974; Bradshaw, 1987). Wittwer et al. (1981) found both N and P as limiting factors in southeastern Kentucky mine spoils. Safaya and Wali (1979) and Iverson and Wali (1982) observed P as a major limiting nutrient during colonization and early succession processes and found that when P was inadequate, plant growth was adversely affected. Power et al. (1978b) reported universal deficiency of P in arid mine spoils of United States. Prasad and Shukla (1985a) reported N, P and K deficiency in coal mine spoils at Dhanpuri, Madhya Pradesh, India; and Prasad and Pandey (1985) reported lack of nutrients and moisture retention capacity in bauxite mine spoils at Amarkantak, MP, India. Jha (1992) observed that the generic mine spoils of Jhingurda coal mines, MP, India were nutritionally sub-optimal for plant growth and responded positively to the addition of NPK fertilizers. Maiti et al. (2005), in a coalfield in Jharkhand, India, found that available N in the top 15 cm. of overburden (OB) was 30 ppm as compared to 45 ppm in topsoil. They observed that the lower value of available N in the OB dumps can be attributed to lack of microbial activity and mineralization of organic matter. Available P was also found lower in comparison to topsoil, which may be due to the fixation of phosphorus by coal shell, and in acidic pH, it forms insoluble aluminium phosphate. Exchangeable potassium was found lower in reclaimed OB dumps than topsoil. According to Severson and Gouch (1983), rapid changes in chemical equilibrium are expected in spoils due to young stage of soil development. Kumar et al. (2005) also reported high pH (8.7) and high Na^+ but low available N and P with no nitrification in lignite mine spoils of Rajasthan, India.

Erosion and Water Conservation

Surface mining drastically changes the topography of the area and transfers relatively stable land forms into more erosive ones. In mine spoils, geomorphic system is in disequilibrium due to destruction of the balance between landforms and processes which accelerate erosion rates (Soulliere and Toy, 1986). Down (1975) reported accelerated erosion even on very old (178 years) spoils (32-36° slope).

Hydrological data collected for eastern U.S. mine spoils showed that the potential for runoff and erosion on long slopes of 9% or greater was rather severe (Gilley et al., 1977). Archibold (1980) observed that slopewash due to rains easily moves seeds to the base of the slope. Fine clay particles are also moved through runoff. Therefore, soil erosion is one of the major problems in rehabilitating mine spoils. Properties of mine-spoils, such as texture, extent of fragmentation of rocks, exchangeable Na content, bulk density, compaction, degree and length of slope, influence infiltration, runoff and evaporation of water (Jha and Singh, 1992). Pederson et al. (1980) found that spoils having high bulk density and low porosity had low infiltration rates. Most of the spoils retained only 25% as much plant available water as the water of a natural soil. The dumping and smoothing of spoils with equipments cause differential packing. The central axis of the piles is markedly packed, while packing is less on the outer slopes where materials are pushed or slide down the slope. Thus, the bulk density in the central axis of the piles becomes higher, and lower towards outer slopes. This causes uneven subsidence and differences in surface drainage (Jha, 1990). Mishra and Hota (1991) have opined that reclamation of mined areas having slope more than 14 degrees is rarely successful.

Formation of benches or terraces is a common method to reduce runoff and soil erosion. Terracing reduces the slope length and creates new base portion farther upslope. Armiger et al. (1976) are of the views that increase in the rate of infiltration, absorption of soil water by root, dry matter yield per unit water used, and decrease in rate of runoff, evaporation, and leaching of soil water depend on conservation and efficient use of limited precipitation.

Mine-Spoil Toxicity

Extensive work on the mine spoils in Northern Great Plains has indicated that these materials are characteristically neutral to alkaline in pH, contain soluble salts with sulphates of Ca, Mg and Na predominating, and appreciable amounts of $CaCO_3$ with very few readily soluble chlorides, carbonates or bicarbonate (Power et al., 1978a). Toxicity problems, usually associated with acidity, nutrient deficiency and physical conditions, have been reported by Rimmer (1982) in reclaimed mine sites. Mine spoils may also increase the amount of trace elements at surface of the spoil to potentially hazardous levels. Water soluble B, Cu, Fe, Li, Ni, Sr and Zn contents were greater in mine spoils compared to unmined sites (Wali and Freeman, 1973). Infiltration and storage of water in a soil is affected by exchangeable Na. Sodium adsorption ratio (SAR) values above 12 influence stability and dispersion of clay minerals, surface sealing, restriction in water infiltration, and increased rate of runoff. Thus, less water is stored in the soil for the vegetation use (Power et al., 1978a). According to Merrill et al. (1985), the low hydraulic conductivity of sodic mine spoil (SAR=30) was the dominant factor limiting plant growth. Sandoval and Gould (1978) reported that spoils had high exchangeable Na contents that resulted in deterioration of soil structure, reduced infiltration and greater crusting. Availability of water for plant growth is reduced by salinity (total dissolved salts) of mine spoils. More energy is required to extract water from saline soil which results in less growth per unit of water used. In sodic spoil, gypsum is used to replace Na on the cation exchange complex. Power et al. (1978a) reported that gypsum additions reduced exchangeable Na content by 30-50 per cent in the upper 30 cm of spoil within a few years.

Due to biochemical oxidation of pyrites, the pH values may decrease from 8.0 to 2.0 within 10 years, which is an important factor for plant growth in several mine spoils (Chadwick, 1973; Doubleday, 1974; Caruccio, 1975; Armiger et al., 1976; Bennett et al., 1976). Acidic spoils contain toxic levels of soluble elements such as Fe, Al, Mn and Cu, and inadequate supply of Mg, Ca and P; and seedling establishment in these spoils is difficult because of the lack of active organic matter (Jha and Singh, 1992). Costigan et al., 1981 reported that dieback of well-established vegetation may result from continued acidification for many years. After a study of active surface mines, Barth and Martin (1984) were of the view that acidic spoil was the most difficult to revegetate, while in the sodic spoil Na migrated subsequently into the replaced topsoil up to a depth of 14 cm. in sodic spoil. An exceptionally high concentration of total B (boron) was reported in limestone mine spoils (Sharma et al., 2000). Chemical analysis of the tin mine tailings indicated that deficiencies of P, K, Ca, Mg, S and trace elements, together with aluminate toxicity associated with high pH, might be responsible for the paucity of natural colonization (Phia et al., 1995).

The ecologically elegant solution for reclamation is to use metal tolerant plants and to supply fertilizers (Smith and Bradshaw, 1979). However, nutrient cycling is so much reduced by the heavy metal toxicity that growth is limited unless repeated fertilizer applications are given (McNeilly et al., 1984). Hence, Bradshaw (1987) is of the view that in case of metal contaminated sites, the nature of the toxicity is such that direct treatment is not completely satisfactory and for successful restoration, vegetation colonizing natural, undisturbed metal-contaminated areas is indicative. Strong negative relationship between tree growth and high concentration of magnesium has been drawn by Richardson and Evans (1987) and Bending and Moffat (1999). In manganese mine spoils in Central India, Juwarkar et al. (2001) informed that spoils had pH in the range of 6.9-7.4 with low soluble salts, N, P, K and organic carbon content (0.104-0.108%). Although, heavy metals like Cr, Pb, Cu, Mn and Zn were present, however, they were not of serious concern. In a study of a coalfield mine spoil in Jharkhand, India, Maiti et al. (2005) also found that acidic nature of mine spoil increases bioavailability of trace elements (micronutrients) but it is not going to cause any toxicity problem for the plants, except zinc. The average concentration of bioavailable zinc in the range of 5.8 to 6.2 mg/kg may cause toxicity problems. In four coalmine sites in Jharkhand, Prasad and Roy (2001) found that metals at high level in spoils could be toxic to plants and among them Ni, Mn, Zn and Cu are prominent. A decrease in total chlorophyll content was reported, in agreement with other similar findings, from various plant species grown on metal enriched soil from all over the world. The physico-chemical properties of the metal-contaminated mine spoils tend to inhibit soil-forming processes and plant growth. Other adverse factors of mine spoil are absence of top-soil, absence of soil forming fine materials, shortage of essential nutrients, etc. (Wong, 2003). However, in an experiment on evaluation of coal mine spoil as a medium for plant growth, Jha (1992) found that mine spoils were non-toxic and suitable, although sub-optimal medium, for plant growth.

MANAGEMENT INTERVENTIONS

Topsoil Replacement

Soils disturbed by surface mining are highly susceptible to erosion and difficult to stabilize by vegetation. For restoring the productivity, and for protection of soil and water resources, many authors have advocated covering the spoil with topsoil of variable depth. In opencast mining, it is accepted that removal, storage and replacement of the original topsoil is essential to the restoration process (Dickie et al., 1988). Topsoil replacement is a common method in enhancing physical and nutrient status of spoils. Several soil properties, e.g., water holding capacity, nutrient supplying capability, buffering capacity, and plant root depth are affected by the thickness of the respread A and B horizon materials. These factors affect plant growth, plant composition and nutritional quality of the vegetation (Jha and Singh, 1992). Topsoil, particularly fresh topsoil, is the major contributor to plant diversity on rehabilitated bauxite mines. The majority of the plant species (72%) on rehabilitated areas come from seed stored in the topsoil. Many of these topsoil species are smaller understory or annual species. Nevertheless, they are an important part of the jarrah forest ecosystem, so rehabilitation procedures must be developed to maximize their establishment (Ward et al., 1996).

Deep ripping brakes up the compacted mine pit floor and improves tree root growth. Double stripping is removing and replacing the soil in two layers to ensure that the valuable top layer (15 cm in case of ALCOA's mining area) is retained at the top of the restored profile and not diluted with the underlying soil. Direct return is the removal of soil and immediate replacement onto a restored site without stockpiling. Double stripping and direct return of topsoil were incorporated into the restoration protocols at ALCOA's mines in Western Australia in the late 1970s (Gardner and Bell, 2007).

In progressive mining operations, it is becoming mandatory for surface soils to be conserved and replaced. In the restoration of small, existing degraded areas, the use of layer of top soil to cover up whatever is wrong beneath is common (Bradshaw, 1987). Parrotta and Knowles (2001) reported that tree basal area, canopy height and crown cover percentages were significantly lower in the mixed native species treatment plots with inadequate topsoil application than in plots in which the prescribed site preparation protocols were followed. In the former plots, rates of litter accumulation and humus development were relatively poor and individuals more than 2 m tall were relatively absent.

Soil handling, stockpiling, and soil mixing can impact on nutrients, biology and organic matter contained in the soil. The concentrations of many nutrients in the forest are greatest in the surface soil and may decrease rapidly with depth. Restoration operations can cause these nutrient rich upper layers of soil to become diluted with soil from deeper in the profile (Ward, 2000; Grant et al., 2007). Carlson et al. (1961) have suggested that P concentration in several crops could be enhanced by returning topsoil or adding manure. Power et al. (1982) have evaluated the effect upon various parameters of crop quality after four years of topsoil and subsoil replacement. Higher plant P concentration was reported where spoil was covered with 20 or 60 cm thick topsoil layer. Gilley et al. (1977) have reported greater runoff and erosion when the smoothed spoils had steep slopes. When such a steep slope (9-10 per cent) was covered with 25 cm topsoil without revegetation the soil loss (36 t ha^{-1}) was greater compared to uncovered spoil (7.8 t ha^{-1}). Thus, the replaced topsoil would be quickly eroded from steep slopes in absence of vegetation. Power et al. (1978b) found that addition of straw reduced the erosion and runoff from the respread topsoil.

From a modeling exercise, Parton et al. (1979) indicated that increasing the topsoil depth from 5 cm to 60 cm increased plant production by 20 to 28 per cent as a result of increased N uptake. Application of N fertilizer caused a sixfold increase in primary production and when combined with irrigation the production increased by 100 per cent compared to fertilization alone. A study of coalmine overburdens in India reveals that stockpiling of soil material has been found to degrade soil quality continually. Topsoil should be removed in a separate layer and stockpiled only when it is not feasible to promptly redistribute on degraded areas. The stockpiled materials should be selectively placed on a suitable area and protected from erosion, compactness and contaminants which will lessen the capability of materials to support vegetation. The dumps height and slope should be maintained properly (Ghose and Kundu, 2001). Other researchers (Moffat and Boswell, 1997; Bending and Moffat, 1999) have also suggested for careful spoil and soil placement in order to prevent compaction during post-mining landform design. Bradshaw (1987) has also opined that it is commonly found that the performances (measured as yield) of ecosystem reconstructed on topsoil are inadequate. Experimentation shows that soil used is unable to sustain vigorous growth unless supplied with extra nitrogen (Bloomfield et al., 1981). This can be due either to deterioration of the soil during storage between gathering and respreading, or to excessive amount of sub-soil being included in 'topsoil'. Sandoval and Gould (1978) and Merrill et al. (1980a; 1983), have reported that since in some spoils sodium moves upwards into topsoil from the spoil causing a decrease in plant yield in later years, a small layer of topsoil replacement cannot maintain the productivity for long. Doll et al. (1984) have suggested more exacting guidelines for soil replacement based upon spoil properties such as texture, electrical conductivity and sodium adsorption ratio. Careful site preparation practices, particularly judicious topsoil handling and re-application prior to tree planting, are essential for the establishment of forest cover, elimination of competing grasses, and acceleration of natural forest succession on reclaimed bauxite mine sites (Parrotta and Knowles, 2001). It is reported that the soil seed bank of the stored topsoil can be maintained throughout the store, so long as the storage period is no longer than a few months (Dickie et al., 1988).

The topsoil in areas of the jarrah forest mined for bauxite commonly contains about two-thirds gravel by weight (Ward, 2000). Approximately 90% of seeds in topsoil pass through a 5-mm screen (Murphy and Loneragan, 2000). Study by Koch and Ward (1994) indicated that 77% of species found in rehabilitated pits germinate from seed of the topsoil rather than from the applied seed, indicating the importance of topsoil in any site rehabilitation. But Grant et al. (1996) reported that the efficiency of use of seed in the topsoil for bauxite mine rehabilitation appears to be low. The highest efficiency of establishment of seedling from topsoil stored seed appears to be about 2%. The loss of seed was attributed to a number of factors, including physical damage during soil handling procedures, loss of seed while soil is stockpiled, dilution of seed when soil is respread, and burial of seed at a depth too deep for emergence.

Top-soil can be collected, the gravel sieved out, and the remaining much smaller volume of seed-rich source can be re-spread on post-mine areas. Spreading a thin layer of this seed-rich soil after ripping ensures that very little of the seed is buried too deeply to germinate. If the soil is collected and sieved in summer, when seed stores are highest, the effectiveness of this seed store is maximized (Koch, 2007a). Study by Koch et al. (1996) indicated that there were large reserves of seeds in the soils of jarrah forest, but substantial losses of seed can occur during soil handling activities, particularly if the soil is stockpiled. For effective use of these seeds, they recommended to respread topsoil to a lesser depth than it was collected from.

Soil Amendments

The sodic and acidic spoils may require chemical amendments in order to speed up the process of rehabilitation and to reduce the cost associated with topsoil replacement. To reach a productively level equivalent to the pre-mining forest, an input of nutrients to the system is required (Ward and Koch, 1996; Ward, 2000). Bradshaw (1983), Kendle and Bradshaw (1992), Moffat and Buckley (1995) and Bending and Moffat (1999) have suggested that without inorganic or organic nitrogen amendment, reclamation of coal spoils to forestry is unrealistic. Merrill et al. (1980b) reported that for the sodic spoils of Northern Great Plains the three requirements associated with chemical reclamation are: (i) Ca or Mg salt, (ii) sufficient water to transport the Ca-ions to the cation exchange sites to displace the Na-ions, and (iii) sufficient hydraulic conductivity so that Na-ions are transported out of the root zone. Power et al. (1978a) have reported that under the climate prevailing in most mine areas of Northern Great Plains, gypsum additions reduce exchangeable Na content by 30-50 per cent in the upper 30 cm of material within a few years after treatment.

The need for nutrient additions, when ecosystems are being reconstructed on skeletal material, is obvious. Ecosystems, even in a juvenile condition, do need considerable amounts of nutrients for growth, and these substances are likely to be limiting in most of the skeletal materials found on severely degraded land. It is easy to provide the necessary nutrients by means of ordinary agricultural fertilizers containing nitrogen, phosphorus and potassium. Practical experience shows that these nutrients must be provided repeatedly or growth is substantially reduced (Bradshaw, 1987). Mine overburdens may be excellent materials for plant growth provided that the deficiencies in organic matter and nitrogen can be remedied (Das et al., 1992).

Application of fertilizer has proven highly successful in replacing nutrient pools lost through the mining and restoration process, particularly P (Ward, 2000; Grant et al., 2007). Day et al. (1979) observed in growth chamber experiments that even with adequate irrigation but without proper fertilization, plant growth in spoils and top soil materials was extremely poor. Jha (1992) found significant difference in shoot growth for NPK treated plants compared to those grown in mine spoil only and in mine spoil with forest soil. Singh et al. (2000a), from a fertilizer treatment experiment in Singrauli Coalfields, India, found that NPK fertilization of mine spoils with seeded ground cover promotes growth, particularly in non-leguminous woody species, consequently leading to potentially increased biomass production. The response of tree species to nutrient amendment is variable, however. The impact of fertilization in most species was more on diameter growth than on height growth. Norman et al. (2006) concluded that fertilizing sites with N is beneficial for increasing growth and establishment in newly rehabilitated sites, but there appears to be no long–term floristic benefits over unfertilized sites. Despite this, fertilizer application is considered a necessary component of the rehabilitation process to increase nutrient stores and encourage nutrient cycling. Although, in the short term, fertilizers can be applied to restored areas, in the long term the demand for nutrients must be met principally from nutrient cycling supplemented by inputs in rain fall, dry deposition, rock weathering, and for nitrogen, by biological fixation of atmospheric N (Grant et al., 2007). Koch (2007) reported use of diammonium phosphate with potassium and micronutrients at the rate of 280 kg/ha to newly restored areas for restoration of bauxite mined areas.

Singh et al. (1996a) studied the effectiveness of amendments on growth of *Pithecellobium dulce* in coal mine overburden while Singh (2004) studied the influence of NPK fertilization on N and P resorption efficiency in nine native tropical tree species planted on coal mine spoil. Of these, *Acacia catechu, Albizia lebbeck, Dalbergia sissoo* and *Pongamia pinnata* were legumes, while *Azadirachta indica, Gmelina arborea, Phyllanthus emblica, Tectona grandis* and *Terminalia bellirica* were non-legumes. The N and P resorption efficiencies dropped in fertilized plots in all the species. Non-leguminous tree species had exhibited greater efficiency for N resorption than leguminous species. The study indicated that nutrient enrichment reduced internal cycling of nutrients thus facilitating greater amount of nutrient return to soil, which in turn would enhance the reclamation process by allowing the colonization of more species due to increased habitat fertility. For the ten tropical tree species, Singh and Singh (2001) and Singh and Singh (2006) reported that the impact of fertilization was comparatively more on non-leguminous tree species than on the leguminous tree species. Nitrogen-fixing species may overcome the nitrogen deficiency by virtue of their nitrogen fixing ability; hence they may remain unaffected by N fertilization but may respond to P fertilization (Singh et al., 2000a; Singh and Singh, 2001). The impact of fertilization in most species was more on diameter growth than on height growth (Singh et al., 2000a).

Seeding and Planting

Seed and propagule dissemination plays a key role in community dynamics during plant succession (Noble and Slatyer, 1980). Forest topsoil, which contains seedbank, has been used to revegetate disturbed lands (Johnson and Bradshaw, 1979; Tacey and Glossop, 1980). The effect of seed size on depth of burial and subsequent emergence has been well documented in the literature. Generally, the larger the seed the greater will be the seedling's capacity to emerge from deeper locations in the soil (Grant et al., 1996). The extent to which one can rely on seed-bank germination (from applied topsoil) to establish species-rich forest cover is largely unknown; although studies of forest succession following natural disturbances and agricultural abandonment can provide some guidance and generally suggest that early successional tree species need not be planted due to their abundance in the soil seed-bank and adaptability to degraded site conditions (Parrotta and Knowles, 1999). Hodder (1977) has reported that the top 5 cm soil layer contains sufficient number of seeds to revegetate the areas beyond their original densities, though the type of vegetation may be altered because most of the seedlings emerging from the seedbank may be of pioneer species. Many viable seeds and propagules from a large number of species were present in forest topsoil associated with surface mining and practically all plants emerging from these propagules survived on typical mine soils and were well established by the end of one growing season (Farmer et al., 1982). In topsoil stored for several years, than months, the overall viable seed population will be significantly reduced except near the surface of the store. In addition, the longer the stores remain, the more will the surface be subjected to invasion by widely dispersed and undesirable ruderal and weed species. These will then add their seeds to the bank unless they are effectively controlled. Sole reliance on buried viable seed from redistributed topsoil would not guarantee rapid establishment of closed vegetation, with further opportunities for undesirable species to enter the consequent gaps. There may be a reduction in total number of viable seeds, and any potential for a reduced diversity would be unwelcome in a conservation context (Dickie et al., 1988). However, contrary to the above findings, from a coalmine, Johnson and West (1989) reported that the greatest number of seeds occurred in the oldest topsoil stockpiles. Inversion of soil layers during soil handling processes cause a reduction in seed numbers and seeds of most prevalent colonizers after reclamation were not present in top soil, rather they emigrated from surrounding areas (Iverson and Wali, 1982). Positive response of the role of soil seed bank in natural restoration of degraded areas has been reported by Liu et al. (2009) also.

Seeding or planting of suitable species speeds up succession that fulfils the revegetation goal (Jha and Singh, 1992). Not applying any seed to rehabilitated sites is an undesirable rehabilitation technique because the resultant flora only reflects the soil seed bank and species that are able to disperse to a site, such as native and exotic ephemeral species. If the soil seed bank were depleted because it had been stockpiled, then few plants would establish from the source (Norman et al., 2006). Direct seeding of tree species as a means of revegetation of coal mine spoils was evaluated by Jha et al. (2000) and reported that direct sowing of a mixture of tree species and grasses and forbs is a technique with potential for rapid stabilization of flat sloping areas of coal mine spoil, and can accelerate natural plant succession processes, as the ground cover acts as a nurse crop and can trap air-borne seeds. Sindelar (1979) reported that seeded species, initial seeding success, cultural practice, and weather influence plant succession on mine spoils. Davidson (1980) favoured direct seeding for revegetating surface mine spoils because this is easier and cost-effective than spot-seeding or drilling. In Northern Great Plains, seeding of native grasses yielded favourable results (Ries et al., 1978; Ries and DePuit, 1984). Barth (1986) recommended sod-forming grasses for surface protection and erosion control in mine spoil. Direct seeding of trees into more normal site is a promising approach (Zarger et al., 1973; Luke et al., 1982) but careful attention is essential. Parrotta and Knowless (2001) found that direct seeding was suitable for species having seeds larger than 2 cm long & broad, however, the seeds of many tropical tree species require mechanical or chemical scarification to speed up their germination.

Wittwer et al. (1981), Creighton et al. (1983) and Cunningham and Wittwer (1984) found that direct spot-seeding at a proper rate is a better and cost-effective method of revegetation for oaks, pine black walnut and black locust, compared to planting of bare-root seeding on mine spoils. Direct seeding results in a more complete occupation of the site. Seeding through drilling or broadcast with judicious inputs of N and P and proper timing with respect to rainfall regime and temperature conditions has proved successful for establishment for mono- and mixed stands of grasses and legumes (Ries and DePuit, 1984). Initially seeded rehabilitated sites had higher native species richness than unseeded rehabilitated sites. The unseeded treatment achieved species richness values equivalent to those of the seeded treatments at 8 and 14 years of age, but still had lower diversity and lower evenness. This is undesirable because although the same number of species may be present, the vegetation community is dominated by large number of individuals of a few species. Although there was no effect of the different seed mixture on exotic species density in 1-year-old rehabilitated site, unseeded sites had significantly higher exotic species density and richness than seeded sites at 5 and 8 years of age, respectively (Norman et al., 2006). Rehabilitated sites had higher exotic species richness, density and cover than forest sites (Ross et al., 2004; Norman et al., 2006). Artificial seeding of grasses and legumes or both has been a commonly used method to stabilize unconsolidated mine tailings and to encourage natural invasion of tree and shrub seedling. This ultimately improves site fertility and moisture retention capacity. Once the abandoned mine lands have vegetation growing on the surface, the regeneration of these areas for productive use has begun and offsite damages are minimized. In addition, establishment of the vegetation on an abandoned mine land also improves the aesthetics of the area (Singh et al., 2002).

The seeding of leguminous herbs, *Stylosanthes hamata,* and grass *Pennisetum pedicellatum* and *Heterpogon contortus* in the experimental plots of flat and sloppy areas enhances the colonization of a large number of plant species. Seeding with native grasses and legumes (nitrogen fixing plants) is an inexpensive method to quickly cover a site. Grasses like Dinanath grass *(Penissetum pedicellatum)* and legumes (*Stylosanthes humulis*) have been successfully used for mine reclamation (Maiti and Singh, 2007). Where phytotoxicity is suspected, it is particularly important to include plant material from populations growing naturally on mine sites and other areas likely to contain similar toxic factors (Phia et al., 1995).

Hossner and Hons (1992) reported that bauxite spoils are highly erodible and a minimum ground cover of 80% is recommended. Because tailings are so variable, those parameters that limit successful reclamation are expected to be site specific and must be evaluated on case-by-case basis. Acid-tolerant and nodulated legume trees were used to revegetate the bauxite dumps in Amazon (Franco and De Faria, 1997).

Floristic Enrichment/Colonization of Species

Natural restoration of mine spoil is a slow process. Floristic enrichment of the reforestation areas through natural regeneration of 'colonizing' tree species (i.e., those not planted or present in the applied soil seed bank) is largely dependent on seed-dispersing wildlife, mainly bats, birds and terrestrial mammals (Parrotta and Knowles, 2001). Numerous studies have demonstrated that land rehabilitation benefits from plantations because it acts as a catalyst to succession. The catalytic effects of plantations are due to changes in understory microclimatic conditions (increased soil moisture, reduced temperature, etc.), increased vegetational-structured complexity, and development of litter and humus layers that occur during the early years of plantation growth. Artificial revegetation is often used to facilitate the generally slow natural rehabilitation process (Bradshaw, 1983).

The rate of succession may also depend on the species planted first. Larson (1984) found low tree invasion rates on white pine plots and significantly higher invasion rates on black locust plots, for 30-yr old strip mine plantations. Schuster and Hutnik (1987), while studying community development on 35-yr old planted mine spoil banks, observed delayed succession on white and red pine plots and accelerated succession on black locust and green and white ash plots. Most importantly, some species can fix and accumulate nitrogen rapidly in sufficient quantities to provide a nitrogen capital, in a drastically nitrogen-deficient condition, more than adequate for normal ecosystem functioning. Once the soil characteristics have been restored, it is not difficult to restore a full suit of plant species to form the required vegetation (Dobson et al., 1997). Other advantages are that establishment of desirable tree species capable of maintaining the site will slow or prohibit invasion of less desirable weedy species, will provide economic returns in the long term, will aid in developing wildlife habitat and will promote hydrologic balance in the watershed (Singh et al., 2002).

Norman et al. (2006) carried an experiment with diverse seed treatments – legume and under-storey mix (major plus minor [MM] mix), under-storey mix only (minor [M] mix), and no seed (NS) – and sites were fertilized with N & P or P only. They found that the species composition of M and NS sites were more similar to that of the forest than to MM sites. The vegetation structure of M sites at 14 years was more similar to the forest structure than were the other treatments. The M treatment produced a lower fire-hazard, and more forest-like appearance and accessibility, due to lack of tall, fast growing legumes in the seed mix. The multivariate analysis of cover showed that seeded sites were more similar to the forest than to unseeded sites due to the establishment of mid-storey and over-storey structure. An MM site with reduced abundance of large legume species appears to be the most appropriate seeding treatment to implement in rehabilitated areas to increase similarity to the forest. Although the M mix alone produced a vegetation structure and floristic composition most similar to the forest, a similar result could be achieved by reducing the abundance of the large legumes in the MM mix, while still retaining some beneficial effects of nitrogen fixation. Reducing the amounts of large legumes can increase the presence of under-storey species in rehabilitated sites and reduce the fire risks. The dominance of annual species in unseeded treatments means that not applying seed is not a viable rehabilitation option. Applying nitrogen fertilizer increased the richness, density and cover of exotic species. They also recommended to immediately establishing all the plant species that are desired in the target community.

Species Selection and Revegetation

Forest plantations are commonly used to ameliorate harsh mine land soil and microclimatic conditions and speed establishment of native temperate and tropical vegetation. Plantings of native forest species combined with topsoil recovery and careful site preparation has been shown to reduce soil compaction, promote mycorrhizal symbiosis, and facilitate colonization of a species-rich forest community on Amazonian bauxite mines (Parrotta et al. 1997b; Parrotta and Knowles 1999). Throughout the tropics, forest plantings containing either native or non-native species have been promoted as “foster ecosystems” that can modulate microclimatic extremes and improve soil nutrient availability.

Plantation is the oldest technology for the restoration of lands damaged by human activity (Filcheva et al., 2000). Plantations can play a critical role in restoring productivity, ecosystem stability and biological diversity to degraded areas. A primary objective for achieving satisfactory rehabilitation of a mined landscape is to establish a permanent vegetation cover. There is increasing evidence that forest plantations can play a key role in harmonizing long-term forest ecosystem rehabilitation or restoration goals with near-term socioeconomic development objectives. Relative to unplanted sites, plantations have a marked catalytic effect on native forest development (succession) on severely degraded sites (Parrotta et al., 1997 a). Plantations have an important role in protecting the soil surface from erosion and allowing the accumulation of fine particles. They can reverse degradation process by stabilizing soils through development of extensive root systems. Once they are established, plants increase soil organic matter, lower soil bulk density, moderate soil pH and bring mineral nutrients to the surface and accumulate them in available from (Chakraborty and Chakraborty, 1989). Their root systems allow them to act as scavengers of nutrients not readily available. The plants accumulate these nutrients and re-deposit them on the soil surface in organic matter, from which nutrients are much more readily available by microbial breakdown (Singh et al., 2002).

The development of a plantation canopy can alter the understory microclimate and soil's physical and chemical environment to facilitate recruitment, survival and growth of native forest species. Otherwise, native species would only very slowly, if ever, regenerate on degraded site. Thus, plantations may act as 'foster ecosystems', accelerating development of genetic and biochemical diversity on degraded site (Singh et al., 2002). However, for reclamation works, proper selection of the species that will adapt with the climatic and local soil condition is a critical step (Maiti et al., 2006). Afforestation of mine spoils with fast growing tree species accelerates the revegetation process and fulfils the restoration goal (Singh, 2008). Follow-up plantings are a possible method of increasing species richness in rehabilitation sites over a period of time (Norman et al., 2006). Enrichment plantings of shade-requiring species, about 5 years after rehabilitation, occur successfully in the Brazilian Amazon (Parrotta and Knowles, 1999). Planting in a 10- to 13-year-old rehabilitated sites in the jarrah forest of Western Australia following thinning and burning operations was successful and may be used to increase the species richness of rehabilitation of this age (Grant and Norman, 2006). However, it is generally believed that introduced species pose a formidable problem in successional management. These species have greater potential for out-competing the less aggressive natives thus resulting in reduction in species and habitat diversity and formation of monospecific stands (Masoodi et al., 2003). The tendency of spread by native species on infertile soils must be an important feature of conservation because it speeds up the direction of change and rate of ecosystem development. The species with such characteristics have a high adaptive efficiency and hasten the process of natural succession on developing lands (Masoodi et al., 2003).

Achievement of an early vegetation cover and high biomass production can be approached though proper selection and planting of early successional native tree species because such species are able to exist under harsh soil conditions and require less long-term maintenance. Preference should be given to local leguminous species because they can survive in unfavourable conditions and will reduce initial nitrogen requirement and also fix the nitrogen required for the subsequent successional species (Singh and Singh, 1998). Developing species selection and establishment guidelines to ensure rapid revegetation of harsh planting site is a universal challenge among mine revegetation projects. Unlike restoration of less severely degraded land, the use of non-native plants remains an acceptable option for mineland revegetation (D' Antonio and Meyerson, 2002; Li 2006) if they fulfill a temporary successional role to colonize and ameliorate severely degraded sites and facilitate colonization and eventual dominance by native flora (Seo et al., 2008).

The choice of plant species depends on the characteristics of spoil, climatic conditions and ultimate land use. The goal of plant species selection model is to provide a method to identify the suitable tree/plant species for reclamation of land contaminated from mining activities. A better understanding of the tree properties on restored mine spoils can lead to beneficial changes in reclamation practice and renewed the prospect of commercial forestry on these restored lands. These plantations can provide fuel, timber, and non-wood forest products to local people (Dutta and Agrawal, 2003). Jefferies et al. (1981) are of the view that legumes are an effective means of improving the nitrogen status of derelict land, with mineralization allowing substantial transfer to companion species. However, Vasistha et al., (1995) recommended *Trema politoria* for mine spoil rehabilitation. This species is a moderate sized tree and is commonly distributed in sub-Himalayan ranges from Punjab to Bengal, Assam, Bihar, Orissa, Rajasthan and Madhya Pradesh. The species mostly comes up in profusion on landslips, mined out areas and other places where the soil is exposed. Owing to its capacity to grow on landslips and mined degraded areas as a primary colonizer, it is suitable for afforestation of such degraded areas. This plant is also used as a source of green fodder during lean period in sub-Himalayan tract. The species is capable of providing above ground protection and underground soil binding to the degraded loose substrate. The species has effective soil binding capacity and thus it is a very suitable soil conserving species for degraded mined lands. Besides being fast growing and of soil binding value it is socio-economically very useful.

A plant species can be selected if it has the similar growth environment with the successful rehabilitation species. During selecting plant species process, the factors of site conditions representing landform, climate, parent rock, soil and vegetation types, vegetation composition, coverage and growth should be considered. The selected plant species should not only include the local plants which can be grown well in the local area, but also the imported plants or new species. Suitable plant species can be identified by the similarity in topo-climate, soil quality, and vegetation cover factors (Chen and Li, 2008). When selecting plants for the reclamation, indigenous species are most preferred over exotic species because they are likely to fit into a fully functional ecosystem and to be climatically adopted but most to the native species in the region are slow growing species. Exotic fast-growing species were therefore tried in view of their large-scale use in afforestation schemes in the country (Parrotta, 1999). Exotic species may be especially recommended for primary rehabilitation on bare coal mine spoil due to their fast growth and establishment as observed in the present study. Care is, however, needed so that introduced exotic species may not become a problematic weed to local flora (Dutta and Agrawal, 2003). Nevertheless, current reclamation and rehabilitation efforts often make use of exotic species. These practices include afforestation or plantation programmes, which include large-scale introduction of fast-growing exotic species such as *Leucaena leucocephala, Acacia auriculiformis*, *Grevillea robusta, Eucalyptus* species, etc., and species of commercial importance such as *Tectona grandis*, *Casuarina equisetifolia,* etc. The success of any biological reclamation depends on; climatic conditions, nature of spoils, types of plant species, nature of dumps, proximity to seed banks (nearby vegetation) and types of amendments used (Maiti and Singh, 2007).

Kuusipalo et al. (1995) suggested using a fast-growing exotic species as a "sacrifice fallow" to aid in restoring degraded forest lands invaded by the large fire-enhancing native grass *Imperata cylindrica* in the humid tropics of Southeast Asia. *Imperata* creates conditions of soil compaction, nutrient deficiency, and hydrologic instability, and the susceptibility of the grass to fire prevents natural succession by destroying propagules and root symbionts. To prevent further cutting of rainforest, Kuusipalo et al. (1995) suggested planting introduced *Acacia* spp. into *Imperata* stands. The rationale behind this is that topsoil can be improved and desired woody vegetation will then be favoured at the expense of more light-demanding grasses and a more heterogeneous light environment is created, favouring a higher degree of biodiversity (D'Antonio and Meyerson, 2002).

The selection of tree species that are capable of rapid early growth under stressful conditions is essential for rapid occupancy of the reforestations site and suppression of grasses that can very often be a severe barrier to regeneration of all but a very limited number of pioneer tree and shrub species. A mixture of both fast-growing pioneer species and slower growing (and longer-lived) native species is recommended to provide greater canopy habitat diversity within the reforestation area, increase structural and biochemical diversity of redeveloping forest floor and soil humus layers. The planting density of nursery-grown seedlings, stumped saplings and/or large seeds should be adequate to minimize the early mortality resulting from competition with grasses and thus facilitate rapid site capture; the planting density of 2500 trees/ha used at Trombetas appears to be an acceptable minimum (Parrotta et al., 1997b).

Evidence presented from the study by Alexander (1989b) suggests that the long-term effect of eucalypt plantations on the spoil soils of the Jos Plateau may be detrimental rather than ameliorative. Although the eucalypts increase both the content of organic matter and some of the bases of the spoil soils, these potentially-beneficial effects have to be balanced against the progressive decrease in base saturation and pH brought about by an increase in cation exchange capacity, a situation, which if continues, will render the soil progressively less productive, rather than more productive as had been the intention of the planting scheme. Restoration managers need to be cognizant of the critical role of wildlife in forest re-development, actively encourage wildlife conservation in the surrounding landscape, and design restoration treatments that will provide suitable habitats for a variety of target wildlife species (Parrotta and Knowles, 2001).

Prasad and Pandey (1985) studied changes in composition of naturally colonizing species in 6-7 yr- old bauxite mine overburdens in Amarkantak and 3-yr-old coal mine overburdens in Dhanpuri (both in central India), planted with different tree species. Among grasses, *Panicum montanum* and among leguminous species *Cassia pumila* were the dominant species on coal mine dumps, whereas *Ageratum conyzoides* was the most dominant species on bauxite mine spoils. In a eucalypt plantation on 3-yr old bauxite and coal mine overburdens, 11 and 18 species colonized, respectively. On 7-yr- old bauxite mine, out of 16 species recorded, grasses, legumes and forbs were, respectively, 4, 2 and 10 species in richness. Coal mine overburdens were comparatively more hospitable than bauxite mine spoils. However, seedlings of *Acacia auriculiformis, Butea monosperma, Grevillea pteridifolia, Kydia calycina, Careya arborea, Eucalyptus sp., Pterocarpus marsupium, Shorea robusta* and *Syzygium cumini* occurred on different ages of plantations on bauxite mine, while on coal mine plantations, only seedlings of *Acaia auriculiformis* were observed. Plantations of *G pteridifolia, Grevillea. robusta* and *A. auriculiformis* on bauxite mine spoils, and *Dalbergia sissoo, A. auriculiformis, Acacia catechu* and *E.* hybrid on coal mine spoils promoted the process of natural plant colonization. Improvement of soil conditions promoted plant succession and sustained stable, productive plant communities (Schafer and Nielsen, 1979). Climax and subclimax native and introduced plant species accelerate soil development processes (Ries et al., 1977). Successional rate may also depend on the species planted first. Larson (1984) found low tree invasion rates on white-pine plots and significantly higher invasion rates on black locust plots in a 30-yr old strip mine plantation. Schuster and Hutnik (1987), while studying community development on a 35-yr- old planted mine spoil banks, observed delayed succession on white- and red-pine plots and accelerated succession on black locust, green- and white-ash plots.

The establishment of a mat of vegetation that is both self-sustaining and amenable to predetermined end uses is usually an important element in a rehabilitation program for waste disposal areas. Silvicultural knowledge is required to select species and establishment techniques appropriate to local site conditions and long-range restoration objectives. In many tropical regions, restorationists lack basic, essential information on seed availability, propagation techniques, growth rates and site adaptability for the hundreds of candidate tree species present in the natural forests (Knowles and Parrotta, 1995). On the other hand, improper selection of plant species is the most common problem associated with revegetation failure. The choice of plantation species is likely to greatly influence both the rate and the trajectory of rehabilitation processes (Parrotta, 1992). Forest restoration programs operated by mining companies in Brazil, Australia and other tropical countries have, therefore, usually relied on the artificial regeneration of either native or exotic forest species to rapidly establish tree cover on reclaimed minesites and thereby facilitate natural forest succession (Parrotta and Knowles, 1999).

In a study of 3 coal mines dumps of eastern India, Singh et al. (1998) observed that *D. sissoo* was the fastest growing species followed by *Leucaena leucocephala* at one site. At another site *Dalbergia sissoo* was followed by *D. regia* and at the third site *L. leucocephala* had the maximum growth and *Acacia nilotica* the lowest. Study by Dadhwal and Singh (1993) revealed that in degraded abandoned limestone mined land *Leucaena leucocephala* and *Bauhinia retusa* among trees, are the plant species suitable for rehabilitation of degraded mine sites because of presence of many root nodules in *Leucaena leucocephala* and soil binding factors of both these species.

Trees can potentially improve soils though numerous processes, including maintenance or increase of soil organic matter, biological nitrogen fixation, uptake of nutrients from below the reach of roots of understory herbaceous vegetation, increase water infiltration and storage, reduce loss of nutrients by erosion and leaching, improve soil biological properties, reduce and leaching, improve soil biological activity. Given time, new self-sustaining topsoils are created by trees (Filcheva *et* al., 2000). However, impact of trees on soil fertility depends on their nutrient-cycling characteristics such as litter chemistry and decomposition. In addition to the nutrient sink function due to mass accumulation, same plantation species exhibit high nutrient use efficiency and may be more effective nutrient sink than the other species (Singh et al., 2002). In temperate environment, slower-growing, broad-leaved native trees are regarded as better for amenity, but less efficient for timber production (Filcheva *et* al., 2000). Lei and Duan (2008) concluded that in general, pioneer plant would be the ideal species for phytostobilization of mine tailings. However, Bendfeldt et al. (2001) observed that after 16 years, there appeared to be no lasting soil quality improvements due to addition of organic amendments to the mine soil. Amendments improved short-term production, but their cost of transport and application may be difficult to justify based on long-term soil quality improvement. Singh and Vashistha (2004) observed that positive impact on hydrological attributes of mined degraded watersheds was mainly due to the effect of silvi-pastoral interventions that may play a vital role in the development of mined degraded watersheds in lower Himalayas.

The role of exotic or native species in rehabilitation needs careful consideration, because we may have to use species combinations (native, exotic or combination thereof) that are capable of surviving into new conditions (Singh et al., 2002; Singh et al., 2004a). Alexander (1989c) found that native *Acacia albida* had more beneficial effect on the spoil soils than the exotic eucalypts in tin-mine spoils. Woody legumes perform best on tin (Sn) mine tailings and pulverized fuel ash waste sites, because, apart from being N-fixing, their roots can penetrate into deeper, moist layers (Phia et al., 1995). In the study by Zhang et al. (2001), root nodules were formed in *Leucaena leucocephala* derived from the soil seed bank. The lead-uptake pattern of *L. leucocephala* revealed that considerable Pb was accumulated in the root, branch, stem bark, and xylem, accounting for more than 80% of total metal in this plant. Therefore, such woody legumes may have a greater advantage in metal-phytoremediation than herbaceous plants. The shoots of herbaceous plants, which contained rather higher Pb concentration, would become litter during their growth (Zhang et al., 2001).

For soil reclamation, the difficulty is not in finding a system that fixes enough N to supply the plant needs, but in managing the system and finding species that will grow under the harsh conditions common in degraded soils. Low soil pH is characteristic of most weathered degraded soils of the humid tropics. It is even more of a problem in mining spoils that generate acidity (Franco and De Faria, 1997).

Many studies are available on selection of suitable species for rehabilitation (Singh et al., 2002; HFRI, 2004; Dubey, 2007). Careful selection of species is needed, as newly introduced exotics may also become pests in other situations. Exotic species are believed to negatively impact site conditions, escape into pristine habitats and displace native species (Lugo, 1997). Therefore, candidate species should be screened for their potential to become problematic weeds in relation to local and regional floristics (Parrotta et al., 1997a). As a part of revegetation efforts, selection of desirable species adapted to the local environment has been emphasized. For artificial introduction, use of species that are well adapted to the local environment should be emphasized (Dobson et al., 1997; Wali and Freeman, 1973; Gibson et al., 1985). Indigenous species are preferable to exotics because they are most likely to fit into a fully functional ecosystem and to be climatically adapted (Phia et al., 1995). Aweto (1981c) suggested that in view of the fact that soil textural composition affects plant growth, the afforestation schemes should be soil specific.

On mine spoils, nitrogen is a major limiting nutrient and regular addition of fertilizer nitrogen may be required to maintain healthy growth and persistence of vegetation, and hence, an alternative approach might be to introduce legumes and other nitrogen-fixing species (Singh et al., 2002). Dobson et al. (1997) emphasized that the use of nitrogen-fixing species requires good knowledge of their biology, both their soil preferences and their interactions with other species. Nitrogen-fixing species can have dramatic effect on soil fertility through the production of readily decomposable, nutrient-rich litter and turnover of fine roots and nodules (Singh et al., 2002). Mineralization of N-rich litter from these species will allow substantial transfer to companion species and subsequent cycling, thus enabling the development of a self-sustaining ecosystem (Jefferies et al., 1981). It should be emphasized that leguminous species also differ in their soil enrichment capabilities.

Many studies document the positive role of grass cover as a nurse crop (Bramble and Ashley, 1955). Grasses may have both positive and negative effects on restoration ofmine lands. They are frequently needed to stabilize soils during the restoration process, but they may compete with woody regeneration. Grasses have fibrous roots that can slow erosion and their sod-forming tendencies eventually produce a layer of organic soil. They are useful in restoration of mined land because they stabilize soil, conserve soil moisture, and may compete with weedy species. This initial cover must allow the development of diverse, self-sustaining plant communities (Singh et al., 2002). In a study done for restoration of damaged coal mine areas in India, Singh et al. (1996b) reported that compared to native non-leguminous species, native leguminous species show greater improvement in soil fertility parameters. Also, native legumes are more efficient in bringing out differences in soil properties than exotic legumes in the short term (Singh et al. 1996b). Jha and Singh (1993) from coal mine rehabilitation site in central India reported good performance of *Bothriochloa pertusa, B. intermedia, Chrysopogon fulvus* and *Cenchrus setigerus.* Comparing the survival and growth of three species on coal mine spoil, Singh et al. (2004c) observed that high survival rate was observed in *Albezia procera* and lowest in *Tectona grandis*. At the age of six years, height was maximum in *Albezia lebbek* and minimum in *Tectona grandis*, whereas diameter was maximum in *A. procera* and minimum in *T. grandis*. Similar findings were recorded for two *Albezia* species by Singh et al. (2004b). Singh and Singh (1999a) have studied the effect of mulches on nutrient uptake of *A. procera* planted in coal mine overburden.

Moreover, vegetative material used in reclamation should consist of trees, which is consistent with site capabilities and should be designed to provide a cover consistent with the stated land-use objective and which does not constitute a health hazard. It should fulfill the demand of the local people (Dubey, 2007). Gupta and Singh (2005) observed that *Albizzia lebbek* was the best-adapted plant species for a mine site, followed by *Dalbergia sissoo*. *Albizzia lebbek* and *Dalbergia sissoo* can be considered as early successional tree species and might have atmospheric nitrogen fixing capacity to cope up with the impoverished, slightly alkaline overburden mine spoil of the area.

Bending and Moffat (1999) have suggested that landform design, selection of suitable soil or soil-forming materials, spoil placement technique and appropriate species choice are central to future success of forestry on restored mined area in South Wales. A suitable species for planting on mine spoils should possess the ability: (i) to grow on poor and dry soils; (ii) to develop the vegetation cover in a short time and to accumulate biomass rapidly; (iii) to bind soil to arrest soil erosion and check nutrient loss; and (iv) to improve the soil organic matter status and soil microbial biomass, thereby enhancing the supply of plant nutrients available (Singh and Singh, 1999b); and in addition, the species should be of economic importance (Singh et al., 2006). Bradshaw (1997) is of the view that current methods of reclamation may lead to the establishment of a limited number of plant species only, resulting in the production of ecosystem of low diversity with restricted land use potential and wildlife conservation value. Phyto-stabilization is the use of metal-tolerant plant species to immobilize heavy metals through absorption and accumulation by roots, adsorption onto roots, or precipitation within the rhizosphere. This process reduces metal mobility and also reduces bioavailability for entry into the food chain. By using metal tolerant plant species for stabilizing mine spoils, it could also provide improved conditions for natural attenuation. The normal practice is to choose drought-resistant, fast-growing species which can grow in metal-contaminated and nutrient-deficient soils (Wong, 2003).

Most investigations on vegetation of man-made mine wastes especially emphasize the selection of metal-tolerant species, and few are related to initial stages of primary succession of plants on these wastelands (Gibson, 1982; Ernst, 1988; Chambers and Siddle, 1991). For reclaimed coal mine spoils, Sanchez and Wood (1989) reported that revegetation species were selected for their ability to withstand environmental stress from severe climatic conditions and shallow, infertile soils; ability to stabilize the ground surface of redistributed soil material; and ability to withstand grazing pressure. Seed mixtures used in the reclamation project were composed of warm season and cold season grasses, forbs and shrubs.

Since spoils usually lack nitrogen, legumes appear to be a logical choice for quick cover (Maiti and Banerjee, 1993; Maiti, 1997; Vajpayee et al., 2000). On the other hand, legumes face severe difficulties both in growth and nodulation on acid spoils since many studies of legumes seeded on acid soils indicate that poor growth under acid condition often can be attributed to toxicities caused by excess manganese and aluminium brought into solution by the acid conditions (Fail Jr. and Wochok, 1977). Legumes are commonly planted, since the bacteria associated with their root nodules have the capacity to fix atmospheric nitrogen. In a study, Macedo et al. (2008) also found that due to their association with arbuscular mycorrhizal fungi, leguminous tree species were selected that not only can establish under harsh conditions, but also produce high biomass yields of low C:N ratio. Soil N increase is very important in degraded land rehabilitation projects, since, it enhances the capacity of the system to support a more complex community.

Doubleday and Jones (1977) pointed out that spoils may be reclaimed by directly establishing grass but the grass swards often deteriorate over a number of years and show signs of moisture stress in summer months and sometimes water-logging in the winter (Doubleday, 1974). Armiger et al. (1976) found that steep slopes (≥60%) in some situations can be stabilized by seeding and transplanting leguminous species with phosphorus application. Stabilization of spoils and reduction in erosion can be effected by seeding the spoil with sod-forming grasses. Grasses have ameliorative effects on both physical and chemical properties of spoils. Alexander (1989 a, b) has compared the beneficial effect of *Acacia albida* and *Eucalyptus camaldulensis* on the tin-mine spoil, and recommended that *A. albida* has an ability to improve both the nutrient status and physical conditions in the top 20 cm of the soil beneath its canopy, whereas *E. camaldulensis* caused a progressive increase in the soil acidity and reduction of base content, although organic C increased. *Robiniapseudoacacia,* a N-fixing legume, raised soil N levels considerably more than the other non-leguminous species (Vimmerstedt et al., 1989).

Along with fixing atmospheric N, fast-growing legumes provide rapidly established vegetative cover (reducing erosion), increase soil organic matter, modify the microclimate at the soil surface, provide a resource for fauna, and favorably affect soil physical properties (Ward et al., 1996; Ward, 2000). However, there are a number of disadvantages including increased competition leading to a restoration in plant species richness, dense vegetation making areas less accessible, and increased rate of fire-fuel load accumulation, and hence, the density of legumes needs to be controlled to limit competitive exclusion of other species. Therefore, low-biomass nitrogen-fixing species should be used in restoration programme to assist in the re-accumulation of micronutrients following initial fertilizer application (Grant et al., 2007). Wong (2003) has reported successful establishment and colonization of several pioneer plant species, e.g., *Vetiveria zizanioides, Sesbania rostrata* and *Leucaena leucocephala* on Pb- Zn mine spoils. Vetiver grass *(Vetiveria zizanioides)* has a massive finely structured and deep root system capable of reaching 3-4 m in the first year. Due to its unique morphological and physiological characteristics, it has been commonly known for its effectiveness in erosion and sediment control, and has also been found to be highly tolerant to extreme soil conditions including prolonged drought, flood, submergence, extreme temperature (-10^0 to 48^0 C), and a wide range of soil acidity and alkalinity (pH from 3 to 10.5). This species is highly tolerant to soil salinity, sodicity, Al, Mn and heavy metal (such as As, Cd, Cr, Ni, Pb, Zn, Hg, Sc and Ce) toxicities in the soil. Several workers from Pb-Zn mine spoils reported that vetiver grass was found to be the best plant spices (in terms of biomass production and coverage) when compared with other three grass species, namely *Paspalum notatum, Cynodon dactylon* and *Imperata cylindrica* used for revegetating Pb-Zn mine tailings (Wong, 2003). *Sesbania rostrata,* which possesses stem as well as root nodules, can be used to modify properties of mine spoils, by supplying the much-needed N and organic matter. Apart from having a very high growth rate, it is also tolerant to toxic metals and low nutrient status, and therefore, will be an ideal pioneer species to accelerate ecological succession of the man-made habitats (Wong, 2003). Ye et al. (2001) also found that *Sesbania rostrata* is a better choice as a pioneer species for revegetation of lead/zinc mine tailings than *Sesbania cannabina.*

A study of three-year-old reclaimed bauxite mine site in central India by Chaturvedi (1983) shows that *Grevillea teridifolia* and *Eucalyptus camaldulensis* had given best results. *Pinus caribaea*, *G. robusta* and *Acacia auriculiformis* had also come up well. Amongst the indigenous species, *Shorea robusta* and *Pongamia pinnata* had shown good survival and healthy look, though spectacular height growth was missing. *Themeda australis* (kangaroo grass) was found suitable for use in large-scale mine rehabilitation and it was planted for mine site restoration (Windsor and Clements, 2001). Calcareous bauxite mining spoils was revegetated by seeding of woody species (Brofas and Karetsos, 2002). From a study of coal mine spoil in central India, Dutta and Agrawal (2002) reported that among different plant species, the physical characteristics are maximally improved by *Eucalyptus* hybrid, *Acacia auriculiformis* and *Casuarina equisetifolia*. In soil texture, maximum silt and clay was found in the plots of *Eucalyptus* hybrid. Bulk density was higher in *Casuarina equisetifolia* followed by *Acacia auriculiformis* and lowest in *Gravellia pteridifolia* plots. Water holding capacity was also highest in the plots of *Eucalyptus* hybrid. Moisture content also followed a similar trend. In chemical characteristics, total N content was higher for *Acacia auriculiformis*, a nitrogen-fixing species, whereas *Casia siamea* showed lowest N content in soil. Organic C and total P also had a similar trend. These characteristics can be due to the lower growth and poor nodulation of *Casia siamea* plants. *Casuarina equisetifolia*, a non-leguminous nitrogen fixing species showed highest available P content and a higher N level in the soil.

Sengupta (1991) reported that for stabilization of a south Indian lignite mine slopes, *Brachiaria mutica* and *Cynodon dectylon* are planted before going for tree plantation. Mathur (1978) and Gupta (1979), for overburden of semiarid to subhumid regions in India, suggested for seeding/planting of *Acacia auriculiformis, Cassia siamea, Albizia lebbek, Dalbergia sissoo, Prosopis juliflora, Terminalia arjuna, Syzygium cumini, Pongamia pinnata, Madhuca latifolia, Alstonia scholaris, Pterocarpus marsupium, Cleistanthus collinus, Vitex negundo, Ipomoea carnea, Tephrosia candida, Sesbania aegyptiaca, Pennisetum pedicellatum, P. purpureum, Cenchrus ciliaris* and *Chrysopogon fulvus*. Saxena (1979) suggested *Cenchrus ciliaris, Cenchrus setigerus, Cynodon dectilon, Dichanthium annulatum, Sehima nervosum, Sporobolus marginatus* and *Eleusine compressa* as grasses, and *Prosopis cineraria, Zizyphus nummularia, Acacia tortilis, A. senegal, Calotropis procera, Salvadora oleoides, Indigofera oblongifolia, Grewia tenax* and *Prosopis juliflora* among shrubs and trees for revegetation of gypsum and bentonite mines of northwest India. Revegetation of acentral Indian bauxite mine spoils was studied by Prasad and Shukla (1985b) and Prasad (1989). They found *Eucalyptus camaldulensis, Grevillea robusta, G. pteridifolia, Acacia auriculiformis, Pinus caribaea* and *P. roxburghii* as successful exotics and *Shorea robusta, Pongamia pinnata, Albizia procera* and *Phyllanthus emblica* as successful indigenous species there. For reclaiming rock-phosphate mine site in northern India, Soni and Vasistha (1986) found *Pennisetum purpureum, Saccharum spontaneum, Acacia catechu, Dalbergia sisoo, Leuceana leucocephala, Rumex hastatus, Salix tetrasperma, Mimosa himalayana* and *Buddleia asiatica* as most successful and suitable for environmental amelioration of that mine site. For another rock-phosphate mine, Soni et al. (1989) found *Pennisetum purpureum, Pragmites sp., Eriophorum comosum* and *Debregea siavelutina* as the most suitable grass and shrub species, and *Pinus roxburghii, Robinia pseudoacacia, Cupressus sp., Bauhania purpurea* and *Cedrus deodara* as the most suitable tree species for reclaiming that area. From a field trial on limestone overburdens yet in norhern India, Soni et al. (1990) have recommended *Acacia catechu, Dalbergia sissoo, Salix tetrasperma, Leucaena leucocephala, Populus deltoides, Pinus roxburghii, Robinia pseudoacacia, Cupressus spp., Cedrus deodara, Wendlandia exserta,* etc. among trees, *Vitex negundo, Dodonia viscosa, Ipomea carnea, Crotalaria sericea,* etc. among shrubs, and *Arundo donax, Saccharum spontanum, Pennisetum purpureum,* etc. among the grasses and forbs for revegetation. After 5 years, the area was free from erosion, and a good number of grasses, shrubs and trees colonized naturally. The total number of plant species on reclaimed spoils was nearly equal to, or more than the native undisturbed forest. *Dalbergia sissoo* and *Acacia catechu* suited well, and natural regeneration of *Buddleja asiatica* and *Trema politoria* was observed. Prasad and Shukla (1985a), from central Indian coal mines, reported that *Eucalyptus territicornis, E. camaldulensis and Acacia auriculiformis* were more successful species in terms of height and diameter growth than *Phyllanthus emblica, Acacia nilotica, A. catechu* and

D. sissoo. D. sissoo and A. nilotica showed better performance than *Pongamia pinnata* and *A. catechu* in terms of diameter growth. *Dendrocalamus strictus* also showed good performance. *Albizia lebbeck* and *A. procera* were also reported as suitable species. Provided the fast growing species are chosen, it is clear that a forest vegetation canopy can be rapidly established on these mine overburdens. Better performing species are nitrogen fixing and this fact is understandable in view of the nitrogen deficiency of the material. Tree growth on unamended overburden materials can be as good as on ordinary soil, provided nitrogen is available (Das et al., 1992).

Biological reclamation on 10-year-old coalfield mine spoil in Eastern India was carried out during 1992. It was observed that after five years of reclamation, out of 14 plant species selected for reclamation, *Acacia auriculiformis*, *Acacia arabica*, *Albizia lebbeck, Leucena leucocephala* and *Gmelina arborea* were successful. Organic carbon, available major nutrients (N, P and K), pH, exchangeable cations and cation exchange capacity increased whereas trace elements decreased. A progressive richness in species diversity in natural vegetation was observed indicating rejuvenation of the ecosystem (De and Mitra, 2002). After study of revegetation program of an eastern India coalfields, Singh et al. (2005) found that *Pterocarpus ascerifolium, Bauhinea variegata, Eucalyptus spp.* And *Zizyphus zuzuba* attenuate sulphur dioxide and nitrous oxide, whereas bamboo spp., *Dalbergia sissoo* and *Cassia siamia* are good attenuaters of sulphur dioxide only. *Syzygium cumini, Mymosops elengi* and *Madhuca indica* attenuate oxides of nitrogen and particulate matters. For restoring gypsum mine spoil, Rao and Tarafdar (1998) found that *Acacia senegal, Acacia tortilis, Azadirachta indica, Cenchrus ciliaris, Colophospermum mopen, Dichrostachys nutans, Pithecellobium dulce, Prosopis juliflora, Salvadora oleoidis* and *Zizyphus nummualaria* can be used. Among these plant species, *Azadirachta indica, C. mopen and Z. nummualaria* showed better exploitation of micro nutrients on mine spoils than normal soil; and *S. oleoidis, C. mopen and P. dulce* were identified as calcium-loving plants. The significantly higher availability of micro-nutrients in gypsum mine spoil compared to the normal cultivated soil of the surroundings may be the reason for higher concentration in plants. Growth of most of the selected plant species in gypsum mine spoil was equal to or more than that observed in normal soil. Gupta et al. (2004 b) reported that *Bambusa pallida, Dendrocalamus hemiltonis, Bambusa nutans, B. multiplex* and *Phyllostachys reticulate* exhibited 60-80% survival on mine spoil slopes, whereas *B. tulda* failed to survive. Growth parameters of the above survived bamboo species were also almost comparable to these species growing in natural habitat.

Muzzi and Fabbri (2007) tested the comparative performance of 24 shrub and tree species on a revegetated mineral clay soil and found that the best performing shrub species were *Spartium junceum, Ligustrum vulgare and Cotinus coggygria* followed by the substrate adaptive but less developed *Rahmnus cathartica, Pinus spinosa and Crataegus monogyna*. The tree species that responded best were *Sorbus domestica, Pyrus pyraster, Fraxinus ornus* and *Fraxinus angustifolia* followed by the adoptive but underperforming *Ulmus minor, Fraxinus excelsior* and *Acer campestre.* Pot experiments were carried out to find out the growth potential of the native plant species and some exotic species, a total of 40 plant species were selected. Nursery raised sapling in polyethylene bags were transported to the site and transplanted in wasteland areas at an eastern India coalfield. The study of plant growth performance with respect to increment in height and girth indicated that the growth potential was maximum in case of *Dalbergia sisoo,* followed by *Gmelina arborea, A. indica, Albezia lebbeck, Tectona grandis, P. enormi, Alstonia scholaris* among the selected plant species (Pal et al., 2005). A series of experiments was conducted on the rehabilitation of mine spoils in a dry tropical region of India for determining the suitability of tree species and impact of the plantations on the restoration of biological fertility of the mine spoils. It was found that the growth of the majority of the species could be improved by amending the mine spoil with NPK fertilizer.

Direct seeding showed the greatest height of *Zizyphus jujuba* and *Pongamia pinnata* on the flat surface and *Terminalia arjuna* on the slope. The plantation of trees significantly accelerated the soil redevelopment process on the mine spoil (Singh and Singh, 2006). Singh et al. (2000 b), from a study of rehabilitated limestone mined land near Mussoorie, reported that agroforestry interventions with establishment of trees, shrubs and grasses on mine spoil have improved the soil conditions significantly. Boulder percentage decreased and the soil fraction below 2mm size increased in rehabilitated mined area in comparison to unrehabilitated control plot. The sand percentage decreased and silt and clay increased significantly after agroforestry interventions. Water holding capacity increased while bulk density decreased after plantation.

For lead/zinc mine tailings, Shu et al. (2005) suggested that artificially accelerating the colonization would be necessary to create self-sustaining vegetation on the tailings. Relatively cost-effective methods for revegetation of the tailings might include: (1) using organic wastes for buffering metal toxicities and improving physicochemical properties; (2) applying top soil from adjacent areas as seed resources and also providing favourable microsites for plant colonization; and (3) introducing seeds of some nature rhizomatous, perennial grasses that might have high colonization potential. For bauxite-mined lands, the most productive treatment in terms of basal area development and height growth was the mixed commercial species treatment, followed by the natural regeneration, direct seeding and mixed native species treatments. Tree basal areas in the restoration treatments ranged from 18% (in the mixed native species plots) to 33% (in the mixed commercial species plots) of that in the primary forest plots (Parrotta and Knowles, 1999). Analyses of basal-area dominance among restoration treatments clearly show that it will take at least several decades before the species that presently dominate the primary forest, though often present in the understory of the restoration stands, will assume a significant structural role (Parrotta and Knowles, 1999).

The mixed commercial species treatment, dominated by *Eucalyptus spp., Sclerolobium paniculatum* and *Acacia mangium* stands out clearly at the most productive as indicated by data on basal area and tree height growth. The abundance, mean height and species richness of trees and shrubs regenerating in the understory were significantly lower, however, than in either the direct seeding or mixed native species treatments. The relatively low regeneration density in this treatment is most likely due in part to the rapid growth rates of the planted trees, to relatively high understory light levels and to increased grass and herb densities all of which result in increased root competition among germinating woody species. The significantly lower understory species richness in treatments also suggests possible limitations on seed-bank germination and/or seed inputs by birds, bats and other mammals that are the main agents of seed dispersal from the primary forest (Parrotta and Knowles, 1999).

Notwithstanding the merits of planting mixed stands of native hardwood species attempts to establish plantations of indigenous species have often been either difficult or unsuccessful owing to the lack of adequate knowledge on their biology, ecology and silviculture (Leopold et al., 2001). Conversely, plantation establishment using exotic species has increased due to (i) readily available information on propagation techniques, silvicultural behaviour and management practices, and (ii) at least initially fast growth rates, and production of wood that can be used for various purposes in a relatively short period of time. However, exotic tree plantations are meanwhile widely considered to have serious adverse effects on the environment, which include (i) harmful changes in the physical, chemical and biological conditions of the soil; (ii) competition with agronomical land use for monetary reasons; (iii) displacement of the local flora, the native vegetation and, in part, the native fauna; and (iv) enhancing problems of susceptibility of the exotic species to epidemic diseases and pests (Feyera et al., 2002).

Despite the various benefits that accrue from reforestation with exotic species, growing concern regarding the disadvantages of such ventures and reluctance or even resistance of local people to the introduction and establishment of exotic species is observed. It is claimed that fast-growing exotic trees, once established in a plantation, may exert higher competitive strength for water, nutrients, and light than indigenous plants. Some may also hamper germination, establishment and growth of other, in particular indigenous, plants by releasing allelopathic chemicals. As a result, environmental changes beneath and around the plantation may take place that are detrimental to the normal functioning of the habitat (Feyera et al., 2002).

However, the impact of plantations of exotic species on the site conditions and on the possible regeneration of indigenous species under that canopy depends not only on the planted species but also on the history of the site, the forest management practices employed, such as planting density and coppicing. Moreover, quality and quantity of light penetrating through the canopy, microclimatic conditions, composition of the seed bank in the soil, and availability of recent seed sources in the vicinity of the plantation are important. Therefore, generalizations are at least risky since they may lead to inappropriate conclusions and recommendations. Potentially, in addition to production of wood as an economic resource, exotic tree plantations may also have advantages, such as improving microclimatic conditions, protecting degraded lands against soil erosion, stabilizing soil development, thereby enhancing soil nutrient status and increasing soil organic matter through enhancement of litter and humus production and supporting water catchment values. In addition, such plantation may foster the regeneration of native species under their canopy and provide a habitat for wildlife (Keenan et al., 1997). Thus, exotic tree plantation potentially may greatly improve physical and biological site conditions catalyzing subsequent succession processes towards a natural forest (Parrotta, 1992; Parrotta et al. 1997b).

Several recent studies in different parts of the world have shown that tree plantations can have a nurse effect and thus be an effective means for rehabilitation of degraded sites and catalyzing the restoration of native vegetation beneath their canopy (Parrotta, 1992; Parrotta et al., 1997a). In many sites and with various exotic species, substantial numbers of indigenous woody species regenerate beneath the canopy of the plantations, thus showing an increase in their biodiversity (Feyera et al., 2002). The establishment of a permanent cover of vegetation involves not only growing plants, but it necessitates bringing into a plant community that will maintain itself indefinitely without attention or artificial aid, and support native fauna. The presence of certain tree species in a productive system can result in better soil structure and increased soil nutrient availability. Such performance could be achieved by selecting species adapted to grow, spread and reproduce under severe conditions provided both by the nature of the dump material and the exposed situation on the dump surface (Singh et al., 2002).

In an experiment with the N_2-fixing legume-tree, *Albezia lebbek* was used as a plantation treeand was found that seedling mortality of the plantation tree itself under the canopy was much larger (47% in a 1.3 year period) than that of secondary forest species (23-26%) underlining the role of the plantation as a "foster ecosystem" for successional development of secondary forest (Parrotta, 1993). While most species appear to act as catalysts for ecosystem rehabilitation, broadleaf species seem to give better results than conifers (Parrotta et al., 1997a). Of these, fast-growing species that represent lower successional stages should have preference, particularly those known to establish and grow well on degraded sites. With the appropriate management, reforestation with exotic trees can foster under-growth. Tree planting may facilitate the process of forest succession by providing a nurse effect for colonizing native species. This nurse effect is attributed to a protective microenvironment under the canopies of tree plantation which rhythmically changes upon coppicing of the nurse tree. Intermittent changes of the light and microclimate obviously increase germination and growth rates as well. In addition, the associated increase in vegetation structural complexity, and the production of multi-component litter and the concomitant enrichment of humus layers bring about improvement of forest floor conditions. In turn, these changes facilitate seedling establishment by suppression of grasses that normally prevent tree seeds germination and growth of the seedlings (Feyera et al., 2002). They (Feyera et al. 2002) concluded that objection against the establishment of exotic species cannot be warranted, especially in countries where, on the one hand, there is a desperate and urgent need of expanding the forest resource base to meet the ever-increasing demand for timber, firewood and non-timber products and, on the other hand, remaining natural forests need protection and regenerative extension of stands of native trees is highly desirable.

D' Antonio and Meyerson (2002) opined that in some degraded sites it may be necessary to introduce an exotic species to assist with the restoration process. At many places fast-growing but sterile exotic grasses were used to quickly establish cover. These grasses do not seed and presumably give way to native species. In other places where land uses, such as mining, have resulted in loss of soil fertility, fast-growing exotic N-fixing trees have been used to ameliorate harsh site conditions. Gupta et al. (2005) reported better growth of *Dalbergia sissoo* and *Melia azedarach* than *Casia fistula* and *Tectona grandis* on coal mine spoils due to atmospheric N-fixation into the soil pool together with higher litter decomposition rate which in turn increase the biomass. Parrotta (1992) found that an Asian N-fixing tree could grow well in degraded pastures and eventually accelerate regeneration of native rain-forest. Lugo (1988) provided observational evidence that non-indigenous trees can ameliorate harsh environmental conditions on barren sites, eventually facilitating the establishment of native tree species.

In a study by Kumar et al. (2004) on vetiver grass (*Vetiveria zizanioides*) planted on coal overburden dumps, it was found that it added nutrients by its biomass and improved the physico-chemical properties of the dumps to support other species to come up. It was concluded that selecting vetiver grass species is the economical bio-reclamation approach for the coal overburden dump. Gupta et al. (2004 a) reported that *Amaranthus caudatus*, regarded as pseudo-cereal, is an under-utilized plant and can be a promising plant species to be used for restoration of coal mine spoils. This species can be used as vegetable, cereal and even as fodder and its establishment will not only help restriction of degraded land but also be useful for people living around mine area. In a study of coal-mine overburden rehabilitation by four species, Singh et al. (2006) concluded that a non-legume (*Dedrocalamus strictus*) can also play a significant role in restoring mine-spoil habitats, in the same way that has been especially reported for leguminous species. Aber (1987) was also of the view that several non-leguminous species, particularly in genera of woody plants associated with wetlands (e.g., *Alnus spp.*) have different but also effective N-fixing microbial associations.

Singh (2006), in a study of mixed plantations raised on mine spoil, found that the effect of N fixing species as neighbour is not always beneficial for the growth of non-N fixing species. *Tectona grandis* exhibited poor performance when grown with legumes *Dalbergia sissoo* and *Leuceana leucocephala* but its growth was significantly greater when planted with the non-legume *Dendrocalamus strictus*. In contrast to result obtained for *T. grandis*, the growth of non-legumes *Gmelina arborea* and *Terminalia bellirica* was greater when these species were grown with the legume *Pongamea pinnata* than when grown with non-legumes. This suggested a positive effect of N fixing legumes on growth of non-legumes. Joshi and Singh (1998) reported better growth in oak in the presence of N fixing alder than when the oak was grown alone. Benefits from N fixation are more likely to occur on sites that are less fertile (Binkley, 1983). This may be the reason behind better growth of non-legume *Gmelina arborea* and *Terminalia bellirica* in combination with N-fixing legume because mine spoils are nutrient poor habitats. On degraded soil, N fixing woody species could have long term effects on improvement in growth of non-N fixing woody species (Singh, 2006).

Singh (2006) found that the legume *Pongamea pinnata* exhibited better growth with non-legume in mixed plantations. This suggests that the legume and non-legume companionship is not only beneficial to non-legume but is also advantageous for legume. The poor performance of *Pongamea pinnata* with *Acacia catechu* suggests that perhaps both the legumes competed for same resource, hence the growth of *Pongamea pinnata* was reduced when planted with *Acacia catechu*. Competition for resource is a dominant process influencing performance of species in mixed plantations (Shainsky and Radosevich, 1992). Pairs of species which have both a large competitive effect and response to one another, probably use the same resources, and hence, are limited by those resources (Miller and Werner, 1987). In mixed plantation, nitrogen may mediate interaction between and within species in at least two different ways. First, because the nitrogen fixer utilizes a different source of nitrogen (the atmosphere) form that of the non-fixer (the soil), potential exists for reduced competition between species through resource partitioning. Second, facilitation may occur through the addition of fixed nitrogen to the system, and enhanced availability to the non-fixer. *Acacia catechu* planted with legumes, showed better growth with *Pongamea pinnata* than with *Albizia lebbeck*. Greater N fixatation rate has been reported in *P. pinnata* than in *A. lebbeck*. The significantly greater N concentration in *G. arborea* and *Terminalia bellirica* grown with leguminous trees suggests that N fixing leguminous species have ability to improve the foliar N status of the companion species in mixed plantation. On the other hand, N fixing legumes *Dalbergia sissoo* and *Leucaena leucocephala*, as neighbouring species, reduced the foliar N concentration in *Tactona grandis* suggesting that N fixers may not always have a positive effect on foliar N status of non- Nfixing companion species in mixed plantation. Therefore, Singh (2006) concluded that in mixed plantation, effect on and response to one another will differ from species to species and this aspect needs to be evaluated before recommending a particular species combination for mixed plantation.

Mine spoils need to be recreated before the mature plant community will function. Altering soil condition through plants has a potential problem that plants that have unusual physiological mechanisms (e.g., symbiotic N fixation) and are, therefore, more likely to survive and to ameliorate soil conditions are often exotics. The role of exotic species in restoration has often been looked with concern due to their negative impact on soil fertility and biodiversity (Lugo, 1997). The available literature, however, do not support this (Parrotta, 1999). Aber (1987) cited an example of reclamation of mine wastes in the humid area where *Robinia pseudoacacia,* a N fixing legume and *Populus sp.* a highly productive cation pump species is often planted first and concluded that once the function and structure are restored, then the exotic community can be removed. Dutta and Agrawal (2003) are of the view that exotic species may be recommended for primary rehabilitation on bare coal mine spoil due to their fast growth and establishment. Care, however, is needed so that introduced exotic species may not become a problematic weed to local flora. They reported that among the tested exotic species, viz. *Acacia auriculiformis, Casurina equisetifolia, Cassia siamea, Eucalyptus* hybrid and *Gravellia pteridifolia,* the most suitable species for coal mine spoils were *E.* hybrid, *A. auriculiformis* and *C. equisetifolia. E.* hybrid showed the highest growth, biomass and net primary production followed by *C. equisetifolia* and *A. auriculiformis. G. pteridifolia* ranked fourth with respect to all the studied parameters while *C. siamea* did not show good growth and biomass accumulation on coal mine spoils.

Among species that may be considered suitable for a given degraded site, there may be considerable variations in their capacity to stabilize soils, increase soil organic matter and available soil nutrients, and facilitate under-storey development. These variables include susceptibility to pests and diseases, patterns of aboveground and root biomass accumulation, nutrient utilization and allocation, nutrient use efficiency, nutrient retranslocation, litter decomposition, and the presence of secondary compounds that may inhibit the activity of decomposing organisms (Singh et al., 2002).

KORAPUT BAUXITE MINE: INITIAL REHABILITATION SUCCESS

The National Aluminium Company started bauxite mining on Panchpatmali hills of Koraput district in Odisha (Latitude 18°45'N to 18°55' N, Longitude 82°55' to 83°05' E) in 1985. Panchpatmali plateau of Eastern Ghats Region of India stands at elevation of 1154 m to 1366 m above mean sea level. The area experiences sub-tropical to tropical climate (temperature between 7.0 degree Celsius to 42.3 degree Celsius and rainfall between 1340 mm to 2100 mm). Forest type of this plateau is Dry Tropical Savannah Forests. This is a fully mechanized opencast mine in operation serving feedstock to Alumina Refinery located on the foothills. Bauxite occurs over the full length of the Panchpatmali plateau spanning over 18 kms. Rehabilitation of mine spoils in the initial phases of mining was done with limited success. However, since the first half of the last decade of the last century, some successful plantations have been raised on mine spoils. One rejuvenated effort is presented along with the photographs taken during the study period 2009 - 2011 by the first author (AKS). The mining Area of the Bauxite-bearing plateau top before mining is naturally carpeted with *Phoenix acaulis*, (looking like grass in Figure 2), and is devoid of tree species. Far background is the relics of some failed plantation (mainly silver oak) of rehabilitation work done long back. Minute observation shows a few stumps of *Terminalia chebula* and *Phyllanthus emblica* plants on the plateau top before mining. Blast Hole Drill Machine (Figure 3) was being used for drilling and blasting the overburden above the bauxite reserves. Trench method of mining is being practiced there. Excavators and dumpers were in operation for cutting and hauling the mineral mix (Figures 4). Figure 5 shows the exposed bauxite profile in a trench during mining. Bauxite occurs as capping of various thicknesses over the khondalite and charnokite rocks. Typical bauxite profile in that area is i) overburden comprising of soil, clay and laterite; up to 12 m. thick; ii) bauxite - up to 25 m. thick; and iii) parent rock.

Figure 2. Mining Area of the plateau top before excavation.

As discussed earlier, in surface mining with inversion of soil horizons, the unweathered 'C' horizon gets exposed which contains unweathered clay with high cation exchange capacity and high level of exchangeable bases, and almost complete absence of organic carbon and nitrogen (Das et al., 1992). To overcome this problem, it is an established procedure to spread the fresh topsoil stockpiled separately. Adopting this principle, stockpiling of topsoil separately and spreading it after filling the mined-out trenches is being practiced at Panchpatmali bauxite mines before going for further plantation activities.

Figure 3. Drilling for blasting the overburden above the bauxite reserves.

Figure 4. Excavation and loading of bauxite.

Figure 5. Bauxite profile in vertical plane with excavator marks on it.

Figure 6. Top soil brought for spreading over filled up area after mining.

Figure 7. Spreading of top soil after filling up the excavated area.

Figure 8. Leveled area ready for plantation after spreading of top-soil.

Figure 9. One year old plantation of miscellaneous species suitable for such area.

After the closure of mining operations, the pits were filled with debris and topsoil was spread over it (Figures 6 and 7). Proper dredging was done to make the ground levelled and ready for plantation (Figure 8). Nursery raised plants of different species were used during rainy season. Till the middle of the first decade of the current century, *Eucalyptus* hybrid, *Grevillea robusta, Acacia auriculiformis, Pinus kesiya, Pongamia pinnata, Casuarina equisetifolia, Syzygium cumini with little Jtropha carcus, Simarouba glauca* and other minor species were plated for rehabilitating the area. However, since 2007 *Acacia nilotica, Aegle marmelos, Anacardium occidentale, Artocarpus heterophyllus, Bambusa spp., Casia fistula, Casia siamea, Casuarina equisetifolia, Madhuca indica, Mangifera indica, Melia azedarach, Phyllanthus emblica, Michellia champaka, Syzygium cumini, Syzygium jambos, Tamarindus indica,* and *Terminalia arjuna* were mainly planted. Figure 9 is depicting the status and initial success of one year old plantation having survival rate of 95%.

Conclusion

Post mined areas may be converted into refractory sites and their restoration becomes highly difficult. Nevertheless, some management interventions based on tried and tested techniques help in bio-rehabilitation and bio-replacement. For this, adoption of a definite path for restoring the soil is critical. At the same time selection of suitable species is extremely important for the development of self- sustaining ecosystem. This must be determined by the fact as to what will grow best in the particular soil and climate of the area. Based on the facts presented in discussion section, the possible strategy for rehabilitating a mine spoil through vegetative cover should include following steps: (i) Survey of vegetation before disturbing the mining site; (ii) Collection of seeds of all the species; grasses, shrubs and tree; (iii) Top soil replacement; (iv) Seeding with the collected seeds, supplemented with extra seeds of the species surveyed on the mining site before disturbance; (v) Chemical amendments, wherever required; and (vi) Selection of species for revegetation.

As discussed, (Figure 1) above, the aim of the management efforts at Panchpatmali plateau is to achieve 'rehabilitation' of the classical Bradshaw model of various options for degraded or derelict ecosystems. Although, detailed survey of the vegetation of Panchpatmali plateau and its adjoining areas were done by the management, collection of seeds of species found were not being done and plantation was being taken up after filling the trenches and spreading topsoil over them. The Mine Management, in a verbal communication informed that no soil amenders were used for plantations earlier than 2007, but afterwards cowdung manure (about two kgs per pit) were being applied.

Although, many grasses and shrubs invaded the rehabilitated sites, only *Woodlandia heynei, Bombax ceiba, Mangifera indica, Terminalia chebula* and *Syzygium cumini* could invade the rehabilitated areas, as far as tree species are concerned. This may be because of infestation of the area by *Lantana camara,*

Chromolaena sp. and *Cymbopogon citratus. Cymbopogon citratus* was introduced there in early stages of rehabilitation to check soil erosion. Recurrent fire due to presence of *Cymbopogon citratus* might be other reason for very insignificant invasion by tree species. Therefore, it can be concluded that these exotic species were acting as pioneer species only and were not going to be the dominant species of the climax community.

From the discussion above, it may be concluded that while selecting suitable species for a rehabilitation package, the basic principles should include that the species: should be adaptive to that adaphic-climatic condition, should create a plantation canopy in a shorter possible time to alter the understorey micro-climate and adaphic condition, and should fulfill the needs of the people dependent on that area for their food, fodder, timber, fuel wood and herbal medicinal requirements. In view of the success of *Eucalyptus* hybrid, *Grevillea robusta, Acacia auriculiformis, Pinus kesiya, Pongamia pinnata, Casuarina equisetifolia, Syzygium cumini, Jatropha curcas, Simarouba glauca,* etc. in nearby plantation sites of the Koraput district, these species were tried. However, *Jatropha curcas and Simarouba glauca* did not perform well as far as growth parameters and survival percentage are concerned. It was observed during the present study that these exotic species (other than *Jatropha curcas and Simarouba glauca*) performed well fulfilling the abovementioned requirement of creating a plantation canopy in a shorter possible time. However, these species are of not much significant need, other than timber requirement, of people dependent on these areas. Further plantations of species more useful to the local highlanders were done in subsequent years and their performances will be a matter of interest.

References

Aber, John D. 1987. Restored forests and the identification of critical factors in species-site interactions. In: Jordan III, William R., Gilpin, Michael E. and Aber, John D. eds. *Restoration Ecology: a synthetic approach to ecological research.* Cambridge University Press. pp. 241-250.

Alexander, M. J. 1989a. A review of the bureaucratic, political and legislative problems encountered in the reclamation of the plateau tin fields of Nigeria. *Landscape and Urban Planning,* 17: 35-45.

Alexander, M. J. 1989b. The long term effect of *Eucalyptus* plantations on tin-mine spoil and its implication for reclamation. *Landscape and Urban Planning,* 17: 47-60.

Alexander, M. J. 1989c. The effect of *Acacia albida* on tin-mine spoil and their possible use in reclamation. *Landscape and Urban Planning,* 17: 61-71.

Anderson, D. W. 1977. Early stages of soil formation on glacial till mine spoils in a semiarid environment. *Geoderma,* 19: 11-19.

Andrews, J. A., Johnson, J. E., Torbert, J. L. Burger, J. A. and Kelting, D. L. 1998. Minesoil and site properties associated with early height growth of eastern white pine. *J. Environ. Qual.* 27: 192-199.

Archibold, O. W. 1980. Seed input as a factor in the regeneration of strip-mine wastes in Saskatchewan. *Can. J. Bot.,* 58: 1490-1495.

Armiger, W. H., Jones, Jr., J. N., and Bennett, O. L. 1976. *Revegetation of land disturbed by strip mining in Appalachia.* ARS-NE-71, USDA, Washington, DC.

Aweto, A. O. 1981. Secondary succession and soil fertility restoration in South-Western Nigeria.III. Soil and Vegetation interrelationships. *J. Ecol.*, 69: 957- 963.

Barth, R. C. 1986. Reclamation technology for tailing impoundments. Part 2: Revegetation. *Mineral and Energy Resources,* 29: 24.

Barth, R. C. and Martin, B. R. 1984. Soil depth requirements for revegetation of surface-mined areas in Wyoming, Montana and North Dakota. *J. Environ. Qual.,* 13: 399-404.

Beaton, J. D. 1974. *Soil-plant relationships*. Land Reclamation Short course Notes. Vancouver, B.C.: Centre for Continuing Education, University of British Columbia.

Bell, T. J. and Ungar, I. A., 1981. Factors affecting the establishment of natural vegetation on a coal strip mine spoil bank in southeastern Ohio. *Am. Midl. Nat.,* 105: 19-31.

Bendfeldt, E. S., Burger, J. A. and Daniels W. L. 2001. Quality of amended mine soils after sixteen years. *Soil Science Society of America Journal,* 65(6): 1736-1744.

Bending, N. A. D. and Moffat, A. J. 1999. Tree performance on mine spoils in the south wells Coal fields. *Journal of Ecology,* 36: 784-797.

Bennett; O. L., Armiger, W. H. and Jones, Jr., J. N. 1976. Revegetation and use of eastern surface mine spoils. In: *Land application of waste materials.* Soil Conservation Society of America, Ankeny, Iowa, pp. 195-215.

Binkley, D. 1983. Ecosystem production in Douglas-fir plantations: interaction of red alder and site fertility. *Forest Ecology and Management*, 5: 215-227.

Bloomfield, H. E., Handley. J. F. and Bradshaw, A. D. 1981. Topsoil quality. *Landscape Design,* 135: 32-34.

Bradshaw, A. D. 1983. *The reconstruction of ecosystems*. Presidential address to the British Ecological Society, December 1982. *J. Appl. Ecol.,* 20: 1- 17.

Bradshaw, A. D. 1984. Land Restoration: Now and in the Future. *Prc. R. Soc. Land. B. Great Britain.* 223: 1-23.

Bradshaw, A. D. 1987. The reclamation of derelict land and the ecology of ecosystems. In: Jordan III, William R., Gilpin, Michael E. and Aber, John D. eds. *Restoration Ecology: a synthetic approach to ecological research.* Cambridge University Press, pp. 53-74.

Bradshaw, A. D. 1997. Restoration of mined lands – using natural process. *Ecological Engineering,* 8: 225-269.

Bradshaw, A. D., Dancer, S. W., Handley, J. F. and Sheldon, J. C. 1975. The biology of land revegetation and the reclamation of the china clay wastes of Cornwall. In: Chadwick, M. J. and Goodman, G. T. eds. *The ecology of resource degradation and renewal.* Symposium of the British Ecological Society. Blackwell Scientific Publications, Oxford. pp. 363-384.

Bramble, W. C. and Ashley, R. H., 1955. Natural revegetation of spoil bank in central Pennsylvania. *Ecology,* 36: 417-423.

Brofas, G. and Karetsos G. 2002. Revegetation of calcareous bauxite mining spoils by seeding of woody species on Ghiona Mountain, Central Greece. *Land Degradation and Development,* 13(6): 461-467.

Carlson, C. W., Grunes, D. L., Alessl, J. and Reichman, G. A. 1961. Corn growth on garden surface and subsoil as affected by applications of fertilizer and manure. *Soil Sci. Soc. Am. Proc.,* 25: 44-47.

Caruccio, F. T. 1975. Estimating the acid potential of coal mine refuse. In: Chadwick, M. J. and Goodman, G. T. eds. *The ecology of resource degradation and renewal.* Blackwell Scientific Publications, Oxford. pp. 197- 206.

Chadwick, M. J. 1973. Methods of assessment of acid colliery spoils as a medium for plant growth. In: Hutnik, R. J. and Davis, G. eds. *Ecology and reclamation of devastated land* (Vol. 1). Gordon and Breech, London. pp. 81-91.

Chakraborty, R. N. and Chakraborty, D. 1989. Changes in soil properties under *Acacia auriculiformis* plantations in Tripura. *Indian Forester,* 115(4), 272-273.

Chambers, J. C. and Siddle, R. C. 1991. Fate of heavy metals in an abandoned lead-zinc tailings pond: I. Vegetation. *Journal of Environmental Quality,* 20: 745-751.

Chaturvedi, J. K. 1983. Afforestation of bauxite mined area in Central India. *Indian Forester,* 105(7): 458-465.

Chen, Yingyi and Li, Daoliang. 2008. Spatial Decision Support System for Reclamation in Opencast Coal Mine Dump. *WSEAS Transactions on Computers,* Issue 5, Vol. 7: 519-531.

Clüsener-Godt, Miguel. 2004. Implementing Agenda 21: Nature conservation and sustainable development in the humid tropics. *Tropical Ecology,* 45(1): 183-186.

Costigan, P. A., Bradshaw, A. D. and Gemmell, R. P. 1981. The reclamation of acidic Colliery spoil. I. Acid production potential. *J. Appl. Ecol.,* 18: 865-878.

Creighton, J. L., Muller, R. N. and Wittewer, R. F. 1983. Biomass and nutrient assimilation of intensively-cultured black locust on eastern Kentucky mine spoil. In: *Symposium on surface mining hydrology, sedimentology and reclamation.* University of Kentucky, Lexington, Kentucky. Proceedings. pp. 503-508.

Crocker, R. L. and Major, J. 1955. Soil development in relation to vegetation and surface age at Glacier Bay, Alaska. *J. Ecol.,* 43: 427-448.

Cunningham, T. R. and Wittwer, R. F. 1984. Direct seeding oaks and black walnut on mine soils in Eastern Kentucky. *Reclam. Reveg. Res.,* 3: 173-184.

D' Antonio, Carla. and Meyerson, L. A. 2002. Exotic Plant Species as Problems and Solutions in Ecological Restoration: A Synthesis. *Restoration Ecology,* 8(4): 703-713.

Dadhwal, K. S. and Singh, Bijendra. 1993. Rooting behavior of different plant science in limestone mined area. *Indian Forester,* 119 (1): 71-74.

Dancer, W. S., Handley, J. F., and Bradshaw, A. D. 1977. Nitrogen accumulation in kaolin mining wastes in Cornwall. II. Forage legumes. *Plant Soil,* 48: 303-314.

Das, P., Panda, P. C., Samantaray, S., Deo, B., Mallick, U. C. and Bradshaw, A. D. 1992. Revegetation of mined land in a Monsoon climate: A preliminary report on chromite and coal overburdens of Eastern India. *Orissa Journal of Horticulture,* 20(2): 32-38.

Davidson, W. H. 1980. Direct seeding for reforestation. In: *Trees for reclamation*, USDA Forest Service General Technical Report NE-61, pp. 93-97.

Day, A. D., Mitchell, G. F., Tucker, T. C. and Thames, J. L. 1979. Emergence and growth of plant species in coal mine spoils. *J. Environ. Qual.,* 8: 110-114.

De, S. and Mitra A. K. 2002. Reclamation of mining-generated wastelands at Alkusha-Gopalpur abandoned open cast project, Raniganj Coalfield, Eastern India. *Environmental Geology,* 43(1-2): 39-47.

Dickie, J. B., Gajjar, Kamini H., Birch, P. and Harris, J. A. 1988. The Survival of Viable Seeds in Stored Topsoil from Opencast Coal Working and its Implications for Site Restoration. *Biological Conservation,* Vol. 43: 257-265.

Dobson, A. P., Bradshaw, A. D. and Baker, A. J. M. 1997. Hopes for the future: restoration ecology and conservation biology. *Science,* 277: 515-522.

Doerr, T. B., Redente, E. F. and Sievers, T. E. 1983. Effect of cultural practices on seeded plant communities on intensively disturbed soil. *J. Range Manage.,* 36: 423-428.

Doll E. C., Merrill, S. D. and Halvorson, G. A. 1984. Soil replacement for reclamation of strip-mined lands in North Dakota. *Agricultural Experimental Station Bulletin.* 24 p.

Doubleday, G. P. 1974. The reclamation of land after coal mining. *Outlook on Agriculture,* 8: 156-162.

Doubleday, G. P. and Jones, M. A. 1977. Soils of reclamation. In: Hackett, B. ed. *Landscape reclamation practice.* IPC Science and Technology Press, Guildford. pp. 85-124.

Down, G. C. 1975. Soil development on colliery waste tips in relation to age. *J. Appl. Ecol.,* 12: 613-635.

Dubey, Kumud. 2007. Bioreclamation of Silica Mining Area, Shankargarh, Allahabad. Ph. D. thesis. F. R .I. Deemed University, Dehradun, 271 p.

Dutta R. K. and Agrawal M. 2002. Effect of tree plantations on the soil characteristics and microbial activity of coal mine spoil land. *Tropical Ecology.* 43(2): 315-324.

Dutta, R. K. and Agrawal, M. 2003. Restoration of opencast coal mine spoil planting exotic tree species; a case study in dry tropical region. *Ecological Engineering,* 21: 143-151.

Ernst, W. H. O. 1988. Response of plants and vegetation to mine tailings and degraded materials. In: Salomons, W. and Forstner, V. eds. *Chemistry and biology of solid waste.* Springer-Verlag, Berlin, Germany. pp. 54-72.

Fail Jr., Joseph L. and Wochok, Zachary S. 1977. Soybean growth on fly ash-amended strip mine spoils. *Plant and Soil,* 48: 473-484.

Farmer, R. E., Jr., Cunningham, M. and Barnhill, M. A. 1982. First-year development of plant communities originating from forest topsoils placed on southern Appalachian mine spoils. *J. Appl. Ecol.,* 19: 283-294.

Feyera, Senbeta, Beck, Erwin and Lüttge, Ulrich. 2002. Exotic trees as nurse-trees for the regeneration of natural tropical forests. *Tree Structure and Function,* 16: 245-249.

Fierro, A., Angers, D. A. and Beauchamp, C. J. 1999. Restoration of ecosystem function in an abundance sandpick: Plant and soil responses to paper de-inking sludge. *J. Applied Ecology,* 36: 244-253.

Filcheva, E., Noustorova, M., Gentcheva-Kostadinova, S. and Haigh, M. J. 2000. Organic accumulation and microbial action in surface coal-mine spoils, Pernik, Bulgaria. *Ecol. Eng.* 15 (1-2): 1-15.

Franco, Avilio A. and de Faria, Sergio M. 1997. The contribution of N2 – fixing tree legumes to land reclamation and sustainability in the tropics. *Soil Bio. Biochem.* Vol. 29, No. 5/6: 897-903.

Gardner, John H. and Bell, D. T. 2007. Bauxite Mining Restoration by Alcoa World Alumina Australia in Western Australia: Social, Political, Historical and Environmental Contexts. *Restoration Ecology,* 15(4) Supplement: S3-S10.

Ghose, M. K. and Kundu, N. K. 2001. Feasibility of storing topsoil in coal mining areas without biological reclamation. *J. of Mines, Metals & Fuels,* July, 2001. 244-249.

Gibson, D. J. 1982. The natural revegetation of lead/zinc mine spoil in northeastern Oklahoma. *The Southwestern Naturalist,* 27: 425-436.

Gibson, D. J., Johnson, F. L. and Risser, P. G. 1985. Revegetation of unreclaimed coal strip mines in Oklahoma. II. Plant communities. *Reclamation and Revegetation Research,* 4: 31-47.

Gilley, J. E., Gee, G. W., Bauer, A., Willis, W. O. and Young, R. A. 1977. Runoff and erosion characteristics of surface-mined sites in western North Dakota. *Agric. Engineers,* 20: 297-704.

Grant, C. D. 2006. State and- Transition Successional Model for Bauxite Mining Rehabilitation in the Jarrah Forest of Western Australia. *Restoration Ecology,* 14(1): 28-37.

Grant, C. D., Bell, D. T., Koch, J. M. and Loneragan, W. A. 1996. Implications of Seedling Emergence to Site Restoration following Bauxite Mining in Western Australia. *Restoration Ecology,* 4(2): 146-154.

Grant, C. D. and Norman, M. A. 2006. Investigating thinning and burning operations in 10- to 13-year old rehabilitated bauxite mines in jarrah forest. *Environmental Department Research Bulletin.* Alcoa World Alumina Australia.

Grant, C. D., Wood, S. C. and Morley, S. C. 2007. Return of Ecosystem Function to Restored Bauxite Mines in Western Australia. *Restoration Ecology,* 15(4) Supplement: S94-S103.

Gupta, R. K. 1979. Reclamation and use of coal-mined ecosystem in India: A review. In: Wali, M. K. ed. *Ecology and Coal Resource Development* (Vol. II). Pergemon Press, N.Y. pp. 510-515.

Gupta, K. C., Mishra, P. K. and Singh, R. S. 2004a. Performance study of *Amaranthus* on mine spoil. *Indian J. Environ. & Ecoplan.* 8(1): 143-145.

Gupta, K. C., Mishra, P. K., and Singh, R. S. 2004b. Performance study of bamboo species for reclamation of mine spoil. *Poll. Res.,* 23(4): 661-663.

Gupta, K. C. and Singh, R. S. 2005. Reclamation of waste/degraded land through afforestation. *Minetech,* Vol. 26 (2): 50-54.

Gupta, K. C., Srivastava, N. K. and Singh, R. S. 2005. Effect of leaf litter on plant growth behavior of four species raised on mine spoil. *Indian J. Forestry,* 28(4): 353-358 p.

Hodder, R. L. 1977. Dryland techniques in the semiarid west. In: Thames, J. L. ed. *Reclamation and use of disturbed lands in the southwest.* University of Arizona Press, Tuscon. pp. 271-278.

Holl, K. D. and Cairns, J. C. Jr. 1994. Vegetation community development on reclaimed coal surface mines in Virginia. *Bulletin of the Torrey Botanical Club,* Vol. 121(4): 327-337.

Hossner, L. R. and Hons, F. M. 1992. Reclamation of mine tailings. In: Lal, R. and Stewart, B. A. eds. *Advances in Soil Science* (Vol. XVII). pp. 311-350.

India. Ministry of Environment and Forests. Indian Council of Forestry Research and Education. Himalayan Forest Research Institute (HFRI). 2004. *Guidelines for eco-rehabilitation of lime stone mined areas in Himachal Pradesh.* GUI: 5, Himalayan Forest Research Institute. pp. 1-5.

India. Ministry of Environment and Forests (MOEF). 2010. *Towards Sustainable Growth: Transport/Chemicals/Waste Management/Mining/SCP.* Ministry of Environment and Forests, Govt. of India, New Delhi. p.128.

Iverson, L. R. and Wali, M. K. 1982. Reclamation of coal mined lands: The role of *Kochia scoparia* and other pioneers in early succession. *Reclam. Reveg. Res.,* 1: 123-160.

Jefferies, R. A., Wilson, K. and Bradshaw, A. D. 1981. The potential of legumes as a nitrogen source for the reclamation of derelict land. *Plant and Soil,* 59: 173-177.

Jefferson, L. V. 2004. Implications of plant density on the resulting community structure of mine site land. *Restoration Ecology,* 12: 429-438.

Jha A. K. 1990. *A study on revegetation of coal-mine spoil.* PhD thesis. Banaras Hindu University, Varanasi, India. 242 p.

Jha, A. K. 1992. Evaluation of coal mines spoils as a medium for plant growth in a dry tropical environment, India. *Indian Forester,* Vol. 118(12): 909-916.

Jha, A. K., Singh, Arvind, Singh, A. N., Singh, J. S. and Singh, A. 2000. Evaluation of direct seeding of tree species as a means of revegetation of coal mine spoils. *Indian Forester,* 126 (11): 1217-1221.

Jha A. K. and Singh, J. S. 1992. Rehabilitation of Mine Spoils. In: Singh, J. S. ed. *Restoration of Degraded Land: Concepts and Strategies.* Rastogi Publications, Meerut, India. pp. 212-254.

Jha, A. K. and Singh, J. S. 1993. Growth performance of certain directly seeded plants on mine spoil in a dry tropical environment, India. *Indian Forester,* 119 (11): 920-927.

Johnson, C. K. and West, N. E. 1989. Seed reserves in stockpiled topsoil on a coal stripmine near Kemmerer, Wyoming. *Landscape and Urban Planning,* 17: 169-173.

Johnson, M. S. and Bradshaw, A. D. 1979. Ecological Principles for the restoration of disturbed and degraded land. *J. Appl. Ecol.,* 4: 141-200.

Joshi, A. and Singh, S. P. 1998. Response of *Quercus leucotrichophora* and *Alnus nepalensis* seedling along a nutrient gradient. *Tropical Ecology,* 39: 117-123.

Joshi, P. K., Kumar, M., Midha, N., Vijayanand and Paliwal, A. 2006. Assessing areas deforested by coal mining activities through satellite remote sensing images and GIS in parts of Korba, Chattisgarh. *J. Indian Soc. Remote Sensing,* Vol. 34 (4): 415-421.

Juwarker, A., Sahni, D. K., Dubey, K., Khobragade, R. and Nimje, M. 2001. Bioremediation and restoration of productivity of mine soil dumps- a case study. In: *Workshop on Reclamation & Rehabilitation of Mined-out Areas*, 16-17 Feb. 2001, Bhubaneshwar, India (Volume of Papers). Proceedings. Bhubaneshwar. pp. 60-68.

Keenan, R., Lamb, D., Woldring, O., Irvine, T., and Jensen, R. 1997. Restoration of plant biodiversity beneath tropical tree plantations in Northern Australia. *For. Ecol. Manage.* 99: 203-214.

Kendle, A. D. and Bradshaw, A. D. 1992. The role of soil nitrogen in the growth of trees on derelict land. *Arboricultural Journal,* 16: 103-122.

Knowles, O. H. and Parrotta, J. A. 1995. Amazonian forest restoration: an innovative system for native species selection based on phenological data and performance indices. *Commonwealth Forestry Review,* 74: 230-243.

Koch, J. M. 2007. Alcoa's Mining and Restoration Process in South Western Australia. *Restoration Ecology,* 15(4) Supplement: S11-S16.

Koch, John M. and Ward, Samuel C. 1994. Establishment of Understory Vegetation for Rehabilitation of Bauxite-mined Areas in the Jarrah Forest of Western Australia. *Journal of Environmental Management,* 41: 1-15.

Koch, J. M., Ward, S. C., Grant, C. D. and Ainsworth, G. L. 1996. Effects of Bauxite Mine Restoration Operations on Topsoil Seed Reserves in the Jarrah Forest of Western Australia. *Restoration Ecology,* 4(4): 368-376.

Kumar, N., Sundararajan M., Kumar, B. and Gupta, A. 2004. Vetiver Grass Technology: An Economical Bio-Reclamation Approach for Coal Overburden Dumps. In: *National Seminar on Environmental Engineering with special species on Mining Environment.* NSEEME. ENVIS. Proceedings.

Kumar, P., Kumar, S., Aggarwal, D. K., Chaudhary, A. and Narain, P. 2005. Returning back to crop and grass production after bauxite mining. In: *5th Mines Environment & Mineral Conservation Week Celebration*, 8-14 Jan. 2005, Udaipur, India. Proceedings. pp. 121-122.

Kuusipalo, J., Adjers, G., Jafarsidik, Y., Otasamo, A., Tuomela, K. and Vuokko, R. 1995. Restoration of natural vegetation in degraded *Imperata cylindrica* grassland: understorey development in forest plantations. *J. Vegetation Science,* 6: 205-210.

Larson, M. M. 1984. *Invasion of volunteer tree species on stripmine plantations in east-central Ohio*. Ohio Agricultural Research and Development Centre, Wooster, OH. p. 10.

Lei, Dongmei and Duan, Changqun. 2008. Restoration potential of pioneer plants growing on lead-zinc mine tailings in Lanping, southwest China. *J. Environmental Science,* Vol. 20: 1202-1209.

Leopold, A. C., Andrus, R., Finkeldey, A. and Knowels, D. 2001. Attempting restoration of wet tropical forests in Coast Rica. *For. Ecol. Manage.* 33: 627-642.

Li, M. S. 2006. Ecological restoration of mine land with particular reference to the metalliferous mine wasteland in China: a review of research and practice. *Science of the Total Environment,* 357: 38-53.

Likens, G. E., Borman, F. H., Johnson, N. M., Fisher, D. W. and Pierce, R. S. 1970. Effect of forest cutting and herbicide treatment on nutrient budgets in the Hubbard Brook watershed ecosystem. *Ecol. Monograph,* 40: 23-47.

Liu, M., Jiang, G., Yu, S., Li, Y. and Li, G. 2009. The Role of soil Seed Banks in natural Restoration of the Degraded Hunshandak Sandsland, Northern China. *Restoration Ecology,* 17(1): 127-136.

Lugo, A. E. 1988. The future of the forest: ecosystem rehabilitation in the tropics. *Environment,* 30: 41-45.

Lugo, A. E. 1997. The apparent paradox of reestablishing species richness on degraded lands with tree monoculture. *Forest Ecol. Manage.,* 99: 9-19.

Luke, A. G. R., Harvey, H. J. and Humphries, R. N. 1982. The creation of woody landscapes on roadsides by seeding- a comparision of past approaches in West Germany and the United Kingdom. *Reclam. Reveg. Res.,* 1: 243-253.

Lyle, E. S. Jr. 1987. *Surface Mine Reclamation Manual*. Elsevier, New York. 33p.

Macedo, M. O., Resende, A. S., Garcia, P. C., Boddey, R. M., Jantalia, C. P., Urquiaga, S., Campello, E. F. C. and Franco, A. A. 2008. Changes in soil C and N stocks and nutrient dynamics 13 years after recovery of degraded land using leguminous Nitrogen-fixing trees. *Forest Ecol. & Manag.,* 25: 1516-1524.

Maiti, S. K. 1997. Nitrogen accumulation in Coalmine spoils by legume (*Stylosanthus humilis*). *Environment and Ecology,* 15 (3): 580-584.

Maiti, S. K. 2003. Ecorestoration problems of coalmine overburden dumps. In: *International Conference on Ecorestoration*, Dehradun and New Delhi, 14-21 October 2003. Proceedings. Dehradun and New Delhi.

Maiti S. K. 2006. Ecorestoration of coalmine overburden dumps in Korba, Gevra and Kusmunda area of SECL, India. In: *Internal Sym. on Environmental issues of Mineral industries*, VNIT, Nagpur, 11-15 Jan. 2006. Proceedings. Nagpur, VNIT.

Maiti, S. K. and Banerjee, S. P. 1993. Coal spoils reclamation with legumes & grass- An experimental field study. In: Banerjee, S. P. ed. In: *2nd National Seminar on Minerals & Ecology.* 1993. Proceedings. Oxford & IBH Pvt. Ltd., Cal. pp. 111-120.

Maiti, S. K. and Das, Debalina. 2004. Ecorestoration of coalmine overburden dumps. *Environment & Ecology,* 22(Spl.-2): 329-333.

Maiti, S. K. and Ghose, M. K. 2005. Ecological restoration of acidic coalmine overburden dumps- an Indian case study. *Land Contamination & Reclamation,* 13(4): 1-10.

Maiti, S. K. Shree, C and Jha, P. C. 2006. Selection of plant species for reclamation of mine degraded land in the Indian Context. *Land Contamination & Reclamation,* 14 (2): 1-14.

Maiti, Subodh Kumar and Singh, Gurdeep. 2007. Ecorestoration scenario of coal mine degraded land in India: Present status and future prospects. In: *Proceedings of Golden Jubilee Seminar on "Present Status of Mining and Future Prospects."* Hyderabad, April 6-8, 2007. Hyderabad, 1-17.

Maitti, S. K., Singh, G. and Nandhini, S. 2005. Biodiversity of micronutrients and heavy metals in reclaimed coalmine overburden dumps and their bioaccumulation in tree species- a case study. In: Sinha, A. and Singh, Satyendra K. eds. *First Indian Mineral Congress & Technological Exhibition: Showcasing the Mineral Industry in the 21st Century*, Dhanbad, 28 Feb. - 1 March 2005. Proceedings. Allied Publishers Pvt. Ltd., N. Delhi. pp. 112-124.

Marrs R. H. Roberts, R. D. and Bradshaw, A. D. 1980. Ecosystem development on reclaimed china clay wastes. I. Assessment of vegetation and capture of nutrients. *J. Appl. Ecol.,* 17: 709-717.

Marrs, R. H., Roberts, R. D., Skeffington, R. A. and Bradshaw, A. D. 1983. Nitrogen and the development of ecosystem. In: Lee, J. A., Neill, S. M. and Ravinson, I. H. eds. *Nitrogen as an Ecological Factor.* Oxford: Blackwell. pp. 113-136.

Masoodi, T. H., Masoodi, N. A., Soni, P. and Zubair, A. 2003. Plant strategies for successful colonization on degraded lands. *Ann. For.* 11 (1): 37-51.

Mathur, H. N. 1978. Afforestation of open cast coal mines. *Journal of the IPHE, India,* Vol. 1978 (2): 43-52.

McFee, W. W., Byrnes, W. R. and Stockton, J. G. 1981. Characteristics of coal mine overburden important to plant growth. *J. Environ. Qual.* 10: 300-308.

McNeilly, T., Williams, S. T. and Christian, P. J. 1984. Lead and zinc in a contaminated pasture at Minera, N. Wales and their impact on productivity and organic matter breakdown. *The Science of the Total Environment,* 18: 183-198.

Merrill, S. D., Doering, E. J. and Power, J. F. 1980a. Changes of sodicity and salinity in soils reconstructed on stripmined land. *Agricultural Experiment Station. Farm Research,* 37: 13-16.

Merrill, S. D., Doering, E. J., Power, J. F., and Sandoval, F. M. 1983. Sodium movement in soil-mine spoil profiles: diffusion and convection. *Soil Sci.,* 136: 308-316.

Merrill, S. D., Sandoval, F. M., Doering, E. J. and Power, J. F. 1980b. The role of gypsum and other amendments in the reclamation of strip-mined lands in semi-arid environments. In: Borris, D. P. and Boody, P. W. eds. *Phosphogypsum: Proceedings of the International symposium on phosphogypsum* (Vol. 1). The Florida Institute of Phosphate Research, Bartow, FL. pp. 218-242.

Merrill, S. D., Smith, S. J. and Power, J. F. 1985. Effect of disturbed soil thickness on soil water use and movement under perennial grass. *Soil Sci. Soc. Am. J.,* 49: 196-202.

Miller, T. E. and Werner, P. A. 1987. Competitive effects and responses between plant species in a first year old-field community. *Ecology,* 68: 1201-1210.

Mishra, S. K. and Hota, J. K. 1991. Mineral Development and Environment. In: *International Conference on Mineral Development and Environment (Volume of Papers)*, Indian School of Mines, Dhanbad, India. Proceedings. Dhanbad, Indian School of Mines. pp. 44-61.

Moffat, A. J. and Boswell, R. C. 1997. The effectiveness of cultivation using the winged tine on restored sand and gravel workings. *Soil and Tillage Research,* 40: 111-124.

Moffat, A. J. and Buckley, G. P. 1995. Soils and Restoration Ecology. In: Ferris-Kaan, R. (ed.). *The Ecology of Woodland Creation.* J. Wiley and Sons, Chichester, U.K. pp. 75-90.

Mohanty, B. N. 2004. *The effects of mining on the biodiversity of iron ore belt of Orissa (Bonai- Keonjhar).* PhD thesis. Utkal University, Orissa. 163+ xi p.

Murphy, B. P. and Loneragan, W. A. 2000. The distribution of seed sizes in concentrated topsoil. In: Asher, C. J. and Bell, L. C. eds. *Third Australian Workshop on Native Seed Biology for Revegetation.* Perth, Australia, 17-18 May, 1999. Australian Centre for Mining Environmental Research. Kenmore, Queensland, Australia. Proceedings. Kenmore, Australian Centre for Mining Environmental Research. pp. 187-192.

Muzzi, E. and Fabbri, T. 2007. Revegetation of mineral clay soils: shrub and tree species compared. *Land Degradation and Development,* 18: 441-451.

Noble, I. R. and Dirzo, R. 1997. Forests as human-dominated ecosystems. *Science,* 277: 522-525.

Noble, I. R. and Slatyer, R. O. 1980. The use of vital attributes to predict successional changes in plant communities subject to recurrent disturbances. *Vegetatio,* 43: 5-21.

Norman, M. A., Koch, J. M., Grant, C. D., Morald, T. K. and Ward, S. C. 2006. Vegetation Succession after Bauxite Mining in Western Australia. *Restoration Ecology,* 14(2): 278-288.

O'Neill, R. V., Ausmus, B. S., Jackson, D. R., Van Hook, R. J., Van Voris, P., Washbourne, C. and Watson, A. P. 1977. Monitoring terrestrial ecosystems by analysis on nutrient export. *Water, Air and Soil Poll.,* 8: 271-277.

Pal, D., Singh, R. S., Singh, S. K. and Gupta, A. 2005. Revegetation-carbon sequestration and pollution attenuation in wasteland areas. In: *National Seminar on Policies, Statutes and Legislation in Mines.* POSTALE. 2005. Proceedings.

Parkinson, D. 1979. Microbes, mycorrhizae and mine spoil. In: Wali, M. K. ed. *Ecology and Coal Resource Development* (Vol. II). Pergamon Press, New York. pp, 634-642.

Parrotta, J. A. 1992. The role of plantation forests in rehabilitating degraded tropical ecosystems. *Agricultural Ecosystem and Environment,* 41: 115-133.

Parrotta, J. A. 1993. Secondary forest regeneration on degraded tropical lands: the role of plantations as 'foster ecosystems'. In: Lieth, H. and Lohmann, M. eds. *Restoration of tropical forest ecosystems.* Kluwer Academic, Dordrecht. pp. 66-73.

Parrotta, J. A. 1999. Productivity, nutrient cycling and succession in single and mixed species plantations of *Casuarina equisetifolia*, *Eucalyptus robusta* and *Leucaena leucocephala* in Puerto Rico. *Forest Ecol. and Management,* 124: 45-77.

Parrotta, J. A. and Knowles, O. H. 1999. Restoration of Tropical Moist Forests on Bauxite Mined Lands in the Brazilian Amazon. *Restoration Ecology,* 7(2):103-116.

Parrotta, J. A. and Knowles, O. H. 2001. Restoring tropical forests on lands mined for bauxite: Examples from the Brazilian Amazon. *Ecological Engineering*, 17: 219-239.

Parrotta, J. A., Turnbull, J. W. and Jones, N. 1997a. Catalyzing native forest regeneration on degraded tropical lands. *Forest Ecology and Management,* 99: 1-7.

Parrotta, J. A., Knowles, O. H. and Wunderle Jr., J. M. 1997b. Development of floristic diversity in 10-year-old restoration forests on a bauxite mined site in Amazonia. *Forest Ecology and Management,* 99: 21-42.

Parton, W. J., Ellis, J. E. and Swift, D. M. 1979. The impacts of strip-mine reclamation practices: A simulation study. In: Wali, M. K. ed. *Ecology and Coal Resource Development* (Vol. II). Pergamon Press, New York. pp. 584-591.

Pederson, T. A., Rogowski, A. S. and Pennock, R. 1980. Physical characteristics of some mine spoils. *Soil Sci. Soc. Am. J.,* 44: 321-328.

Phia, M. I., Vallack, H. W., Reeler, B. M. and Michael, N. 1995. A low input approach to vegetation establishment on mine and coal ash wastes in semi-arid regions. I. Tin mine tailings in Zimbabwe. *J. Appl. Ecol.,* 32: 372-381.

Power, J. F. 1978. Reclamation research on strip-mined lands in dry regions, In: F. W. Schaller and P. Sutton. eds. *Reclamation of drastically disturbed lands. American Society of Agronomy*, Madison, Wise. pp. 521-536.

Power, J. F. and Bennett, O. L. 1977. Protection of soil and water resources on land disturbed by mining. In: *2nd National Conference on Energy, Environment*, Washington, DC. Proceedings. pp. 195-201.

Power, J. F., Bond, J. J., Sandoval, F. M. and Willis, W. O. 1974. Nitrification in Paleocene shale. *Science,* 183: 1077-1079.

Power, J. F., Ries, R. E. and Sandoval, F. M. 1978a. Reclamation of coal mined land in the Northern Great Plains. *J. Soil Water Conserv.,* 33: 69-74.

Power, J. F., Ries, R. E. and Sandoval, F. M. 1982. Quality parameters of spring wheat and forage crops related to topsoil and subsoil thickness. *Soil Sci. Soc. Am. J.,* 46: 381-386.

Power, J. F., Sandoval, F. M. and Ries, R. E. 1978b. Restoration of productivity to disturbed land in the Northern Great Plains. In: R. A. Wright. ed. *The reclamation of disturbed arid lands.* U. of N. Mexico Press, Albuquerque. pp. 33-49.

Prasad, R. 1989. Removal of forest cover through mining and technology for revival. Proc. National Seminar on depletion of soil and forest cover. *J. Trop. Forestry,* 5: 109-116.

Prasad, R. and Pandey, R. K. 1985. Natural plant succession in the rehabilitated bauxite and coal mine overburdens of Madhya Pradesh. *J. Trop. Forestry,* 1: 310-320.

Prasad, A. R. K. and Roy, B. K. 2001. Heavy Metal Accumulation and Growth Responses of *Cajanus cajan* L. on Coal Mine Spoils. *Inter. J. Ecol. Environ. Sciences,* 27:155-162.

Prasad, R. and Shukla, P. K. 1985a. Reclamation and revegetation of coal-mine overburden in Madhya Pradesh. *J. Trop. For.* 1: 79-84.

Prasad, R. and Shukla, P. K. 1985b. Restoration of ecological balance to the bauxite mined out areas of Madhya Pradesh. *J. Tropical For.* 1: 236-245.

Rao, A. V. and Tarafdar, J. C. 1998. Selection of plant species for rehabilitation of gypsum mine spoil in arid zone. *J. Arid Environ.* 39: 559-567.

Reeder, J. D. and Berg, W. A. 1977. Nitrogen mineralization and nitrification in a Cretaceous shale and coal mine spoils. *Soil Sc. Soc. Am. J.* 41: 922-927.

Reeder, J. D. and Sabey, B. 1987. Nitrogen. In: Williams, R. D. and Schuman, G. E. eds. *Reclaiming mine soils and overburden in the western United States: Analytic parameters and procedures.* Soil and Water Conservation Society, Ankeny. IA. pp. 155-184.

Richardson, J. A. 1958. The effect of temperature on the growth of plants on pit heaps. *J. Ecol.,* 46: 537-546.

Richardson, J. A. and Evans, M. E. 1987. Establishing trees and shrubs on magnesium limestone waste. *J. Environmental Management,* 24: 157-167.

Richardson, J. A. and Greenwood, E. F. 1967. Soil moisture tension in relation to plant colonization of pit heaps. *Proc. Univ. Newcastle Phil. Soc.,* 1: 129-136.

Ries, R. E. and Depuit, E. J. 1984. Perennial grasses for mined land. *J. Soil Water Conserv.,* 39: 26-29.

Ries, R. E., Sandoval, F. M. and Power, J. F. 1977. Reclamation of disturbed lands in the lignite area of the Northern Plains. In: *Ninth annual lignite symposium,* Grand Forks, North Dakota. Proceedings. pp. 309-327.

Ries, R. E., Sandoval, F. M. and Power, J. F. 1978. Re-establishment of grasses on disturbed land. In: *Proceedings of the 1st International Rangeland Congress.* pp. 700-703.

Rimmer, D. L. 1982. Soil physical conditions on reclaimed colliery spoil heaps. *J. Soil Sci.,* 33: 567-579.

Ripley, E. A., Redmann, R. E. and Crowder, A. A. 1996. *Environmental effects of mining.* St. Lucie Press, Florida. 113-138 p.

Roberts, R. D., Marrs, R. H. and Bradshaw, A. D. 1980. Ecosystem development on reclaimed china clay wastes. II. Nutrient compartmentation and nitrogen mineralization. *J. Appl. Ecol.,* 17: 719-726.

Roberts, R. D., Marrs, R. H., Skeffington, R. A. and Bradshaw, A. D. 1981. Ecosystem development on naturally colonized china clay wastes. I. Vegetation changes and overall accumulation of organic matter and nutrients. *J. Ecol.,* 69: 153-161.

Rodrigue, J. A. and Burger, J. A. 2004. Forest soil productivity of mined land in the midwestern and eastern coalfield regions. *Soil Sc. Soc. Am. J.,* 68: 833-844.

Ross, K. A., Taylor, J. E., Fox, M. D. and Fox, B. J. 2004. Interaction of multiple disturbances: Importance of disturbance interval in the effects of fire on rehabilitating mined areas. *Austral. Ecology*, 29: 508-529.

Russell, M. J. and Roberts, B. R. 1986. Revegetation of coal mine spoil using pasture on the Darling Downs of Queensland, Australia. *Reclamation and Revegetation Research,* 5: 509-519.

Russell, R. S. 1977. *Plant root systems.* McGraw Hill, London. 177p.

Saffaya, N. M. and Wali, M. K. 1979. Growth and nutrient relations of a grass-legume mixture on sodic coal mine spoil as affected by some amendments. *Soil Sc. Soc. Am. J.,* 43: 747-753.

Sagar, R., Raghubanshi, A. S. and Singh, J. S. 2003. Tree species composition, dispersion and diversity along a disturbance gradient in a dry tropical forest region of India. *Forest Ecology and Management,* 186: 61-71.

Sanchez, C. E. and Wood, M. Karl. 1989. Infiltration rates and erosion associated with reclaimed coal mine spoils in Western Central New Mexico. *Landscape and Urban Planning,* 17: 151-168.

Sandoval, F. M. and Gould, W. C. 1978. Improvement of saline and sodium-affected disturbed lands. In: Schaller, F. W. ed. *Reclamation of drastically disturbed lands.* American Society of Agronomy, Madison, Wisconsin. pp. 485-504.

Saxena, S. K. 1979. Degradation of vegetation in the surface-mined area of western Rajasthan. In: Wali, M. K. ed. *Ecology and Coal Resource Development* (Vol. II). Pergamon Press, New York. pp. 626-633.

Schafer, W. M. and Nielsen, G. A. 1979. Soil development and plant succession on 1- to 50-year-old strip mine spoils in southern Montana. In: Wali, M. K. ed. *Ecology and Coal Resource Development* (Vol. II). Pergamon Press, New York. pp. 541-549.

Schafer, W. M., Nielsen, G. A. and Nettleton, W. D. 1980. Mine soil genesis and morphology in a spoil chronosequence in Montana. *Soil Sci. Soc. Am. J.,* 44: 802-807.

Schuster, W. S., and Hutnik, R. J. 1987. Community development on 35-yr-old planted mine spoil banks in Pennsylvania. *Reclam. Reveg. Res.,* 6: 109-120.

Sengupta, S. R. 1991. Mineral development and Environment. In: *International Conference on Mineral Development and Environment.* Indian School of Mines, Dhanbad (Volume of Papers). Proceedings. Dhanbad, Indian School of Mines. pp. 86-95.

Seo, Kyung Won, Son, Yowhan, Rhoades, Charles C., Noh, Nam Jin, Koo, Jin Woo and Kim, Jeong-Gyu. 2008. Seedling Growth and Heavy Metal Accumulation of Candidate Woody Species for Revegetating Korean Mine Spoils. *Restoration Ecology,* Vol. 16 (4): 702-712.

Severson, R. C. and Gouch, L. P. 1983. Rehabilitation material from surface-coal mines in western U.S.A. I. Chemical characteristics of spoil and replaced cover soil. *Reclam. Reveg. Res.,* 2: 83-102.

Shainsky, L. J. and Radosevich, S. R. 1992. Mechanism of competition between Douglas- fir and Red alder seedlings. *Ecology,* 73: 30-45.

Sharma K. D, Kumar S. and Gough L. P. 2000. Rehabilitation of lands mined for limestone in the Indian desert. *Land Degradation and Development,* 11 (6): 563-574.

Sheil, D. 1999. Tropical forest diversity, environmental change and species augmentation; after intermediate disturbance hypothesis. *J. Veg. Sci.* 10: 851-860.

Shu, W. S., Ye, Z. H., Zhang, Z. Q., Lan, C. Y. and Wong, M. H. 2005. Natural Colonization of Plants on Five lead/zinc Mine tailings in Southern China. *Rest. Ecology,* Vol. 13 (1), 49-60.

Shukla, M. K. and Lal, R. 2005. Temporal changes in soil organic carbon concentration and stocks in reclaimed mineoils of Southeastern Ohio. *Soil Science,* Vol. 170(12): 1013-1021.

Sindelar, B. W. 1979. Successional development of vegetation on surface mined land in Montana. In: Wali, M. K. ed. *Ecology and Coal Resource Development* (Vol. II). Pergamon Press, New York. pp. 550-556.

Singh, A. 2004. Effect of fertilization on N and P resorption efficiency of selected leguminous and non-leguminous tropical trees planted on coal mine spoil. *J. Indian Inst. Sci.,* 84: 173-182.

Singh, Arvind. 2006. Growth and leaf nutrient status of companion species as influenced by neighbouring species in mixed plantations raised on mine spoil. *Tropical Ecology,* 47 (2): 259-269.

Singh, Arvind. 2008. Influence of tree canopy on dry matter yield of seedlings of *Leucaena leucocephala* on mine spoil in a dry tropical environment. *Ind. J. For.,* 31 (2): 201-208.

Singh, Arvind, Jha, A. K. and Singh, J. S. 2000a. Effect of Nutrient Enrichment on Native Tropical Trees Planted on Singrauli Coalfields, India. *Restoration Ecology,* 8(1): 80-86.

Singh, A. K., Lepande, C. L. and Singh, R. B. 1996a. Effectiveness of amendments to growth and biomass production of Pithecellobium dulce in coal mine overburden. In: Ram Prakash. ed. *Advances in Forestry Research in India,* Vol. XIV: pp. 52-67.

Singh, Arvind and Singh, J. S. 2001. Comparative Growth Behaviour and Leaf Nutrient Status of Native Trees Planted on Mine Spoil With and Without Nutrient Amendment. *Ann. Bot.,* 87: 777-787.

Singh, A. K. and Singh, R. B. 1999a. Effect of mulches on nutrient uptake of *Albizia procera* and subsequent nutrient enrichment of coal mine overburden. *Journal of Tropical Forest Science,* 11(2): 345-355.

Singh, A. N., Raghubanshi, A. S. and Singh, J. S. 2002. Plantations as a tool for mine spoil restoration. *Current Science,* 82(12): 1436-1441.

Singh, A. N., Raghubanshi, A. S. and Singh, J. S. 2004a. Impact of native tree plantations on mine spoil in a dry tropical environment. *Forest Ecology & Management,* 187: 49-60.

Singh, A. N., Raghubanshi, A. S. and Singh, J. S. 2004b. Comparative performance and restoration potential of two *Albizia* species planted on mine spoil in a dry tropical region, India. *Ecological Engineering,* 22: 123-140.

Singh, A. N., Raghubanshi, A. S. and Singh, J. S. 2004c. Survival and growth pattern of three tropical forest plantations raised on coal-mine spoils of central India. *Indian Forester,* 130: 376-383.

Singh, A. N. and Singh, J. S. 1999b. Biomass, net primary production and impact of bamboo plantation on soil development in a dry tropical region. *Forest Ecology & Management,* 119: 195-207.

Singh, A. N. and Singh, J. S. 2006. Experiments on ecological restoration of coal mine spoil using native trees in a dry tropical environment, India: a synthesis. *New Forest,* 31: 25-39.

Singh, A. N., Zeng, D. H., and Chen, F. S. 2006. Effect of young woody plantations on carbon and nutrient accretion rates in a redeveloping soil on coal mine spoil in a dry tropical environment, India. *Land Degradation & Development,* 17: 13-21.

Singh, Dhan and Vasistha, H. B. 2004. Rehabilitation of mined degraded lands in the Himalayas through silvi-pastoral models. Indian Forester, Vol. 130 (4). 398-404.

Singh, Dhan, Vasistha, H. B. and Sharma, S. D. 2000b. Changes in soil physical attributes under age series agroforestry interventios in limestone mined lands of Garhwal Himalaya. *Annals of Forestry,* Vol. 8 (1): 120-130.

Singh, J. S., Singh, K. P. and Jha, A. K. 1996b. An integrated ecological study on revegetation of mine spoil: Concepts and Research Highlights. *Final report of a S&T project sponsored by Ministry of Coal, Government of India.* Department of Botany, BHU, Varanasi.

Singh, P. K. and Singh, K. P. 1998. Biomass production in selected tree species raised on coal mine spoil in a dry tropical region in India. *Tropical Ecology,* 39 (2): 289-292.

Singh, R. S., Chaulya, S. K. and Singh, S. K. 1998. Eco-friendly technology for waste dumps management in India. In: Pasamehmetoğlu, A. G. and Őzgenoğlu, A. eds. *The Fifth International Symposium on Environmental Issues and Waste Management in Energy and Mineral Production-SWEMP' 98.* Ankara/Turkey. 18-20 May, 1998. Proceedings. A. A. Balkema, Publishers, Rotredam, Netherland. pp. 463-468.

Singh, R. S., Pal, D., Singh, S. K. and Gupta, A. 2005. Revegetation- a pollution attenuation perspective. In: Sinha, A. and Singh, Satyendra K. eds. *First Indian Mineral Congress & Technological Exhibition: Showcasing the Mineral Industry in the 21st Century.* 28 Feb.-1 Mar., 2005, Dhanbad. Proceedings. Allied Publishers Pvt. Ltd. New Delhi. pp. 125-130.

Smith, R. A. H. and Bradshaw, A. D. 1979. The use of metal tolerant plant population for the reclamation of metalliferous wastes. *J. Applied Ecology,* 16: 595-612.

Smith, R. M. Tryon, E. H. and Tyner, E. H. 1971. Soil development on mine spoil. West Virginia Agricultural Experiment Station Bulletin, 604 p.

Soni, P. and Vasistha, H. B. 1986. Reclamation of mine spoils for environmental amelioration. *Indian Forester,* 112(7): 621-632.

Soni, P., Vasistha, H. B. and Kumar, O. 1990. Ecorestoration model for surface mined lands. In: *Fifth Billings symposium on planning, rehabilitation, and treatment of disturbed lands.* Billings. Proceedings. Billings. pp. 10-12.

Soni, P., Vasistha, H. B., Kumar, O., Bhatt, V. and Negi, M. 1989. Revegetation and ecological monitoring of an open cast rock phosphate mine. In: Walker, D. G., Powler, C. B. and Pole, M. W. eds. *Conference on Reclamation: a global perspective. Alberta Land Conservation and Reclamation Council Report.* # RRTAC 98-2, 2 vols. Proceedings. pp. 327-332.

Soni, P., Vasistha, H. B. and Om Kumar 1992. Surface mined lands: Problems and prospects. Report on Ecorestoration of rock phosphate mined area. ICFRE Publication, pp. 18-30.

Soulliere, E. J. and Toy, T. J. 1986. Rilling of hillslopes reclaimed before 1977 surface mining law, Dave Johnston mine, Wyoming. *Earth Surface Processes and Landforms,* 11: 293-305.

Stark, N. M. 1977. Fire and nutrient cycling in a Douglas fir/larch forest. *Ecology.* 58: 16-30.

Tacey, W. H. and Glossop, B. L. 1980. Assessment of topsoil handling techniques for rehabilitation of sites mined for bauxite within the Jarrah Forest of Western Australia. *Austr. J. Ecol.,* 4: 195-205.

Torbert, J. L., Tuladhar, A. R., Burger, J. A. and Bell, J. C. 1988. Mine soil property effects on the height of ten-year-old white pine. *J. Environ. Qual.* 17: 189-192.

Toy, T. J. and Shay, D. 1987. Comparison of some soil properties on natural and reclaimed hillslopes. *Soil Sci.,* 143: 264-277.

Treweek, J. 1996. Ecology and environmental impact assessment. *J. Appl. Ecol.,* 33:191-99.

Vajpayee, P., Rai. U. N., Choudhary, S. K., Tripathi, R. D. and Singh, S. N. 2000. Management of fly ash landfills with *Cassia surattensis* Burm: a case study. *Bulletin of Environmental Contamination and Toxicology,* 65(5): 675-682.

Vasistha, H. B., Soni, P. and Kumar Om. 1995. Trema Politoria- A primary colonizer species suitable for reclamation of mined lands in lower Himalayas. *Ann. For.* Vol. 3 (1): 21-25.

Vimmmerstedt, J. P., House, M. C., Larson, M. M., Kasile, J. D. and Bishop, B. L. 1989. Nitrogen and carbon accretion on Ohio coal mine soils: influence of soil forming factors. *Landscape and Urban Planning,* Vol. 17: 99-111.

Visser, S., Zak, J. and Parkinson, D. 1979. Effects of surface mining on soil microbial communities and processes. In: Wali, M. K. ed. *Ecology and Coal Resource Development* (Vol. 2). Pergamon Press, New York. pp. 643-651.

Wali, M. K. 1987. The structure, dynamics, and rehabilitation of drastically disturbed ecosystems, In: Khoshoo, T. N. ed. *Perspectives in environmental management.* Oxford Publications, New Delhi. pp. 163-183.

Wali, M. K. and Freeman, P. G. 1973. Ecology of some mined areas in North Dakota. In: M. K. Wali. ed. *Some environmental aspects of strip mining in North Dakota.* ND Geol. Surv., Educ. Ser. 5, Grand Forks. pp. 25-47.

Wali, M. K. and Kollman, A. L. 1977. Ecology and mining or mining ecology? In: Thames, J. L. (ed.). *Reclamation and use of disturbed land in southeast.* The University of Arizona Press, Tucson. pp. l08-115.

Ward, S. C. 2000. Soil development on rehabilitated bauxite mines in south-west Australia. *Australian Journal of Soil Research,* 38: 453-464.

Ward, S. C., and Koch, J. M. 1996. Biomass and nutrient distribution in a 15.5 year old forest growing on a rehabilitated bauxite mine. *Australian Journal of Ecology,* 21: 309-315.

Ward, S. C., Koch, J. M. and Ainsworth, G. L. 1996. The Effect of Timing on Rehabilitation Procedures on the Establishment of a Jarrah Forest after Bauxite Mining. *Restoration Ecology,* Vol. 4 (1): 19-24.

Wieder, R. K., Carre1, J. E., Rapp, J. K. and Kucera, C. L. 1983. Decomposition of fescue (*Festuca elator* var. arundinaceae) and cellulose litter on surface mines and a tallgrass prairie in central Missouri, *U.S.A. J. Appl. Ecol.,* 20: 303-321.

Windsor, D. M. and Clements, A. M. 2001. A germination and establishment field trial of *Themeda australis* (kangaroo grass) for mine site restoration in the Central Tablelands of New South Wales. *Restoration Ecology,* 9(1): 104-110.

Wittwer, R. F., Carpenter, S. B. and Graves, D. H. 1981. Survival and growth of oaks and Virginia pine three years after direct seeding on mine spoils. In: *Symposium on Surface Mining Hydrology, Sedimentology and Reclamation.* University of Kentucky, Lexington, Kentucky. Proceedings. Kentucky, University of Kentucky. pp. 1-4.

Wong, M. H. 2003. Ecological restoration of mine degraded soil, with emphasis on metal contaminated soil. *Chemosphere,* 50: 775-780.

Ye, Z. H., Yang, Z. Y., Chan, G. Y. S. and Wong, M. H. 2001. Growth response of Sesbania rostrata and S. cannabina to sludge-amended lead/zinc mine tailings: a greenhouse study. *Environment International,* 26(5-6): 449-455.

Young, T. P. 2000. Restoration Ecology and Conservation Biology. *Biological Conservation,* 92: 73-83.

Zarger, T. G., Curry, J. A. and Allen, J. C. 1973. Seeding of pine on coal spoil banks in the Tennessee Valley. In: Hutnik, R. J. and Davis, J. eds. *Ecology and devastated land.* Gordon and Breech, New York. pp. 509-523.

Zedler, J. B. 2005. *Ecological restoration: guidance from theory*. San Francisco Estuary and Watershed Science, Vol. 3(2).

Zeleznik, J. D. and Skousen, J. G. 1996. Survival of three tree species on old reclaimed surface mines in Ohio. *Journal of Environmental Quality,* 25:1429-1435 p.

Zhang, Z. Q., Shu, W. S., Lan, C. Y. and Wong, M. H. 2001. Soil Seed Bank as an Input of Seed Source in Revegetation of Lead/ Zinc Mine Tailings. *Restoration Ecology,* 9(4): 378-385.

In: Critical Research Techniques in Animal …
Editors: Kaushalendra Kumar Jha et al.
ISBN: 978-1-53619-720-4

Chapter 10

WHAT COULD BE THE PRESENT AND FUTURE HABITATS OF RESIDENTS AND WINTERING VULTURES FOR CONSERVATION IN THE INDIAN SUBCONTINENT

***Radhika Jha*[1] *and Kaushalendra Kumar Jha*[2,3*]**
[1]Zoology Department, Lucknow University, India
[2]Indian Institute of Forest Management, Bhopal, India
[3]Department of Forest, Environment and Climate Change, Uttar Pradesh, Lucknow, India

ABSTRACT

The Indian subcontinent is one of the three strongholds of Old-World vultures which provide invaluable ecosystem goods and services. However, the abundance of a few species, especially residents and endemics has plummeted due to past factors. These include anthropogenic changes which, if left unchecked, along with climate change may continue to change the habitat. The objective of this study is to predict the potential habitat expanse of vulture species in the Indian subcontinent with varying edaphoclimatic and physiographic conditions across different countries. This research-based chapter, enriched with a review of literature, documents the research background, methods, and results, and evaluates these findings in comparison with existing published research findings. One of the most popular species distribution models, MaxEnt has been used in this study with the inputs of bioenvironmental covariates and species occurrences. The presence data of nine vulture species found in the subcontinent was procured from published literature and recently developed citizen science observations. For future predictions, CCSM4 (Community climate system model) Global circulation model along with lowest (RCP2.6) and highest

* Corresponding Author's E-mail: jhakk1959@gmail.com.

(RCP8.5) emission scenarios were selected. The models developed were robust enough (0.704-0.729; residents and 0.812-0.826; migrants) and were replicable for conservation. Our results indicated that across all the categories and temporal scenarios, the most important variables influencing the prediction were bio1 (Mean annual temperature), bio12 (Annual precipitation), bio15 (Precipitation seasonality and LULC (Land use land cover). The expanse of suitable areas for vulture conservation was lower and confined to certain regions as compared to the unsuitable area. Migratory vultures had a lower expanse of suitable landcover, though overlapping, as compared to resident ones. The future scenarios indicated habitat dynamism, with losses of suitable habitats and gains in unsuitable areas of the habitat. Except for some inconsistencies, an observed trend was a loss of area for resident vultures in different regions and gains in areas for migratory vultures. Some overall conservation measures based on the results of this study have been suggested, which could be adapted to local conditions.

WHAT IS THE BACKGROUND

Vultures are categorized as the only avian obligatory scavengers which are an invaluable component of the ecosystem, providing goods like faith-based medical care and meat (McKeen et al., 2013; Gbogbo et al., 2016; Boakye et al., 2020) and sanitary, cultural services, etc. (Markandya et al., 2008, Ogada et al., 2012, Moleon et al., 2014) to the benefit of mankind. However, their conservation status is of serious concern as they have been reported to have dwindled in the past. The existence of nine of the sixteen Old-World vultures viz., five *Gyps*, *Gypaetus*, *Sarcogyps*, *Neophron*, and *Aegypius* vultures is under threat in one of their 'Shangri-las', the Indian subcontinent. Several factors have caused their numbers to decline and fall into the categories of threatened to endangered as per the latest IUCN red list conservation status (*Aegypius monachus* Linnaeus 1766 = Near threatened to least concerned, decreasing; *Gypaetus barbatos* Linnaeus 1768 = Near threatened to vulnerable, decreasing; *Gyps bengalensis* Gmelin 1788 = Critically endangered, decreasing; *Gyps fulvus* Hablizl 1783 = Least concerned, increasing; Gyps himalayensis *Hume 1869* = Near threatened, stable; *Gyps indicus* Scopoli 1786 = Critically endangered, decreasing; *Gyps tenuirostris* Gray 1844 = Critically endangered, decreasing; *Neophron percnopterus Linnaeus 1758* = Endangered, decreasing; *Sarcogyps calvus* Scopoli 1786 = Critically endangered, decreasing) (IUCN 2021). Even as anthropogenic activities have driven many wildlife species towards extinction, the geographic distribution of many species is poorly documented, which can limit the effectiveness of conservation (Sharma et al., 2020). Some prominent causes recorded in the case of vultures were indirect poisoning (Pain et al., 2003, Oaks et al., 2004), pesticides (Plaza et al., 2019), food loss (Fakhar-i-Abbas et al., 2013), habitat degradation (Baral et al. 2005) among others. Recent reports from the Indian subcontinent indicate the recovery of few species (Chaudhry et al., 2012, Prakash et al., 2012) which are expected to fill the existing habitat and expand in future due to its territorial nature. Nonetheless, the current climate crisis is expected to change the habitat of most of the biodiversity including predatory raptors and vultures (Sala et al., 2000; Parmesan, 2006; Zhang et al., 2019a; Ilanloo et al., 2020). Therefore, it is pertinent to fill the knowledge gap in this field to manage and conserve these precious species sustainably.

This study identifies the present and future habitat suitability of vultures, using Species Distribution Modelling. Species distribution models (SDM) can help fill knowledge gaps for poorly understood species by predicting suitable habitats within an area of interest, providing information about habitat variables important to the species' distribution, and predicting responses to future climate conditions (Elith and Leathwick, 2009; Pearce and Boyce, 2006; Coxen et al., 2017). Species distribution models predict species distribution by relating occurrence or abundance records to environmental, demographic, climatic and satellite derived predictor variables (Elith et al., 2006). The inputs for SDM are mainly species' occurrence data and environmental data, which are both proliferating with advancements in technology and sampling efforts (Feng et al., 2017). The increase in species' occurrence data is due in large part to the digitization of museum specimens (Page et al., 2015; Rogers, 2016) and citizen science projects (Sullivan et al., 2009; Jackson et al., 2015). For the purpose of the present study, habitat prediction modelling has been done using MaxEnt, bioclim data, land use landcover, and citizen science data and published records. The different vulture species were divided into two groups, resident and migratory, since the former need intensive management throughout the year while the latter require only winter care in the study area. However, the conceptual model of habitat prediction, habitat change due to changing climate and management intervention for conservation is presented in Figure 1.

How it was Done

The study area descends from the heights of Hindukush-Himalaya-Arakans towards the peninsular south in the Indian ocean. It is interspersed by plains, plateaus and hills, which are variably covered with vegetation. Politically known as Indian subcontinent, one of the largest geopolitical regions of Asia, it is divided into India, Pakistan, Bangladesh, Sri Lanka, Nepal, Bhutan and other small islands of Indian ocean, prominently Maldives (Kumar 2012). The Land use Land cover (LULC), Elevation along with the main climatic features which form an important component of the habitat are presented in Figure 2.

Habitat area assessment was done by adopting species distribution modelling, using MaxEnt software. The inputs for MaxEnt run were: a) Occurrence data, b) Environmental layers and c) Bias file. The occurrence coordinates were sourced from published literature (Table 1) as well as websites which collect information from citizen science (eBird and iNaturalist), elaborated in next subsection. To reduce spatial bias within this data, it was first subjected to the remove duplicate tool from MS Excel and then spatially rarefied at a distance of 10 km by the spatially rarefy occurrence data tool in ArcGIS (Angelieri et al., 2016; Ahmad et al., 2019; Zhang et al., 2019b; Garza et al., 2020). Environmental variables comprising climate data, elevation, aspect and land use land cover (LULC), which are presented in Table 2, were used. A suite of climatic variables described with acronyms and/or codes in the literature (Tytar and Nekrasova, 2016; Prockow et al., 2018) are also detailed in this table. The climate data was downloaded from worldclim.org and then converted to MaxEnt friendly ascii format using ArcGIS. Of the different GCM models available for future predictions, the globally accepted CCSM4 was selected since, this model is one of the most efficient global climate projections to predict the influence of future climatic changes on the distribution of animal and plant species (Al-Qaddi et al., 2017; Sanjerehei and Rundel, 2017; Farashi and Erfani, 2018; Liang et al., 2018; Salas et al., 2018; Abdelal et al., 2019; Mohammadi et al., 2019). Within this model, the two extreme emission scenarios RCPs 2.6 (minimum emission hypothesis) and 8.5 (maximum emission hypothesis) were chosen (Abdelal et al., 2019). The land cover data was accessed from Copernicus Global Land Service (Buchhorn et al., 2020). SRTM 1 Arc second global data from earthexpoler.usgs.gov was used to assemble the elevation layer by mosaicking the available tiles. This layer was then used to prepare the aspect layer in ArcGIS. All the environmental layers were subjected to Pearson's correlation tests to remove the collinearity impacts, as spatial autocorrelation and multicollinearity may lead to false ecological conclusions in modelling spatial distribution of species (de Frutos et al., 2007). Although literature survey for threshold value indicated use of various values between 0.5 and 0.9, many studies chose 0.7 or >0.7 (Ahmad et al., 2019; Farrel et al., 2019; Zhang et al., 2019b; Perkins-Taylor and Frey, 2020) in different species. Anoop et al. (2020) and Phipps et al. (2017) selected 0.7 and <0.7 in the case of vultures. The threshold value in this study was fixed at ±0.7 and the remaining non-corelated variables were used for modelling with MaxEnt. To correct for latitudinal bias as well as incorporate absence points, a bias file was prepared using the SDM toolbox (Brown et al., 2017).

Twelve models were run at default setting with the exception of the replicated run type set as bootstrap, the replicate value set at 10 and random test percentage at 25. After checking for robustness of the model, the output generated was processed in ArcGIS. The MaxEnt heatmap was reclassified twice, once into 3 categories for generating suitability classes (0-0.3: Unsuitable; 0.3-0.6: Moderately suitable; 0.6-10: Highly suitable) and again into two categories only (0-0.3: Unsuitable and 0.3-1.0: Suitable) for finding out changes in area in future terms. The area corresponding to these categories was calculated for each model. Using the latter classification, the present-day layer was compared with the projected future layers and again the area was calculated under the following categories: Unsuitable, Gain, Loss and Suitable. The flow chart of the modelling methodology adopted from input to output is presented in Figure 3. The percent contribution tables and Jackknife test bar charts were used for finding variables contribution in prediction.

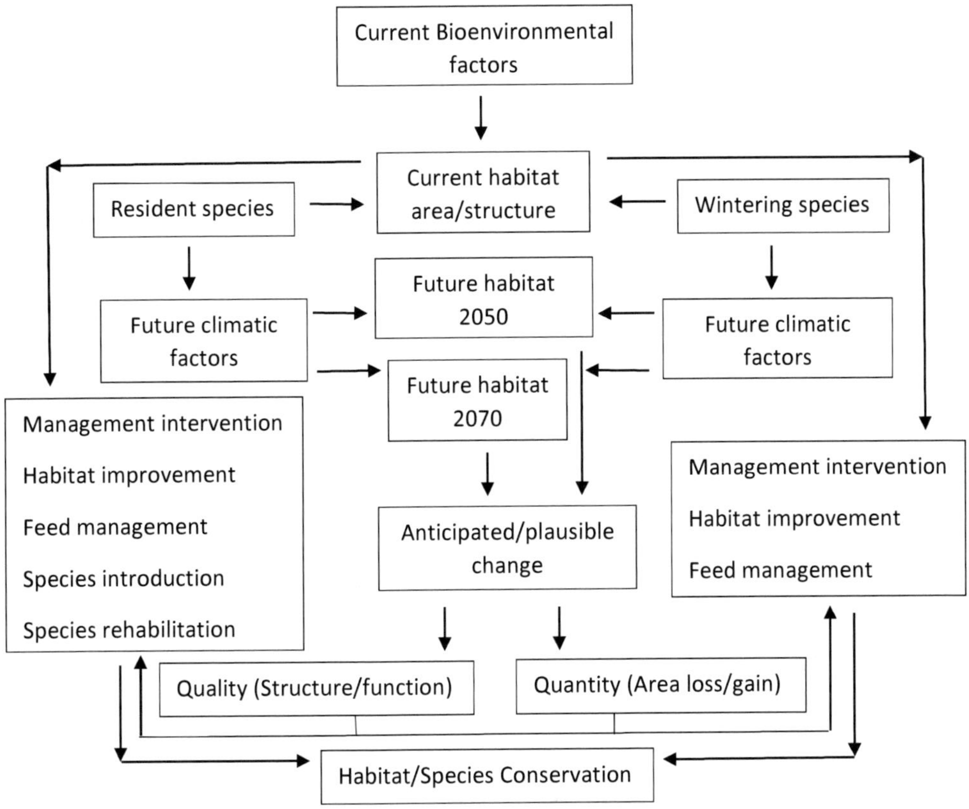

Figure 1. Conceptual model of climate change impact and habitat conservation for Vultures. Boxes showing habitats (current and future), habitat impacting factors (bioenvironmental and bioclimatic), climate impacted change (habitat quality and quantity) and management interventions for resident (left) and wintering (right) species. One-way arrows show links among different items.

Citizen Science

Vulture locations across the sub-continent presented in Figure 2 (vulture occurrence points) were collected using published literature and citizen science repository. Although, citizen science has its merits and demerits, as discussed in the following paragraphs, it was used to offset the stupendous, impractical and exorbitant task of surveys on temporal as well as spatial scales owing to a vast expanse of the sub-continent.

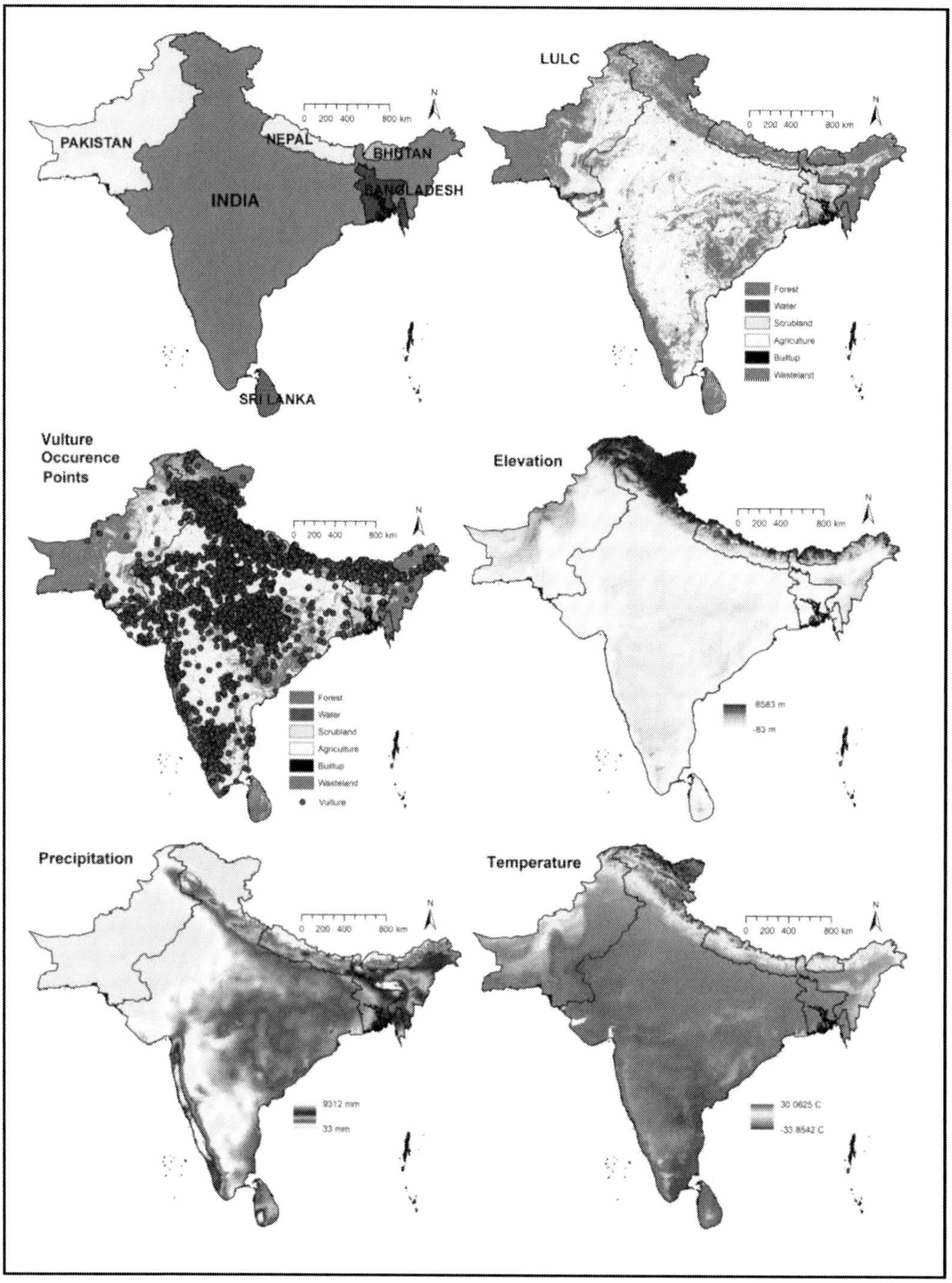

Figure 2. Maps showing features of Indian subcontinent, clockwise from top left: (a) Political boundaries of countries, (b) Land use landcover, (c) Elevation ranges, (d) temperature ranges, (e) rainfall ranges and (f) distribution of vulture locations.

Table 1. Published records used for collection of vulture occurrence data

Country	Authors	Publication
India	Ahmad et al., 2020	Ahmad, S., Rehman, E.U., Muhammad, K. & Fathul, B. 2020. Population Status and Nesting Activity of White-Rumped Vulture *Gyps bengalensis* in Kotli District, Azad Jammu and Kashmir, Pakistan Ornithological Science, 19(1): 55-61. https://doi.org/10.2326/osj.19.55.
	Anoop et al., 2020	Anoop, N.R., Babu, S., Nagarajan, R. & Sen S. 2020. Identifying suitable reintroduction sites for the White-rumped Vulture (*Gyps bengalensis*) in India's Western Ghats using niche models and habitat requirements. Ecological Engineering, 158 (2020) 106034. https://doi.org/10.1016/j.ecoleng.2020.106034.
	Ansari, 2015	Ansari, N.A. 2015. Status and distribution of Vultures in Gautam Budh Nagar District, Uttar Pradesh, India. International Journal of Advanced Research, 3(5): 506-511.
	Chakraborty, 2010	Chakraborty, R. 2010. Vultures are again in the sky. Records of the Zoological Survey of India, 110: 119-123.
	Chhangani, 2009	Chhangani, A. K. 2009. Status of vulture population in Rajasthan, India. Indian Forester, 135(2): 239-251.
	Das et al., 2011	Das, S.K., Dashahare, A., Marathe, S., Kundu, N. & Kesharwani, R. 2011. Status of raptors with special reference to vultures in and around Rajaji National Park, India. World Journal of Zoology, 6: 350–356.
	Jha, 2015	Jha, K. K. 2015. Distribution of vultures in Uttar Pradesh, India. Journal of Threatened Taxa, 7(1): 6750–6763.
	Jha, 2017	Jha, K. K. 2017. Vulture Atlas Central India – MP. Indian Institute of Forest Management, Bhopal, India.
	Jhala et al., 2017	Jhala, Y. V., Qamar, Q. & Saini, S. 2017. Status of vultures in tiger landscapes of India. National Tiger Conservation Authority, New Delhi and Wildlife Institute of India, Dehradun.
	Joshi, 2012	Joshi R. 2012. The Return of the Nature's Guard: Endangered Vulture's Population on Rise in Rajaji National Park, North India. International Journal of Environmental Protection, 2(8): 28-32.
	Kamboj, 2016	Kamboj, R.D., Tatu, K. & Munjpara, S.B. 2016. Status of vultures in Gujrat - 2016. Gandhinagar, Gujrat Ecological Education and Research (GEER) Foundation.
	Kaushik, 2014	Kaushik, M. 2014. Influence of habitat structure and anthropogenic disturbances on diurnal raptor community in Rajaji National Park, India. (Raptor Research & Conservation Foundation project completion report).
	Khatri, 2013	Khatri, P.C. 2013. Home Range Use of Winter Migratory Vultures in and around Jorbeer, Bikaner (Rajasthan) India. Bioscience Discovery, 4(1): 96-99.
	Khatri, 2015	Khatri, P.C. 2015. The king is disappearing: A study of socio-ecological aspects of King Vulture or Red-Headed Vultures (*Sarcogyps calvus*) at Jorbeer, Bikaner. International Journal of Geology Earth & Environmental Sciences, 5(1): 22-28.
	Kumar et al., 2014	Kumar, S., Meena, H., Jangid, P.K. & Nama, K.S. 2014. Current Status of Vulture Population in Chambal Valley of Kota, Rajasthan. Int. J. Pure App. Biosci., 2(5): 224-228.
	Majgaonkar et al., 2018	Majgaonkar, I., Bowden, C.G.R. & Quader, S. 2018. Nesting Success and Nest-Site Selection of White-rumped Vultures (*Gyps bengalensis*) in Western Maharashtra, India. J. Raptor Res., 52(4): 431–442.
	Misher, 2017	Misher, C., Bajpai, H., Bhattarai, B., Sharma, P., Sharma, R. & Kumar, N. 2017. Observations on the breeding of Indian long-billed vultures *Gyps indicus* at Gapernath, Chambal River in Rajasthan, India. Vulture News, 72: 14-21.
	Mishra et al., 2017	Mishra, S., Kumar, A. & Kanaujia A. 2017. Egyptian Vultures: The Unfortunate Victims. G- Journal of Environmental Science and Technology 5(3): 140-142.

Table 1. (Continued)

Country	Authors	Publication
	Navneethan, 2015	Navaneethan, B., K. Sankar, Q. Qureshi & M. Manjrekar 2015. The status of vultures in Bandhavgarh Tiger Reserve, Madhya Pradesh, central India. Journal of Threatened Taxa, 7(14): 8134–8138; http://dx.doi.org/10.11609/jott.2428.7.14.8134-8138
	Pande et al., 2013	Pande, S., Mestri, P., Deshpande, P., Warange, A. & Mahabal, A. 2013. Promising trend of *in situ* breeding of Oriental White-rumped Vulture *Gyps bengalensis* in Raigad District, Maharashtra, India: conservation implications for re-introduction of ex situ populations. Journal of Threatened Taxa, 5(7): 4106–4109.
	Purohit and Saran, 2013	Purohit, A. & Saran, R. 2013. Population Status and Feeding Behavior of Cinereous Vulture (*Aegypus monachus*): Dynamics and Implications for the Species Conservation in and Around Jodhpur. World Journal of Zoology, 8(3): 312-318. DOI: 10.5829/idosi.wjz.2013.8.3.74148.
	Rajkumar, 2015	Rajkumar, K. 2015. Recovery Plan for Vultures in Tamil Nadu. Report for Arulagam Society.
	Ramakrishnan et al., 2012	Ramakrishnan, B., Kannan, G., Samson, A., Ramkumar, K. & Ramasubramanian, S. 2014. Nesting of White-Rumped Vulture (Gyps Bengalensis) in the Segur Plateau of the Nilgiri North Forest Division, Tamilnadu, India. Indian Forester, 140(10): 1014-1018. http://www.indianforester.co.in
	Samson and Balasundaran, 2018	Samson, A. & Ramakrishnan, B. 2018. Long-billed vultures: Additional new breeding site of Long-billed vultures (*Gyps indicus*) in Moyar Valley, Tamil Nadu, Southern India. Bird-o-soar #09. In: *Zoo's Print* 33(1): 23:25.
	Samson et al., 2015	Samson, A., Baladundharam, R., Arunachalam, V. & Pandiyan, R. 2015. Occupation of Indian Giant Squirrel nests by White-rumped Vultures *Gyps bengalensis* in India. Podoces, 10(2): 35-36.
	Saran and Purohit, 2012	Saran, R. P. & Purohit, A. 2012. Eco-transformation and electrocution: A major concern for the decline in vulture population in and around Jodhpur. International Journal of Conservation Science, 3(2): 111-118.
	Saran, 2017	Saran, R. 2017. Population monitoring and annual population fluctuation of migratory and resident species of vultures in and around Jodhpur, Rajasthan. Journal of Asia-Pacific Biodiversity, 10: 342-348.
	Sasikumar and Vishnudas, 2016	Sasikumar, C. & Vishnudas, C.K. 2016. Conservation of the Critically Endangered Vultures in Wayanad and The Neighbouring Areas of Kerala as Part of Establishing a Vulture Safe Zone in South India. A Report of the CEPF-ATREE Western Ghats Small Grants Program 2013 – 2014.
	Singh and Bisht, 2017	Singh V. & Bisht S. S. 2017. Records of Critically Endangered Species of Avian fauna: Vultures *Gyps indicus* (Scopoli, 1786) and *Sarcogyps calvus* (Scopoli, 1786) from Surai Reserved Forest at Foot Hills of Kumaun Himalaya 140-147. In Arya M.K. & Bharti P.K. (2017) Experimental Animal Science - Birds and Insects. Discovery Publishing House Pvt. Ltd., New Delhi (India).
	Thakur, 2014	Thakur M.L. 2014. Breeding records and recent population trends of Himalayan Griffon (*Gyps himalayensis* Hume) in Himachal Pradesh, India. American Journal of Resource Communication, 2(3): 141-152.
	Thakur and Narang, 2012	Thakur, M.L. & S.K. Narang 2012. Population status and habitat-use pattern of Indian White-backed Vulture (*Gyps bengalensis*) in Himachal Pradesh, India. Journal of Ecology and the Natural Environment, 4(7): 173–180.
	Wani et al., 2020	Wani, H.M., Fazili, M.F., Charoo, S.A. & Ahmad R. 2020. Systematic survey on population of *Gyps himalayensis* in Hirpora Wildlife Sanctuary, Jammu and Kashmir, India. Eco. Env. & Cons, 26: S357-S359.
Bangladesh	Khan, 2013	Khan, M.M.H. 2013. Population, breeding and threats to the White-rumped Vulture *Gyps bengalensis* in Bangladesh. Forktail, 29: 52–56.
Nepal	Acharya et al., 2010	Acharya, R., Cuthbert, R., Baral, H.S., Chaudhary, A. 2010. Rapid decline of the Bearded Vulture *Gypaetus barbatus* in Upper Mustang, Nepal. Forktail, 26: 117–120.
	Bhusal and Paudel, 2016	Bhusal, K.P. & Paudel, K. 2016. Distribution and breeding ecology of Red-headed Vulture *Sarcogyps calvus* in Nepal. Danphe, 25:1-4.

Country	Authors	Publication
	Dhakal, 2014	Dhakal, H., Baral, K.M., Bhusal, K.P. & Sharma H.P. 2014. First record of nests and breeding success of Red-headed Vulture *Sarcogyps calvus* and implementation of Vulture Conservation Programs in Nepal. Ela Journal, 3(3): 9-15.
	Joshi et al., 2015	Joshi, M.K., Chalise, M.K., Chaudhary, A., Katuwal, H.B. 2015. Himalayan vultures in Khodpe far west Nepal: Is there any threat? Journal of Threatened Taxa, 7(14): 8128-8133.
	Khadka, 2013	Khadka K.K. & Mandal R.K. 2013. Understanding the Efficacy of Diclofenac Use-ban and its Aftermath Effect on Gyps Vulture in Koshi Tappu Wildlife Reserve, Nepal. Journal of Entomology and Zoology Studies, 1(4): 20-22.
	Paudel et al., 2016	Paudel, K., Galligan, T.H., Bhusal, K.P., Thapa, I., Cuthbert, R.J., Bowden C.G.R., Shah, R. & Pradhan, N.M.B. 2016. A decade of vulture conservation in Nepal. Conference Presentaion.
	Phuyal et al., 2016	Phuyal, S., Ghirmire, H.R., Shah, K.B. & Baral, H.S. 2016. Vultures and people: Local perceptions of a low-density vulture population in the eastern mid-hills of Nepal. Journal of Threatened Taxa, 8(14): 9597–9609.
Pakistan	Awan et al., 2017	Awan, M.S., Minhas, R.A., Ahmad, B. & Khan, A.A. 2017. Impact of Habitat Qualities on the Breeding Activities of Himalayan Griffons (*Gyps himalayensis* Hume, 1869): A Case Study from Azad Jammu and Kashmir, Pakistan. The Journal of Animal & Plant Sciences, 27(2): 627-641.
	Iqbal, 2011	Iqbal M.M. 2011. Distribution and Status of Oriental White-backed Vulture *Gyps bengalensis* in Nagar Parkar, Sindh, Pakistan.
	Mirbahar et al., 2016	Mirbahar N., Firdous F. & Ghalib S.A. 2016. National Vulture Conservation Strategy Project, Pakistan. A Baseline Report.
	Murn et al., 2014	Murn, C., Saeed, U., Khan, U., Iqbal, S. 2014. Population and spatial breeding dynamics of a Critically Endangered Oriental White-backed Vulture *Gyps bengalensis* colony in Sindh Province, Pakistan. Bird Conservation International, Available on CJO 2014. doi:10.1017/S0959270914000483.

Table 2. Variables used in the current vulture habitat prediction

Climatic				Environmental
Temperature		**Precipitation**		
Variable	**Acronym**	**Variable**	**Acronym**	
(i) Annual Mean Temperature	bio1	(xii) Annual Precipitation	bio12	LULC (1. Forest, 2. water, 3. scrubland, 4. agriculture, 5. built-up area, 6. wasteland)
(ii) Mean Diurnal Range (Mean of monthly (max temp - min temp))	bio2	(xiii) Precipitation of Wettest Month	bio13	Elevation
(iii) Isothermality (bio2/bio7) (* 100)	bio3	(xiv) Precipitation of Driest Month	bio14	
(iv) Temperature Seasonality (standard deviation *100)	bio4	(xv) Precipitation Seasonality (Coefficient of Variation)	bio15	
(v) Max Temperature of Warmest Month	bio5	(xvi) Precipitation of Wettest Quarter	bio16	
(vi) Min Temperature of Coldest Month	bio6	(xvii) Precipitation of Driest Quarter	bio17	
(vii) Temperature Annual Range (bio5-bio6)	bio7	(xviii) Precipitation of Warmest Quarter	bio18	
(viii) Mean Temperature of Wettest Quarter	bio8	(xix) Precipitation of Coldest Quarter	bio19	
(ix) Mean Temperature of Driest Quarter	bio9			
(x) Mean Temperature of Warmest Quarter	bio10			
(xi) Mean Temperature of Coldest Quarter	bio11			

Species distribution models (SDM) have become increasingly important in the last few decades for the study of biodiversity, macro ecology, community ecology and the ecology of conservation (Botella et al., 2017). Modelling the distribution of species usually requires a large amount of information collected over multiple years of standardized fieldwork (Stockwell and Peterson, 2002; Phillips et al., 2009; Bled et al., 2013). However, the long-term collection of broad-scale information on a wide range of species is prohibitively expensive (Peterson et al., 2008; Feldman et al., 2021). In the past decade, public participation in science (a.k.a. "citizen science") has gained widespread support and recognition (Bonney et al., 2014; Miller-Rushing et al., 2012; Jackson et al., 2015).

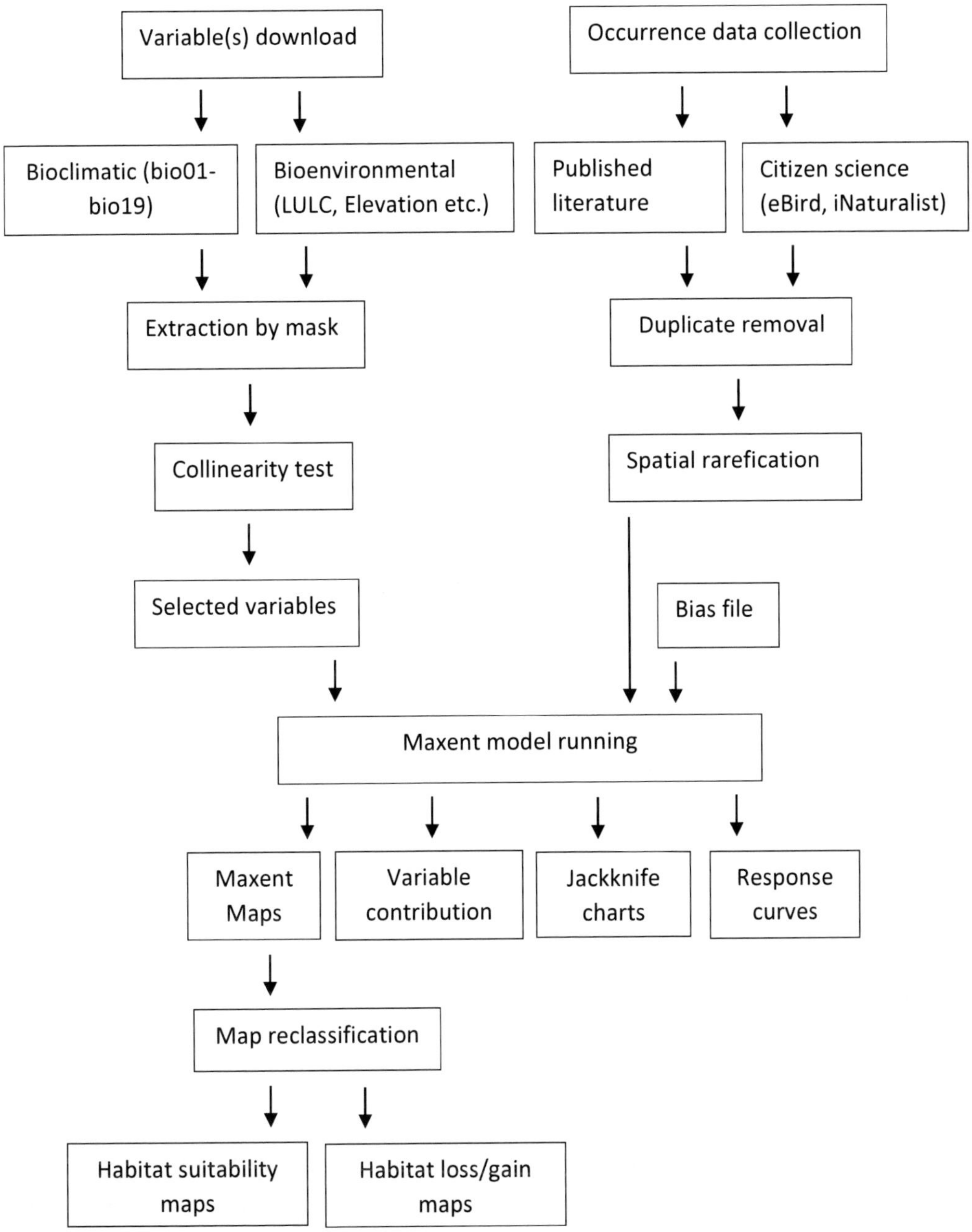

Figure 3. Flow chart of the methodology adopted in the study right from data acquisition and data preparation, running through modelling and getting desired output. Boxes indicate events and one way arrows the linkages and flow.

However, the most common contribution of citizen science is typically through wildlife observations (Tulloch et al., 2013). The field of ornithology provides an exemplary case of the same, with thousands of citizens across the world collecting information. Combined with advanced analytical approaches, the large amount of information derived from technology and citizen science offers unprecedented opportunities to fill the existing gaps in basic ecological knowledge (La Sorte and

Somveille, 2020; Santangeli et al., 2020). Opportunistic observations of species submitted by citizen scientists can be useful for monitoring distributions and abundance of elusive species (Bonney et al., 2009) and have been critical for documenting or predicting climate-driven range shifts for numerous species worldwide (Parmesan et al., 1999; Hickling et al., 2006; Breed et al., 2012; Abolafya et al., 2013; Jackson et al., 2015). However, despite the apparent benefits of citizen science monitoring, biases exist in citizen science data that may be minimized in field survey data. These include higher variability among data collected by citizen scientists compared to experts (Ericsson and Wallin, 1999; Genet and Sargent, 2003), over- or underestimates of species abundance (Bray and Schramm, 2001; Galloway et al., 2006) and misidentification of species (Brandon et al., 2003; Genet and Sargent, 2003; Jackson et al. 2015).

With the increasing availability of open access high-resolution land cover maps, environmental variables (e.g., temperature, precipitation), and species occurrence data from citizen science databases, the application of SDMs has become even more nuanced of late, and thus more accurate in estimating species distributions (Ramesh et al. 2018). Though, opportunistic occurrence data of the type collected through citizen science is known to be subject to reporting bias (Callcutt et al., 2018; Dickinson et al., 2010) but is often ignored (Croft et al. 2019; Wheeler et al., 2019). Moreover, the benefits of citizen science programs may outweigh their limitations (Crall et al., 2010; Danielsen et al., 2005) and newer, advanced statistical tools are continually being developed that can address potential biases in citizen science datasets (Kelling et al., 2009; Hegel et al., 2010; Jackson et al. 2015).

Maxent Model Selection

Some published literature identifies a trend of using single model and multiple model ensemble for prediction of habitat or potential species distribution in current and future scenario with varied emissions. At its early stage of development, SDM mainly relied on the application of a single technique or algorithm, like Artificial Neural Network, Bioclim, Domain, Generalised Additive Model, General Linear Model, Flexible Decrement Analysis, Support Vector Machine, Mahalnabis, Maximum Entropy, Multivariate Adaptive Regression Spline, Random Forest, etc. (Aguillar et al., 2016, Rew et al., 2020). The choice of the modelling tool depended on the author's involvement in its development, the availability of environmental and species occurrence data, tool performance, computing processing costs and training of users (Stohlgren et al., 2010). It was observed that different models gave varying results for the same species due to differences among algorithms. However, it was emphasised that ensemble prediction may be used in order to tap advantages of different algorithms used on one hand, and on the other to address the weakness of each individual algorithm (Araujo & New, 2007; Marmion et al., 2009). Ensemble modelling was thus used in several studies for single- and multi-species distribution analyses (Fitzpatrick et al., 2011), invasive species (Stohlgren et al., 2010; Poulos et al., 2012; Roura-Pascual et al., 2008), rare and endangered species prediction (Evangelista et al., 2008; Bombi et al., 2009), and forecasting distributions with climate change scenarios (Jarnevich & Stohlgren, 2009).

However, the modern trend has reflected the use of single Maxent model in tens of thousands of studies on various aspects of conservation biology in plants and animals (Feng et al., 2019). Maxent in particular, among other machine languages, is heavily favoured in the scientific community (Kaky et al., 2020) simply because it is one of the most reliable and statistically robust modelling softwares among well-established SDMs (Phillips et al., 2004, 2006; Elith et al., 2006; Wisz et al., 2008). This is also the most frequently used method for SDM/ENM and has been shown to perform better than many other models (Phillips et al., 2006; Elith et al., 2006; Banag et al., 2015). This has become the most popular modelling tool for both its ease of use and functionality (Morales et al., 2017; Mohammadi et al., 2019). This could be due to its higher predictive accuracy than any other method (Elith et al., 2006; Summers et al., 2012). For details chapter 3 of this book and Jha and Jha (2021) may be referred.

WHAT WAS ACHIEVED

The Pearson correlation test at ±0.7 threshold value extracted nine variables (LULC, aspect, bio1, bio2, bio3, bio12, bio14, bio15, bio19) for modelling and provided reduced collinearity effect on prediction. Duplicate removal and spatial rarefication of occurrence points at 10 km setting resulted in screening of 2357 (resident) and 1082 (migrant) presence locations to further improve the modeling. Model evaluation values of AUC ranged between 0.704-0.729 (residents) and 0.812-0.826 (migrants).

Model Determinants

As suggested by the variable contribution table, out of seven non-colinear bioclimatic covariates, top three (bio15 > bio12 > bio1) contributed > 75% and top five (bio15 > bio12 > bio1 > bio19 > bio3) contributed >90% in current habitat determination in resident vultures. In bioenvironmental model LULC (25%) took second place. Within LULC, built-up area, water, scrubland, forest, wasteland and agriculture contributed in decreasing order. Another bioenvironmental covariate 'aspect' contributed insignificantly (1%). In the case of migrant vultures and bioclimatic scenario, first three variables contributing 75% were bio1, bio14, bio15 while first five contributing 90% were bio1, bio14, bio15, bio12 and bio03. In bioenvironmental models LULC was placed first contributing 30%. Within LULC, built up, water, scrubland, forest, wasteland and agriculture contributed in decreasing order. In this case also the aspect contributed very low (<1%).

In the future scenarios (short and long term both) for resident vultures first three and five variables' average contribution was in following order: bio15, bio12, bio01 and bio15, bio12, bio01, bio 19 and bio14, respectively. First four variables were in similar order when compared with current climatic scenario of resident vultures. In the case of short-term and long-term scenarios for migrant vultures the average contribution of the first four variables decreased in the following order (bio1>bio14>bio15>bio12). The fifth variable was bio03 in short term and bio19 in long term.

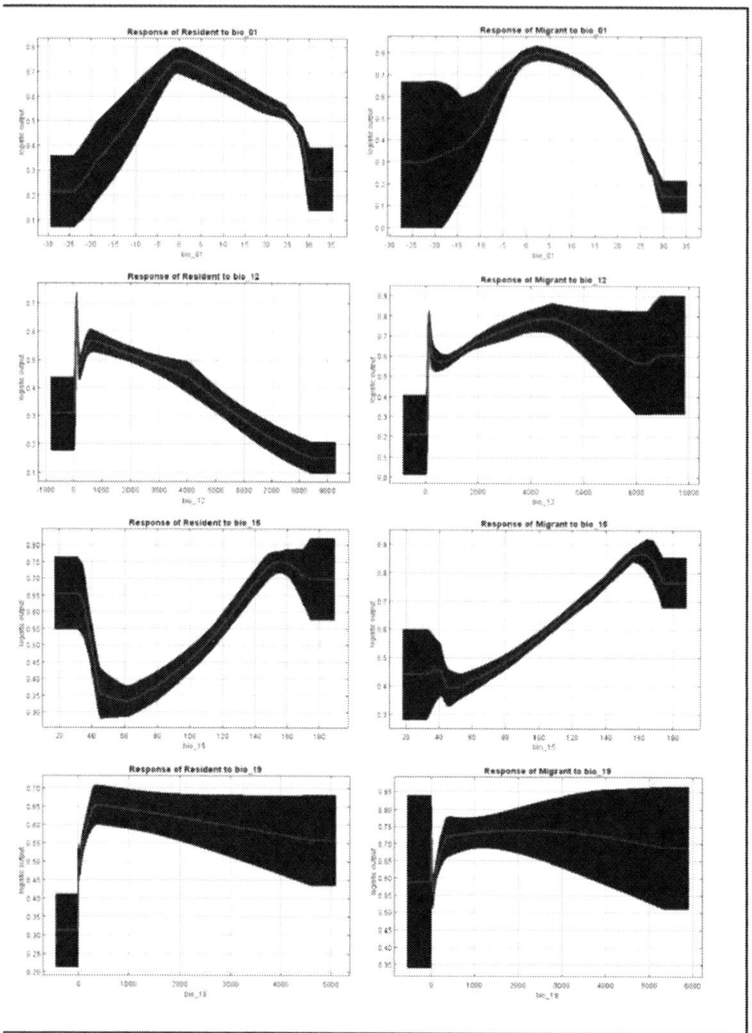

Figure 4. Response curves (bio01, bio12 and bio15) of three most important climatic variables showing probability of presence of vultures. Bio19 is also included from migratory vultures' point of view. Left column pertains to residents while right to migrants.

The results indicate that across all the categories and temporal scenarios, the most important variables influencing the prediction are bio1, bio12, bio15 and LULC. The subcontinental ranges of these three most important habitat determinants are

24.61-176.18 mm (precipitation seasonality, bio15), 33-9312 mm (annual mean precipitation, bio12) and -33.85-30.06°C (annual mean temperature, bio1). Since winter months are important for migratory vultures bio19 (precipitation of coldest quarter), only noncorrelated available variable, was considered as important for prediction. The subcontinental range of this variable (bio19) is 0-6604 mm. The jackknife test showed some variation from contribution table results. The response curves produced by the model along with Jackknife charts are presented in Figure 4 and 5.

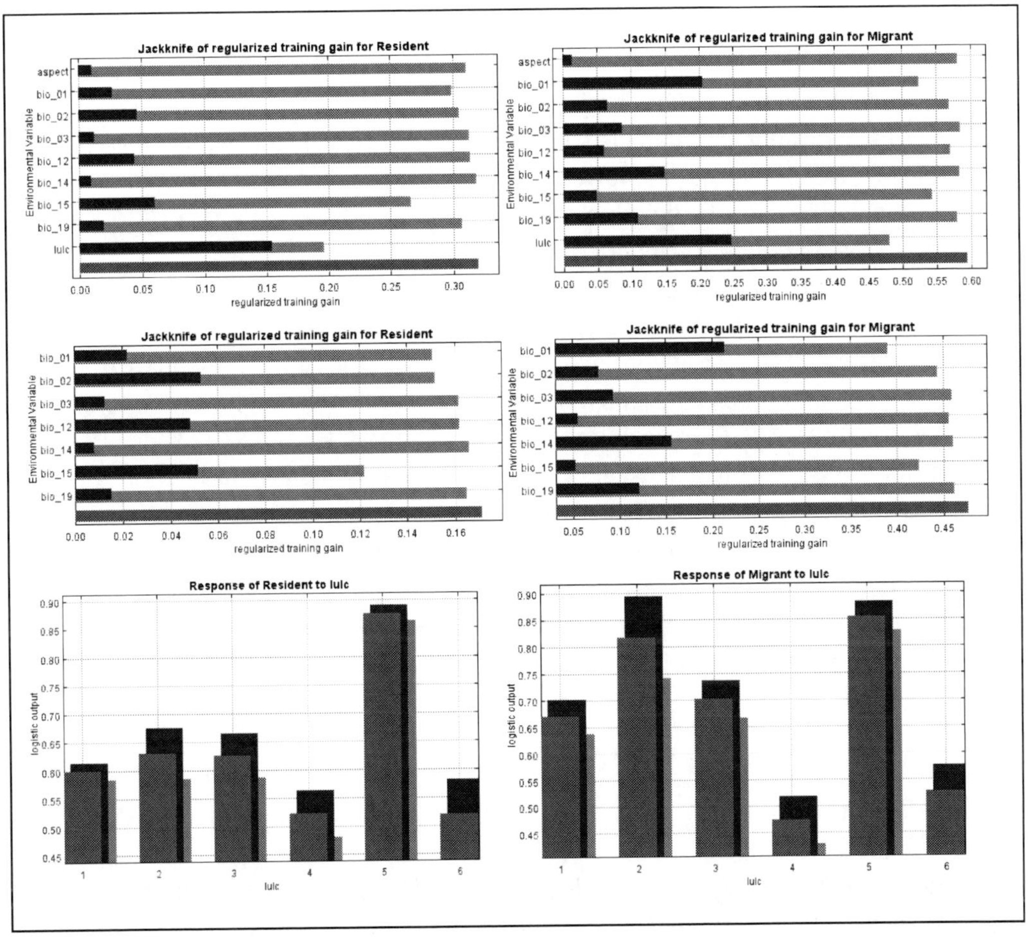

Figure 5. The Jackknife charts showing variables' importance in model prediction. Vertical bar charts depict importance of LULC components (1. Forest, 2. Water, 3. Scrubland, 4. Agriculture, 5. Builtup area and 6. Wasteland). Left column pertains to resident vultures while right to migrant vultures.

Habitat Suitability

Geographical area wise, the biggest country is India (71.9%) followed by Pakistan (19.3%), Bangladesh (3.2%), Nepal (3.2%), Sri Lanka (1.4%, and Bhutan (0.8%). Maldives being a very small country without available vulture records has been ignored for this study. The model predicted area availability under different scenarios and vulture groups are detailed in Table 3 and pictorially depicted in Figure 6 and 7. The total area of the subcontinent (4569483 km^2) is distributed as 2905650 km^2 (63.5%) and 1663833 km^2 (36.4%) suitable and unsuitable area, respectively for resident vultures in the current climatic projection. Corresponding figures for migratory vultures are 1427325 km^2 (31.2%) and 3142158 km^2 (68.7%), respectively. However, environmental models projected lower suitable area by 6.2% and 4.2% in resident and migratory vultures, respectively. India, Nepal and Bhutan followed a similar pattern of prediction for lower area in the case of resident vultures while Pakistan, Sri Lanka and Bangladesh showed the opposite pattern. In the case of migrant vultures this pattern was not evident.

On the subcontinent scale, the suitable area projected in the current scenario in the case of resident vultures decreased while it increased for migratory vultures except RCP2.6/2050. However, while comparing the future scenario among themselves, suitable area of the residents in the subcontinent in short term as well as long term decreased as emission scenario changed from RCP2.6 to RCP8.5. However, in the case of migratory vultures' suitable area increased in short term but decreased negligibly (2595 km^2) in long term. India, the largest contributor of the subcontinent, followed a similar pattern of decrease in residents from current to future but increase in migratory except RCP2.6 of 2050 and 2070. Within the future scenarios also, the resident vultures' area decreased but in the case of migratory vultures it contrasted (increased) in both the terms. Other countries occupying lesser area had mixed trends.

Habitat Dynamics

Projected suitable and unsuitable areas of the subcontinent or the different countries were found dynamic in the future scenarios as there was a loss from suitable areas or gain from unsuitable areas. This is presented in Table 4 and depicted in Figure 8 and 9. At the subcontinent level, in the case of resident vultures, loss is always greater than gain in all the four scenarios (RCP2.6/2050, RCP8.5/2050, RCP2.6/2070 and RCP8.5/2070) while in the case of migratory vultures, gain is more than loss except RCP2.6/2050. India followed a similar pattern in the case of resident vultures and migratory vultures except RCP2.6/2070. In other countries, the area dynamism response was mixed.

However, the quantum of loss at the level of the subcontinent, across all the four scenarios in the case of residents ranged between 83193 km^2 – 127758 km^2 while gain ranged between 34710 km^2 - 57961 km^2. India had a much larger share in change dynamics as compared to smaller countries.

How Did it Fare

Nine different species, although they may compete among themselves for resources, were clubbed in two groups for the purpose of this study – residents (Bearded Vulture *Gypaetus barbatus*, Egyptian Vulture *Neophron percnopterus*, Indian Vulture *Gyps indicus*, Red-headed Vulture *Sarcogyps calvus*, Slender-billed Vulture *Gyps tenuirostris* and White-rumped Vulture *Gyps bengalensis*) and wintering or migratory (Cinereous Vulture *Aegypius monachus*, Himalayan Griffon Vulture *Gyps himalayensis* and Eurasian Griffon Vulture *Gyps fulvus*). Moreover, Marshal et al., (2008) observed that competition is not clear among species if resources are abundant. Management requirements and conservation strategies are different for these two groups. Residents which form the bulk of the abundance, require habitat improvement, feed management, introduction and rehabilitation through the year while the migratory species, comparatively much lower in number, need only habitat improvement and feed management for a few winter months. However, the ultimate objective of the study was to find out coarse grained information needed by the managers for conservation of these species which are more alike than different (Jha and Jha, 2021). Keeping this in view the species and ecological differences were not required to be scaled.

Model Robustness

The AUC of ROC value is the preferred technique to test the accuracy of models (Stockwell and Peters, 1999; Bosch et al., 2014). Our models' performance is worth usable in achieving conservation objective as the AUC value was well within the range of 0-1 and fell in the category of "good performance" in AUC classification (AUC: >0.9 = very good, AUC: 0.7–0.9 = good, AUC: <0.7 = uninformative) (Swets 1988, Baldwin et al. 2009, Lv et al. 2012, etc.). Good model prediction category is well accepted in plants as well as animals (Songer et al., 2012; Khanum et al., 2013; Bosch et al., 2014; Banag et al., 2015; Jafari et al., 2018; Mohammadi et al., 2019; etc.) and considered suitable for conservation planning (Pearce and Ferrier 2000, Lobo et al. 2008, Zeng et al. 2015).

Table 3a. Current, short term (30 years) and long term (50 years) distribution of habitat category area (km^2) of vulture groups in varied scenarios in different countries

Political area	Suitability category	Current				Future 2050				Future 2070			
		Resident		Migrant		Resident		Migrant		Resident		Migrant	
		Environment	Climate	Environment	Climate	RCP2.6	RCP8.5	RCP2.6	RCP8.5	RCP2.6	RCP8.5	RCP2.6	RCP8.5
Sub- continent	Not suitable	1949561	1663833	3335935	3142158	1692234	1723802	3192387	3113371	1732765	1756882	3136763	3139358
	Moderate	2332408	2554565	999059	1191577	2548405	2494268	1141342	1215275	2493280	2469124	1183422	1197524
	High	287515	351085	234489	235747	328844	351412	235754	240836	343438	343477	249297	232601
	Suitable	2619923	2905650	1233548	1427325	2877249	2845680	1377096	1456111	2836717	2812601	1432720	1430125
	Total	4569483	4569483	4569483	4569483	4569483	4569483	4569483	4569483	4569483	4569483	4569483	4569483
India	Not suitable	1113968	837302	2412702	2236409	855250	872940	2280786	2222435	902823	918377	2238896	2230112
	Moderate	1910608	2120732	724520	898201	2126028	2083752	858894	909069	2064618	2059743	893147	909422
	High	263014	329556	150368	152981	306312	330898	147910	156086	320149	309470	155548	148055
	Suitable	2173622	2450288	874888	1051181	2432340	2414650	1006804	1065155	2384767	2369213	1048694	1057478
	Total	3287590	3287590	3287590	3287590	3287590	3287590	3287590	3287590	3287590	3287590	3287590	3287590
Pakistan	Not suitable	695624	697678	717888	707164	702511	712765	712154	697120	695325	708657	701799	712221
	Moderate	180110	182597	147305	161620	178586	168581	159504	172660	184115	168370	165564	158264
	High	6179	1638	16720	13129	816	567	10254	12133	2473	4886	14550	11429
	Suitable	186289	184235	164025	174749	179402	169148	169758	184793	186588	173256	180114	169692
	Total	881913	881913	881913	881913	881913	881913	881913	881913	881913	881913	881913	881913
Bangladesh	Not suitable	135073	140888	142160	142105	147925	147477	142636	140554	148460	148235	141683	142140
	Moderate	13387	7572	6358	6355	535	983	5824	7906	0	225	6777	6320
	High	0	0	0	0	0	0	0	0	0	0	0	0
	Suitable	13387	7572	6358	6355	535	983	5824	7906	0	225	6777	6320
	Total	148460	148460	148460	148460	148460	148460	148460	148460	148460	148460	148460	148460

Table 3b. Current, short term (30 years) and long term (50 years) distribution of habitat category area (km^2) of vulture groups in varied scenarios in different countries

Nepal	Not suitable	12045	5538	16884	9472	3803	6669	9592	5776	7068	6970	7401	7613
	Moderate	123990	128888	73575	81133	128714	126768	76339	81549	126170	119219	75524	78571
	High	11481	13090	57057	56911	14999	14078	61585	60191	14278	21326	64591	61333
	Suitable	135471	141978	130632	138044	143713	140846	137924	141740	140448	140546	140115	139903
	Total	147516	147516	147516	147516	147516	147516	147516	147516	147516	147516	147516	147516
Sri Lanka	Not suitable	9149	9343	58877	61348	10210	10801	61414	61927	4133	625	61366	62120
	Moderate	52758	53704	6418	3187	52795	53341	3452	3683	59104	61412	4244	3490
	High	3702	2563	315	1075	2605	1468	743	0	2373	3573	0	0
	Suitable	56460	56267	6733	4262	55400	54809	4195	3683	61477	64985	4244	3490
	Total	65610	65610	65610	65610	65610	65610	65610	65610	65610	65610	65610	65610
Bhutan	Not suitable	3793	1184	791	863	1265	1808	684	709	2275	1579	668.0996	771.2242
	Moderate	34578	37210	28900	30526	37129	36586	27685	30269	36119	36815	28403.44	30775.31
	High	23	0	8703	7006	0	0	10024	7417	0	0	9322.458	6847.469
	Suitable	34601	37210	37603	37532	37129	36586	37709	37686	3619	36815	37725	37662
	Total	38394	38394	38394	38394	38394	38394	38394	38394	38394	38394	38394	38394

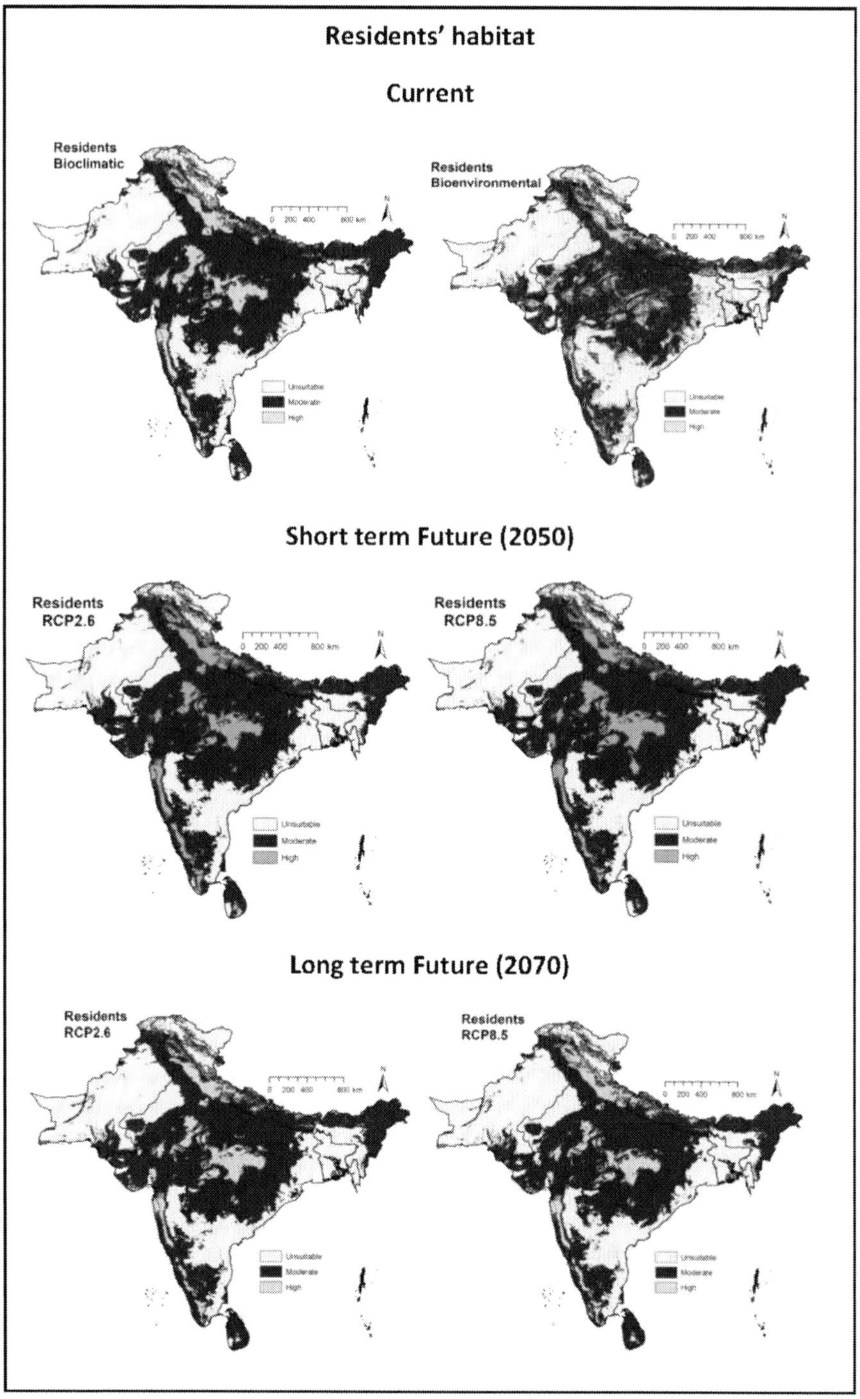

Figure 6. Maps showing expanse of different suitability categories for resident vultures in different countries of the subcontinent. Current climatic and environment models could be compared to note area reduction in the latter (top two maps). Changes in future from the current climatic model (top left) could be compared to future scenarios for area change (bottom left). Even two scenarios could be compared within themselves for change.

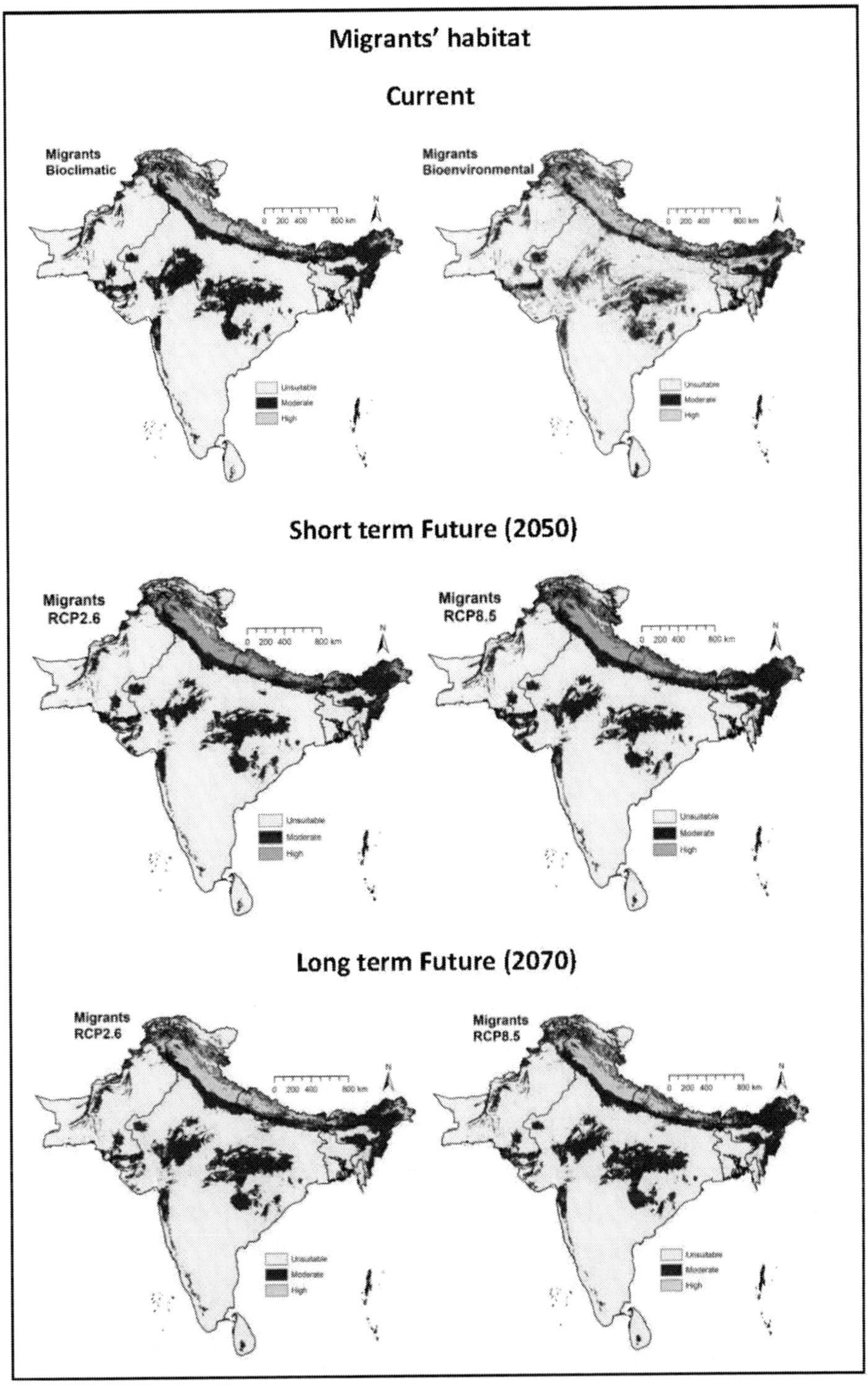

Figure 7. Maps showing expanse of different suitability categories for migratory vultures in different countries of the subcontinent. Current climatic and environment models could be compared to note area reduction in the latter (top two maps). Changes in future from the current climatic model (top left) could be compared to future scenarios for area change (bottom left). Even two scenarios could be compared within themselves for change.

Table 4a. Current, short term (30 years) and long term (50 years) distribution of stable (unsuitable and suitable) and unstable (gain and loss) habitat area (km^2) of vulture groups in varied scenarios in different countries

Political area	Stability category	Residents					Migrants				
		Current	Future (2050)		Future (2070)		Current	Future (2050)		Future (2070)	
			RCP2.6	RCP8.5	RCP2.6	RCP8.5		RCP2.6	RCP8.5	RCP2.6	RCP8.5
Subcontinent	Unsuitable	1663833	1609041	1616953	1605872	1629123	3142158	3091335	3038488	3052213	3069769
	Gain	0	54792	46880	57961	34710	0	50823	103670	89946	72389
	Loss	0	83193	106849	126894	127758	0	101052	74883	84550	69589
	Suitable	2905650	2822457	2798800	2778756	2777891	1427325	1326273	1352441	1342774	1357735
	Total	4569483	4569483	4569483	4569483	4569483	4569483	4569483	4569483	4569483	4569483
India	Unsuitable	837302	793365	796663	795842	816426	2236409	2194572	2154607	2164675	2174857
	Gain	0	43938	40639	41461	20876	0	41836	81802	71733	61552
	Loss	0	61886	76277	106981	101950	0	86214	67828	74221	55255
	Suitable	2450288	2388402	2374011	2343306	2348337	1051181	964968	983353	976961	995926
	Total	3287590	3287590	3287590	3287590	3287590	3287590	3287590	3287590	3287590	3287590
Pakistan	Unsuitable	697678	691099	695680	687216	692204	707164	700749	691394	693066	699387
	Gain	0	6579	1997	10461	5473	0	6415	15770	14098	7776
	Loss	0	11412	17085	8109	16453	0	11405	5726	8733	12833
	Suitable	184235	172823	167151	176127	167783	174749	163344	169024	166016	161916
	Total	881913	881913	881913	881913	881913	881913	881913	881913	881913	881913
Nepal	Unsuitable	5538	3758	5375	5464	5527	9472	7917	5728	7170	7602
	Gain	0	1780	163	74	11	0	1555	3744	2302	1870
	Loss	0	45	1294	1604	1444	0	1675	49	231	11
	Suitable	141978	141932	140684	140373	140534	138044	136369	137996	137813	138034
	Total	147516	147516	147516	147516	147516	147516	147516	147516	147516	147516

Table 4b. Current, short term (30 years) and long term (50 years) distribution of stable (unsuitable and suitable) and unstable (gain and loss) habitat area (km²) of vulture groups in varied scenarios in different countries

Bhutan	Unsuitable	1184	1009	1109	1179	1176	863	644	644	654	751
	Gain	0	175	75	6	9	0	219	219	208	112
	Loss	0	256	698	1097	404	0	65	65	14	21
	Suitable	37210	36953	36511	36113	36806	37531	37467	37467	37517	37511
	Total	38394	38394	38394	38394	38394	38394	38394	38394	38394	38394
Bangladesh	Unsuitable	140888	140888	140875	140888	140888	142105	141926	140532	141461	141801
	Gain	0	7037	12	0	0	0	178	1572	643	304
	Loss	0	0	6602	7572	7348	0	710	22	222	340
	Suitable	7572	535	971	0	225	6355	5646	6334	6133	6016
	Total	148460	148460	148460	148460	148460	148460	148460	148460	148460	148460
Srilanka	Unsuitable	9343	7281	5553	3185	575	61348	61010	61272	60775	61084
	Gain	0	2061	3790	6158	8768	0	338	76	573	264
	Loss	0	2929	5248	948	51	0	404	655	591	1036
	Suitable	56267	53339	51019	55319	56216	4262	3858	3607	3671	3226
	Total	65610	65610	65610	65610	65610	65610	65610	65610	65610	65610

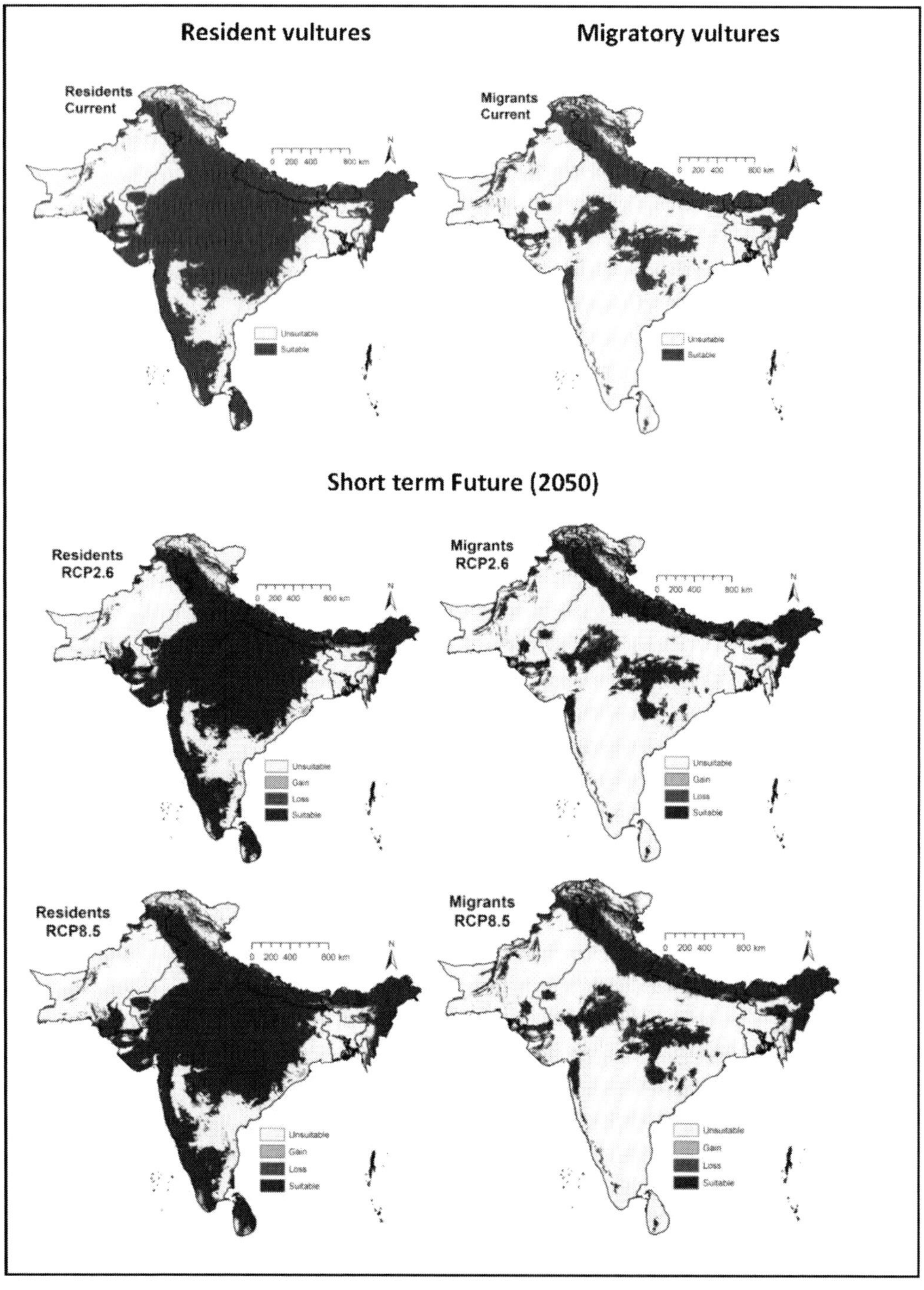

Figure 8. Maps showing stable and unstable habitats. Stable area of current and future are shown in yellow and dark blue while unstable area is shown in red and light blue. Changes may be noted in low (RCP2.6) and high (RCP8.5) emission scenario of resident vultures in left column and the same for migrants in right column. Residents and migrants may also be compared for similar scenario in horizontal rows.

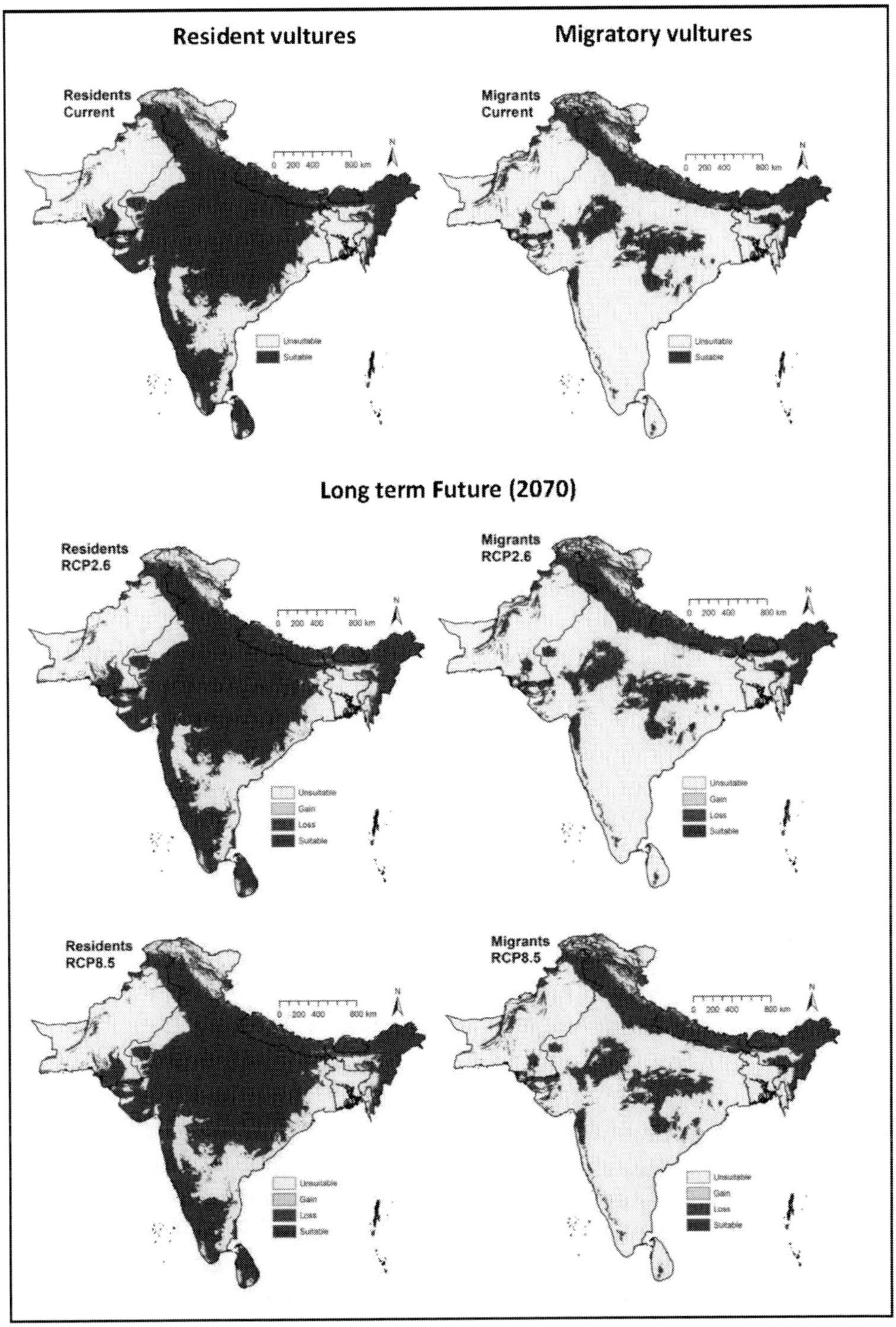

Figure 9. Maps showing stable and unstable habitats. Stable area of current and future are shown in yellow and dark blue while unstable area is shown in red and light blue. Changes may be noted in low (RCP2.6) and high (RCP8.5) emission scenario of resident vultures in left column and the same for migrants in right column. Residents and migrants may also be compared for similar scenario in horizontal rows.

Habitat Drivers

The finding of the present study is in accordance with previous reports suggesting rainfall, temperature, water and shelter influence the habitat suitability of vultures (Tyrberg, 2010; Howard et al. 2012; Hernandez-Baz et al. 2016; Thakur and Narang, 2012; Joshi et al., 2015; Kanaujia and Kushwaha 2014, Kushwaha, 2016, Jha et al., 2020). Even the importance of environmental components (e.g., LULC, Precipitation Seasonality, Annual Precipitation, and Annual Mean Temperature) in habitat prediction is in line with earlier works on vulture (Phipps et al., 2017; Anoop et al., 2020; Ilanloo et al. 2020; Jha and Jha 2020, 2021). However, within LULC, role of builtup area was most important than water or forest as suggested by Jha and Jha (2020) in a study at smaller scale.

Since the habitat of any species is influenced by a mix of different environmental factors, focusing on the response of an individual variable may hamper the efficiency of prediction. Therefore, all response curves generated by a machine learning interface should be interpreted with caution. On the whole, the response curve for bio01 (annual mean temperature) for residents indicates a suitable range of -20°C to 25°C whereas for migratory species the range is -10°C to 25°C. Upon further analysis of field conditions, it can be noted that the species preferring higher elevation (Himalayan Griffon among migratory and Bearded Vulture among resident species) occupy the negative side of the range whereas species dwelling in the plains occupy the positive side of the range. The covariant bio12 (annual mean precipitation) shows a decline in the probability of presence for resident species from 1000 mm onwards. Such decline from 5000 mm in migratory species makes it intriguing since, the visit of the migratory species to the Indian subcontinent coincides with the period which sees little to no rainfall (November to March) especially in the regions north of the Tropic of Cancer (23.5°N) where a bulk of the migratory species are seen. This reinforces the hypothesis of the impact of multi-covariates on modelling and a careful interpretation of any covariates in isolation and anyone could be following Liebig's Law of the Minimum (Golterman 1975).

Habitat Suitability

As evident from the results (Table 3 and 4) India, having maximum stable area (suitable area) for resident and migrant vultures currently as well as in the future, could be concentrated upon for vulture management by opting for different strategies such as, *in situ* conservation, introduction and habitat improvement. Nepal, Bhutan and Sri Lanka also have similar possibilities while those in Pakistan are not so favourable, though there are fair chances of conservation in limited areas. Bangladesh seems to be most unsuitable for vulture conservation efforts as the impact of the change in climate seems to be the maximum here. Additionally, the results indicate the importance of conserving shelter and foraging areas within

Bangladesh. Contrary to other countries and earlier reporting (Jha and Jha 2020, 2021) habitat predicted was little to no suitable area within Bangladesh with climatic factors, but with the incorporation of shelter determining factors, suitable area increased. Sri Lanka, where there are no reports of vulture locations, showed large swaths of suitable area which also remain stable in the future. This can be kept in mind for any future relocation plans especially if any urgent conservation measures need to be taken. Unstable area i.e., conversion of suitable to unsuitable and vice versa in future, requires proactive management of the habitat to have minimum impact on the shelter in almost all the countries of the subcontinent.

Management Implication

This study provides baseline information on the subject for vulture management which could be improved further by refining the input and methodology, as no model is perfect for want of all possible biological considerations. Future prediction models could be marginally overestimation since the LULC is not incorporated as a variable. Jha and Jha (2020, 2021) have shown that the current habitat models with LULC and without LULC project different area, which is lower in the case of models with LULC. Some improvement in prediction accuracy may be achieved by incorporating projected LULC in the models of future scenarios. However, the change in habitat suitability projected in the current study is negative for resident but positive for migrant vultures as the unsuitable area will increase in the case of the former while it will decrease in the case of the latter. These results highlight the importance of shelter and foraging area which are themselves dynamic in nature.

The availability of suitability classes and dynamism in habitat expanse in the future suggest proactive management is needed to compensate for the ill effects of climate change and anthropogenic activities on the habitat and possibly the species in future. Within forests which are home for most of the vultures in the subcontinent, the nesting/roosting cliff or potential nesting/roosting cliff (for *Gyps indicus*, *Neophron percnopterus*) must be kept undisturbed or unaffected from mining or other development projects to provide the vultures territory for expansion (Purohit and Saran, 2013). Similarly, the nesting/roosting trees (for *Gyps bengalensis*, *Gyps tenuirostris*, *Sarcogyps calvus*) or the potential nesting/roosting trees (known species with 8m-10m+ height in Table 5) should be kept out of any potential logging plans. In agriculture or foraging areas, smaller trees known for roosting/nesting should also be protected or rotation may be maximized. All the sheltering or foraging territory of vultures in Protected Areas, reserve forests and agriculture areas must be kept inviolate of diclofenac use (Jha et al., 2021). This must be done with the utmost urgency and strictness. It is worth mentioning at this point that in spite of the best efforts and intentions of banning the use of diclofenac by the government for several years now in the subcontinent, at least in India the ban has not been foolproof (Cuthbert et al., 2017; Galligan et al., 2020; Jha et al. 2021).

Table 5a. Different nesting/roosting trees reported in different regions of the subcontinent

Country	Tree species	Reference*
North India	*Pinus roxburghii*	Ahmad et al., 2020
	Pinus roxburghii	Thakur and Kataria, 2012; Thakur and Narang, 2012; Sehgal et al., 2020
	Haldunia longifolia, *Tectona grandis*, *Bombax ceiba*, *Termanalia tomentosa*, *Dalbergia sissoo* and *Thespesia populnea*	Das et al., 2011
	Ficus bengalensis, *F. religiosa* and *Spondias mangifera*, *Azadirachta indica*, *Tamarindus indica*, *Terminalia arjuna*, *Bombax ceiba*, *Dalbergia sissoo* or *D. latifolia*, *Mitragyna parvifolia*, *Balanitesa egyptica*, *Albizzia lebbeck*, *Acacia nilotica*, *A. leucophloea*, *Crateva nurvala*, and *Cocos nucifera*	Naoroji, 2006
	Haldina cordifolia, *Ficus racemose*, *Tamarindus indica*, *Dalbergia sissoo*, *Tectona grandis*, *Bombax ceiba*	Jha, 2015
South India	*Terminalia arjuna*, *Spondias mangifera*	Ramakrishnan et al., 2014
	Dalbergia lanceolaria, *Terminalia arjuna*	Ramesh et al., 2011
	Terminalia arjuna, *Spondias mangifera*, *Ficus benghalensis*, *Terminalia bellirica*, *Manilkara hexandra*	Samson et al., 2018
	Terminalia arjuna, *Spondias mangifera*	Samson and Ramakrishnan, 2020
Central India	*Acacia catechu*, *Acacia nilotica*, *Adina cordifolia*, *Aegle marmelos*, *Albizia lebbeck*, *Albizia procera*, *Anogeisus latifolia*, *Anogeisus pendula*, *Anthocephalus kadamba*, *Azadirachta indica*, *Bombax ceiba*, *Boswelia serrata*, *Butea monosperma*, *Casearia tomentosa*, *Chloroxylon swietenia*, *Dalbergia paniculata*, *Diospyros melanoxylon*, *Dipterocarpus trubinatus*, *Ficus benghalensis*, *Ficus glomerata*, *Ficus religiosa*, *Gmelina arborea*, *Hardwickia binate*, *Holoptelia integrifolia*, *Lagerstromia parviflora*, *Lannea coromandelica*, *Madhuca indica*, *Mallotus philippinensis*, *Mangifera indica*, *Miliusa tomentosa*, *Mytragyna parviflora*, *Pometia pinnata*, *Prosopis juliflora*, *Pterocarpus marsupium*, *Schleichera oleosa*, *Shorea robusta*, *Sterculia urens*, *Sygygium cuminii*, *Tamarindus indica*, *Tectona grandis*, *Terminalia arjuna*, *Terminalia belerica*, *Terminalia chebula*, *Terminalia tomentosa*, *Wrightia tinctoria*	Jha et al., 2020
	Terminalia arjuna, *Ficus religiosa*, *Terminalia tomentosa*, *Mytragyna parviflora*, *Madhuca latifolia*, *Terminalia belerica*, *Shorea robusta*, *Tectona grandis*, etc.	Jha et al., 2021
	Shorea robusta	Navaneethan et al., 2015
	Mangifera indica, *Cocos nucifera*, *Terminalia cuneata*, *Terminalia bellirica*, *Alstonia scholaris*, *Calophyllum inophyllum*, *Tetramelis nudiflora* *Thespesia populnea* and *Caryota urens*	Pande et al., 2013

*See Table 1 and for remaining Reference section.

Table 5b. Different nesting/roosting trees reported in different regions of the subcontinent

West India	*Cocos nucifera*	Gadhvi and Dodia, 2006
	Cocos nucifera, Terminalia, Mangifera	Majgaonkar et al. 2018
	Cocos nucifera, Prosopis cineraria, Azadiracta indica, Ficus	Paul, 2011
	Prosophis cinerara, Tecomela undulata	Purohit and Saran, 2013
	Prosopis cineraria, Acacia nilotica, Tecomella undulata, Ficus benghalensis, Lannea coromandelica	Chhangani, 2007
	Prosopis cineraria	Khatri, 2015
Nepal	*Bombax ceiba, Mangifera indica, Alstonia scholaris, Acacia catechu, Ficus lacor*	Baral et al., 2005
	Daubanga grandifolia	Dhakal et al., 2014
	Terminalia tomentosa, Pinus roxburghii, Shorea robusta, Bombax ceiba	Rana et al., 2019
	Terminalia tomentosa, Shorea robusta, Bombax ceiba	Subedi, 2009
Bangladesh	*Albizia lebbeck, Albizia procera, Anthocephalus chinensis, Bombax ceiba, Borassus flabellifer, Cocos nucifera, Ficus benghalensis, Ficus religiosa, Mangifera indica, Swietenia mahagoni.*	Khan, 2013
Pakistan	*Prosopis cineraria*	Mirbahar et al., 2016
	Prosopis cineraria, Azadirachta indica, Tecomella undulata, Tamarindus indica	Murn et al., 2014

*See Table 1 and for remaining Reference section.

Three different categories of habitat suitability area and the habitat likely to change in the future may also be strategized differently for conservation of species. Habitat improvement, carcass management in general and *in situ* conservation of species in highly suitable areas in particular, habitat maintenance in moderately suitable area for territory expansion, advance planning of unstable area and development of agroforestry model to provide trees for nesting and roosting and animal husbandry for carcass availability (Hiraldo et al., 1991; Chhangani 2007; Kambale, 2011; Khatri, 2013) are some of the strategies that may be considered. Thus, specialized as well as plastic vulture species may be provided favourable conditions to survive well (Arrondo et al. 2018; Genero et al., 2020). Above all, all the vulture landscape should be provided high security against anthropogenic activities.

CONCLUSION

No biological models are understood to be perfect since it is not possible to imitate nature through algorithms. Similarly, MaxEnt, like other models, may not be a close biological reality (Merow et al., 2013; Coxen et al., 2017), though this climate prediction model is the basic reference. However, this can be further improved by incorporating some more variables like, food availability, future LULC, etc. Incorporating a layer for future LULC in the model may overcome the overestimation in future projection (Van Asselen and Verburg, 2013; Smith et al., 2016; Thapa et al., 2021).

The present study provides comprehensive details of current vulture habitat suitability in the Indian subcontinent and its constituent countries. It also identifies climate vulnerable or unstable areas in the future in the form of loss and gain in suitable and unsuitable areas. It further confirms the hypothesis that precipitation and temperature along with vegetation influence the distribution of vulture species.

Some management interventions for vulture conservation namely, no anthropogenic disturbance in cliff nesting areas, protection of nesting trees, promotion of agroforestry may be suggested. Nevertheless, it would be practical to consider local conditions for implementation of specific management.

References

Abdelaal, M., Fois, M., Fenu, G. & Bacchetta G. (2019). Using MaxEnt modeling to predict the potential distribution of the endemic plant *Rosa arabica* Crép. in Egypt. *Ecological Informatics*, 00 (2019) XX–XX. https://doi.org/10.1016/j.ecoinf.2019.01.003.

Abolafya, M., Onmus, O., Sekercioglu, C. H. & Bilgin, R. (2013). Using citizen science data to model the distributions of common songbirds of Turkey under different global Climatic change scenarios. *PLoS ONE*, 8, e68037.

Aguilar, G. D., Waqa-Sakiti, H. & Winder, L. (2016). Using predicted locations and an ensemble approach to address sparse data sets for species distribution modelling: Long-horned beetles (*Ceram bycidae*) of the Fiji islands. Unitec ePress. Retrieved from http://www.unitec.ac.nz/epress/.

Ahmad, R., Khuroo, A. A., Hamid, M., Charles, B. & Rashid, I. (2019). Predicting invasion potential and niche dynamics of *Parthenium hysterophorus* (Congress grass) in India under projected climate change. *Biodiversity and Conservation*, https://doi.org/10.1007/s10531-019-01775-y.

Al-Qaddi, N., Vessella, F., Stephan, J., Al-Eisawi, D. & Schirone, B. (2017). Current and future suitability areas of kermes oak (*Quercus coccifera* L.) in the Levant under climate change. *Regional Environmental Change*, *17*, 143–156.

Angelieri, C. C. S., Adams-Hosking, C., Ferraz, K. M. P. Md. B., de Souza, M. P. & McAlpine, C. A. (2016). Using Species Distribution Models to Predict Potential Landscape Restoration Effects on Puma Conservation. *PLoS ONE*, *11*(1), e0145232. doi:10.1371/journal.pone.0145232.

Anoop, N. R., Babu, S., Nagarajan, R. & Sen, S. (2020). Identifying suitable reintroduction sites for the White-rumped Vulture (*Gyps bengalensis*) in India's Western Ghats using niche models and habitat requirements. *Ecological Engineering*, *158* (2020), 106034. https://doi.org/10.1016/j.ecoleng.2020.106034.

Araujo, M. & New, M. (2007). Ensemble forecasting of species distributions. *Trends in Ecology & Evolution*, *22*(1), 42–47. doi:10.1016/j.tree.2006.09.010.

Arrondo, E., Moleon, M., Cortes-Avizanda, A., Jimenez, J., Beja. P., Sanchez-Zapata, J. A. & Donazar, J. A. (2018). Invisible barriers: differential sanitary regulations constrain vulture movements across country borders. *Biological Conservation*, *219*, 46–52.

Baldwin, R. A. (2009). Use of maximum entropy modeling in wildlife research. *Entropy*, *11*, 854-866. doi:10.3390/e11040854.

Banag, C., Thrippleton, T., Alejandro, G. J., Reineking, B. & Liede-Schumann, S. (2015). Bioclimatic niches of selected endemic *Ixora* species on the Philippines: Predicting habitat suitability due to climate change. *Plant Ecology*, *216*, 1325–1340.

Baral, N., Gautam, R. & Tamang, B. (2005). Population status and breeding ecology of White-rumped Vulture *Gyps bengalensis* in Rampur Valley, Nepal. *Forktail*, *21*, 87–91.

Bled, F., Nichols, J. D. & Altwegg, R. (2013). Dynamic occupancy models for analyzing species' range dynamics across large geographic scales. *Ecology and Evolution*, *3*(15), 4896–4909. https://doi.org/10.1002/ece3.858 PMID: 24455124.

Boakye, M. K., Wiafe, E. D. & Ziekah, M. Y. (2019). Ethnomedicinal use of vultures by traditional medicinal practitioners in Ghana. *Ostrich*, June 2019. https://doi.org/10.2989/00306525.2019.1578834.

Bombi, P., Salvi, D., Vignoli, L. & Bologna, M. A. (2009). Modelling Bedriaga's rock lizard distribution in Sardinia: An ensemble approach. *Amphibia-Reptilia*, *30*(3), 413–424. doi:10.1163/156853809788795173.

Bonney, R., Shirk, J.L., Phillips, T.B., Wiggins, A. & Ballard, H.L. (2014). Next steps for citizen science. *Science*, *343*, 1436–1437.

Bosch, J., Mardones, F., Perez, A., Torre, A.L. & Munoz, M.J. (2014). A maximum entropy model for predicting wild boar distribution in Spain. *Spanish Journal of Agricultural Research*, *12*(4), 984-999. http://dx.doi.org/10.5424/sjar/2014124-5717.

Botella, C., Joly, A., Bonnet, P., Monestiez, P. & Munoz, F. (2017). A deep learning approach to Species Distribution Modelling. In: A. Joly; S. Vrochidis; K. Karatzas; A. Karppinen; P. Bonnet (Eds). *Multimedia Tools and Applications for Environmental & Biodiversity Informatics*, Springer, pp.169-199, 2018, Multimedia Systems and Applications Series, 978-3-319-76444- 3. ff10.1007/978-3-319-76445-0_10ff. ffhal-01834227. Accessed 29 May, 2021.

Brandon, A., Spyreas, G., Molano-Flores, B., Carroll, C. & Ellis, J. (2003). Can volunteers provide reliable data for forest vegetation surveys? *National Area Journal*, *23*, 254–261.

Bray, G. S. & Schramm Jr., H. L. (2001). Evaluation of a statewide volunteer angler diary program for use as a fishery assessment tool. *North American Journal of Fisheries Management*, *21*, 606–615.

Breed, G. A., Stichter, S. & Crone, E. E. (2012). Climate-driven changes in northeastern US butterfly communities. *Nature Climate Change*, 1–4. http://dx.doi.org/10.1038/NCLIMATE1663.

Brown, J., Bennett, J. & French, C. (2017). SDMtoolbox: a python-based GIS toolkit for landscape genetic, biogeographic, and species distribution model analyses. Methods in *Ecology and Evolution*, doi:PeerJ 5:e4095; DOI 10.7717/peerj.4095.

Buchhorn, M., Smets, B., Bertels, L., De Roo, B., Lesiv, M., Tsendbazar, N. E., Herold, M. & Fritz, S. (2020). *Copernicus Global Land Service: Land Cover* 100m: collection 3: epoch 2019: Globe 2020.

Callcutt, K., Croft, S. & Smith, G. C. (2018). Predicting population trends using citizen science data: Do subsampling methods produce reliable estimates for mammals? European *Journal of Wildlife Research*, *64*(3), 28. https://doi.org/10.1007/s10344-018-1189-7.

Chaudhry, M. J. I., Ogada, D. L., Malik, R. N., Virani, M. Z. & Giovanni, M. D. (2012). First evidence that populations of the critically endangered Long-billed Vulture *Gyps indicus* in Pakistan have increased following the ban of the toxic veterinary drug diclofenac in south Asia. *Bird Conservation International*, *22*, 389–397. doi:10.1017/S0959270912000445.

Chhangani, A. K. (2007). Sightings and nesting sites of Red-headed Vulture *Sarcogyps calvus* in Rajasthan, India. *Indian Birds*, *3* (6), 218–221.

Coxen, C. L., Frey, J. K., Carleton, S. A. & Collins D. P. (2017). Species distribution models for a migratory bird based on citizen science and satellite tracking data. *Global Ecology and Conservation*, *11*, 298e311. http://dx.doi.org/10.1016/j.gecco.2017.08.001.

Croft, S., Ward, A. I., Aegerter, J. N. & Smith, G. C. (2019). Modeling current and potential distributions of mammal species using presence-only data: A case study on British deer. *Ecology and Evolution*, 2019, 00:1–12. DOI: 10.1002/ece3.5424.

Cuthbert, R. J., Dave, R., Chakraborty, S. S., Kumar, S., Prakash, S., Ranade, S. P. & Prakash, V. (2017). Assessing the ongoing threat from veterinary non-steroidal anti-inflammatory drugs to Critically Endangered Gyps vultures in India. *Oryx*, *45*(3), 420–426. DOI:10.1017/S0030605311000135.

de Frutos, A., Olea, P. P. & Vera, R. (2007). Analyzing and modelling spatial distribution of summering lesser kestrel: The role of spatial autocorrelation. *Ecological Modelling*, *200*(1–2), 33–44.

Dickinson, J., Zuckerberg, B. & Bonter, D. (2010). Citizen science as an ecological research tool: Challenges and benefits. *Annual Review of Ecology*, *Evolution*, *and Systematics*, *41*, 149–172. https://doi.org/10.1146/annurev-ecolsys-102209-144636.

Elith, J. & Leathwick, J. R. (2009). Species distribution models: ecological explanation and prediction across space and time. *Annual Review of Ecology*, *Evolution*, *and Systematics*, *40*, 677-697.

Elith, J., Graham, C. H., Anderson, R. P., Dudík, M., Ferrier, S., Guisan, A., Hijmans, R. J., Huettmann, F., Leathwick, J. R., Lehmann, A., Li, J., Lohmann, L. G.,

Loiselle, B. A., Manion, G., Moritz, C., Nakamura, M., Nakazawa, Y., Overton, J. M. M., Peterson, A. T., Phillips, S. J., Richardson, K., Scachetti-Pereira, R., Schapire, R. E., Soberón, J., Williams, S., Wisz, M. S. & Zimmermann, N. E. (2006). Novel methods improve prediction of species' distributions from occurrence data. *Ecography*, *29*, 129–151.

Evangelista, P. H., Norman, J., Berhanu, L., Kumar, S. & Alley, N. (2008). Predicting habitat suitability for the endemic mountain nyala (*Tragelaphus buxtoni*) in Ethiopia. *Wildlife Research*, *35*(5), 409-416. doi: 10.1071/WR07173.

Fakhar-i-Abbas, Rooney, T. P., Haider, J. & Mian, A. (2013). Food limitation as a potentially emerging contributor to the Asian vulture crisis. *The Journal of Animal & Plant Sciences*, *23*(6), 1758-1760.

Farashi, A. & Erfani, M. (2018). Modeling of habitat suitability of Asiatic black bear (*Ursus thibetanus gedrosianus*) in Iran in future. *Acta Ecologica Sinica*, *38* (1), 9–14.

Farrell, A., Wang, G., Rush, S. A., Martin, J. A., Belant, J. L., Butler, A. B. & Godwin, D. (2019). Machine learning of large-scale spatial distributions of wild turkeys with high-dimensional environmental data. *Ecology and Evolution*, *2019*, 9, 5938–5949. DOI: 10.1002/ece3.5177.

Feldman, M. J., Imbeau, L., Marchand, P., Mazerolle, M. J., Darveau, M. & Fenton, N. J. (2021). Trends and gaps in the use of citizen science derived data as input for species distribution models: A quantitative review. *PLoS ONE*, *16*(3), e0234587. https://doi.org/10.1371/journal.pone.0234587.

Feng, X., Park, D. S., Liang, Y. Pandey, R. & Papeş, M. (2019). Collinearity in ecological niche modeling: Confusions and challenges. *Ecology and Evolution*, *2019*, 00:1–12. DOI: 10.1002/ece3.5555.

Feng, X., Walker, C. & Gebresenbet, F. (2017). Shifting from closed-source graphical-interface to open-source programming environment: a brief tutorial on running Maxent in R. *Peer J* Preprints, https://doi.org/10.7287/ peerj. preprints.3346v1.

Fitzpatrick, M. C., Sanders, N. J., Ferrier, S., Longino, J. T., Weiser, M. D. & Dunn, R. (2011). Forecasting the future of biodiversity: A test of single- and multi-species models for ants in North America. *Ecography*, *34*(5), 836–847. doi:10.1111/j.1600-0587.2011.06653.x.

Gadhvi, I. R. & Dodia, P. P. (2006). Indian White-backed Vultures *Gyps bengalensis* nesting in Mahuva, Bhavnagar district, Gujarat, India. *Indian Birds*, Vol. *2*(2), 36.

Galligan, T. H., Mallord, J. W., Prakash, V. M., Bhusal, K. P., Alam, A. B. M. S., Anthony, F. M., Dave, R., Dube, A., Shastri, K., Kumar, Y., Prakash, N., Ranade, S., Shringarpure, R., Chapagain, D., Chaudhary, I. P., Joshi, A. B., Paudel, K., Kabir, T., Ahmed, S., Azmiri, K. Z., Cuthbert, R. J., Bowden, C. G. R. & Green, R. E. (2020). Trends in the availability of the vulture-toxic drug, diclofenac, and other NSAIDs in South Asia, as revealed by covert pharmacy surveys. *Bird Conservation International*, (2020), page 1 of 17. DOI:10.1017/ S0959270920000477.

Galloway, A. W. E., Tudor, M. T. & Haegen, W. M. V. (2006). The reliability of citizen science: a case study of Oregon White Oak stand surveys. *Wildlife Society Bulletin*, *34*, 1425–1429.

Garza, G., Rivera, A., Barrera, C. S. V., Martinez-Ávalos, J. V., Dale, J. & Arroyo, T. P. F. (2020). Potential Effects of Climate Change on the Geographic Distribution of the Endangered Plant Species *Manihot walkerae*. *Forests*, *11*, 689; doi:10.3390/f11060689.

Gbogbo, F., Roberts, J. S. T. & Awotwe-Pratt, V. (2016). Some Important Observations on the Populations of Hooded Vultures *Necrosyrtes monachus* in Urban Ghana. *International Journal of Zoology*, *2016*, ID 7946172. http://dx.doi.org/10.1155/2016/7946172.

Genero, F., Franchini, M., Fanin, Y. & Filacorda S. (2020). Spatial ecology of non-breeding Eurasian Griffon vultures *Gyps fulvus* in relation to natural and artificial food availability. *Bird Study*, *67*(1), 53–70.

Genet, K. S. & Sargent, L. G. (2003). Evaluation of methods and data quality from a volunteer-based amphibian call survey. *Wildlife Society Bulletin*, *31*, 703–714.

Golterman, H. L. (1975). Physiological Limnology: An approach to the physiology of lake ecosystem (Eds.). *Developments in Water Science Series*. Elsevier Publishing company. Doi:10.1016/s0167-5648(08)71058-X.

Hegel, T. M., Cushman, S. A., Evans, J. & Huettmann, F. (2010). Current state of the art for statistical modelling of species distributions. In: Cushman, S.A., Huettmann, F. (Eds.), *Spatial Complexity*, *Informatics*, *and Wildlife Conservation*. Springer, Tokyo, pp. 273–311.

Hernandez-Baz, F., Romo, H., Gonzalez, J. M., Hernandez, M. J. M. & Pastrana R. G. (2016). Maximum entropy niche-based modeling (Maxent) of potential geographical distribution of *Coreura albicosta* (Lepidoptera: Erebidae: Ctenuchina) in Mexico. *Florida Entomologist*, *99*(3), 376-380. https://doi.org/10.1653/024.099.0306.

Hickling, R., Roy, D. B., Hill, J. K. & Fox, R. (2006). The distributions of a wide range of taxonomic groups are expanding polewards. *Global Change Biology*, *12*, 450–455.

Hiraldo, F., Blanco, J. C. & Bustamante, J. (1991). Unspecialised exploitation of small carcasses by birds. *Bird Study*, *38*(3), 200–207.

Hollings, T., Robinson, A., van Andel, M., Jewell, C. & Burgman, M. (2017). Species distribution models: A comparison of statistical approaches for livestock and disease epidemics. *PLoS ONE*, *12*(8), e0183626. https://doi.org/10.1371/journal.pone.0183626

Howard, A. M., Bernardes, S., Nibbelink, N., Biondid, L., Presotto, A., Fragaszy, D. M. & Madden, M. (2012). A maximum entropy model of the bearded capuchin monkey habitat incorporating topography and spectral unmixing analysis. ISPRS *Annals of the Photogrammetry, Remote Sensing and Spatial Information Sciences*, Volume *I-2*, 2012 XXII ISPRS Congress, 25 August – 01 September 2012, Melbourne, Australia.

Ilanloo, S. S., Khani, A., Kafash, A., Valizadegan, N., Ashrafi, S., Loercher, F., Ebrahimi, E. & Yousefi, M. (2020). Applying opportunistic observations to model current and future suitability of the Kopet Dagh Mountains for a Near Threatened avian scavenger. *Avian Biology Research*, *00*(0). https://doi.org/10.1177/1758155920962750.

IUCN, (2021). The IUCN Red List of Threatened Species. Accessed June 01, 2021. https://www.iucnredlist.org/species/22729731/117875047.

Jackson, M. M., Gergel, S. E. & Martin, K. (2015). Citizen science and field survey observations provide comparable results for mapping Vancouver Island White-tailed Ptarmigan (*Lagopus leucura saxatilis*) distributions. *Biological Conservation*, *181*, 162–172.

Jafari, A., Mirzaei, R. & Zamani-Ahmadmahmoodi, R. (2016). Species distribution modeling of wild sheep based on improving bias of occurrence records and selecting appropriate environmental predictors using MaxEnt. *Iranian Journal of Applied Ecology*, *5*(15), 39–49.

Jarnevich, C. S. & Stohlgren, T. J. (2009). Near term climate projections for invasive species distributions. *Biological Invasions*, *11*(6), 1373–1379. doi:10.1007/s10530-008-9345-8.

Jha, K. K. & Jha, R. (2020). Habitat suitability mapping for migratory and resident vultures: A case of Indian stronghold and species distribution model. *Journal of Wildlife and Biodiversity*, *4*(3), 91–111.

Jha, K. K. & Jha, R. (2021). Study of Vulture Habitat Suitability and Impact of Climate Change in Central India Using MaxEnt. *Journal of Resources and Ecology*, *12*(1), 30-42. DOI: 10.5814/j.issn.1674-764x.2021.01.004.

Jha, K. K., Campbell, M. O. & Jha, R. (2020). Vultures, their population status and some ecological aspects in an Indian stronghold. *Notulae Scientia Biologicae*, *12*(1), 124–142.

Jha, K. K., Jha, R. & Campbell, M. O. (2021).The Distribution, Nesting Habits and Status of Threatened Vulture Species in Protected Areas of Central India. Ecological Questions, *32,* (2021), 2 http://dx.doi.org/10.12775/EQ.2021.020.

Joshi, M. K., Chalise, M. K., Chaudhry, A. & Katuwal, H. B. (2015). Himalayan vultures in Khopde, far west Nepal: Is there any threat? *Journal of Threatened Taxa*, *7*(14), 8128–8133.

Kaky, E., Nolan, V., Alatawi, A. & Gilbert, F. (2020). A comparison between Ensemble and MaxEnt species distribution modelling approaches for conservation: A case study with Egyptian medicinal plants. *Ecological Informatics 60* (2020), 101150. https://doi.org/10.1016/j.ecoinf.2020.101150.

Kambale, A. A. (2011). *A study on breeding behaviour of oriental white backed vulture (Gyps bengalensis) in Anjarle and Deobag, Maharashtra*. Wildlife Institute of India, Dehradun.

Kanaujia, A. & Kushwaha, S. (2014). Relationship between Water Bodies and Vulture Populations in Bundelkhand, India. *Species*, *11*(27), 6-19.

Kelling, S., Hochachka, W. S. M., Fink, D., Riedewald, M., Caruana, R., Ballard, G. & Hooker, G. (2009). Data-intensive science: a new paradigm for biodiversity studies. *Bioscience, 59*, 613–620.

Khanum, R., Mumtaz, A. S. & Kumarm S. (2013). Predicting impacts of climate change on medicinal asclepiads of Pakistan using Maxent modelling. *Acta Oecologica*, *49*, 23-31.

Khatri, P. C. (2013). Home range use of migratory vultures in and around Jorbeer, Bikner (Rajasthan) India. *Bioscience Discovery*, *4*(1), 96–99.

Kumar, D. (2012). *The Indian subcontinent and its people. In: Genomics and health in developing world*. Oxford University Press., 1536 p.

Kushwaha, S. (2016). Utilisation of green plant material in nests of Long-billed Vultures *Gyps indicus* in Bundelkhand Region, India. *Vulture News*, 7–21.

La Sorte, F. A. & Somveille, M. (2020). Survey completeness of a global citizen-science database of bird occurrence. *Ecography*, *43*, 34–43.

Liang, Q., Xu, X., Mao, K., Wang, M., Wang, K., Xi, Z. & Liu, J. (2018). Shifts in plant distributions in response to climate warming in a biodiversity hotspot, the Hengduan Mountains. *Journal of Biogeography*, *5* (6), 1334–1344.

Lobo, J. M., Jiménez-Valverde, A. & Real, R. (2008). AUC: a misleading measure of the performance of predictive distribution models. *Global Ecology and Biogeography*, *17*, 145–151.

Lv, W., Li, Z., Wu, X., Ni, W. & Qv, W. (2012). Maximum entropy niche-based modeling (Maxent) of potential geographical distributions of *Lobesia botrana* (Lepidoptera: Tortricidae) in China. *International Conference on Computer and Computing Technologies in Agriculture: Computer and Computing Technologies in Agriculture.* V.D. Li and Y. Chen (Eds.): CCTA 2011, Part III, IFIP AICT 370, pp. 239–246, 2012. doi:10.1007/978-3-642-27275-2_26.

Markandya, A., Taylor, T., Longo, A., Murty, M. N., Murty, S. & Dhavala, K. (2008). Counting the cost of vulture decline: An appraisal of the human health and other benefits of vultures in India. *Ecological Economics*, *67*, 194–204.

Marmion, M., Parviainen, M., Luoto, M., Heikkinen, R. K. & Thuiller, W. (2009). Evaluation of consensus methods in predictive species distribution modelling. *Diversity and Distributions*, *5*, 59–69.

Marshal, J. P., Bleich, V. C. & Andrew, N. G. (2008). Evidence for interspecific competition between feral ass *Equus asinus* and mountain sheep *Ovis canadensis* in a desert environment. *Wildlife Biology*, *14*, 228–236.

McKean, S., Mander, M., Diederichs, N., Ntuli, L., Mavundla, K., Williams, V. & Wakelin, J. (2013). The impact of traditional use on vultures in South Africa. *Vulture News*, *65*, 15-36. http://dx.doi.org/10.4314/vulnew.v65i1.2.

Merow, C., Smith, M. J. & Silander, J. A. (2013). A practical guide to Maxent for modeling species' distributions: what it does, and why inputs and settings matter. *Ecography*, *36*, 1058e1069.

Miller-Rushing, A., Primack, R. & Bonney, R. (2012). The history of public participation in ecological research. *Frontiers in Ecology and the Environment*, *10*, 285–290.

Mohammadi, S., Ebrahimi, E., Shahriari, M. M. & Bosso, L. (2019). Modelling current and future potential distributions of two desert jerboas under climate change in Iran. *Ecological Informatics*, *52*, 7–13. https://doi.org/10.1016/j.ecoinf.2019.04.003.

Moleon, M., Sanchez-Zapata, J. A., Margalida, A., Carrete, M., Donazar, J. A. & Owen-Smith, N. (2014). Humans and scavengers: The evolution of interactions and ecosystem services. *BioScience*, *64*, 394–403.

Morales, N., Fernández, I. C. & Baca-González, V. (2017). MaxEnt's parameter configuration and small samples: Are we paying attention to recommendations? A systematic review. *PeerJ*, *5*, e3093. DOI: 10.7717/peerj.3093.

Oaks, J. L., Gilbert, M., Virani, M. Z., Watson, R. T., Meteyer, C. U., Rideout, B. A., Shivaprasad, H. L., Ahmed, S., Chaudry, M. J. I., Arshad, M., Mahmood, S., Ali, A. & Khan, A. A. (2004). Diclofenac residues as the cause of population decline of vultures in Pakistan. *Nature*, *427*, 630–633.

Ogada, D. L., Torchin, M. E., Kinnaird, M. F. & Ezenwa, V. O. (2012). Effects of vulture declines on facultative scavengers and potential implications for mammalian disease transmission. *Conservation Biology*, *26*, 453–60.

Page, L. M., MacFadden, B. J., Fortes, J. A., Soltis, P. S. & Riccardi, G. (2015). Digitization of biodiversity collections reveals biggest data on biodiversity. *BioScience*, *65*, 841-842.

Pain, D. J., Cunningham, A. A., Donald, P. F., Duckworth, J. W., Houston, D. C., Katzner, T., Parry-Jones, J., Poole, C., Prakash, V., Round, P. & Timmins, R. (2003). Causes and effects of temporospatial declines of *Gyps* vultures in Asia. *Conservation Biology*, *17*, 661–71.

Parmesan, C. (2006). Ecological and evolutionary responses to recent climate change. *Annual Reviews of Ecology*, *Evolution and Systematics*, *37*, 637-669.

Parmesan, C., Ryrholm, N., Stefanescu, C., Hill, J. K., Thomas, C. D., Descimon, H., Huntley, B., Kaila, L., Kullberg, J., Tammaru, T., Tennent, W. J., Thomas, J. A. & Warren, M. (1999). Poleward shifts in geographical ranges of butterfly species associated with regional warming. *Nature, 399*, 579–583.

Paul, V. (2011). *Role of Scavengers in Ecosystem*. Masters Thesis. Faculty of Landscape studies CEPT University, Ahmedabad.

Pearce, J. & Ferrier, S. (2000). Evaluating the predictive performance of habitat models developed using logistic regression. *Ecological Modelling*, *133*(3), 225–245.

Pearce, J. L. & Boyce, M. S. (2006). Modelling distribution and abundance with presence-only data. *Journal of Applied Ecology*, *43*, 405e412.

Perkins-Taylor, I. E. & Frey, J. K. (2020). Predicting the distribution of a rare chipmunk (*Neotamias quadrivittatus oscuraensis*): comparing MaxEnt and occupancy models. *Journal of Mammalogy*, *XX*(X), 1–14. DOI:10.1093/jmammal/gyaa057.

Peterson, A. T., Navarro-Siguenza, A. G. & Benıtez-Dıaz, H. (2008). The need for continued scientific collecting; a geographic analysis of Mexican bird specimens. *Ibis*, *140*, 288–294. https://doi.org/10.1111/j.1474-919x.1998.tb04391.x.

Phillips, S. J., Anderson, R. P. & Schapire, R. E. (2006). Maximum entropy modelling of species geographic distribution. *Ecological Modelling*, *190*, 231–259.

Phillips, S. J., Dudík, M., Schapire & R. E. (2004). A maximum entropy approach to species distribution modeling. In: Greiner R, Schuurmans D (eds.). *Proceedings of the 21st international conference on machine learning.* Alberta, Canada, 655–662.

Phillips, S. J., Dudik, M., Elith, J., Graham, C. H., Lehmann, A., Leathwick, J. & Ferrier, S. (2009). Sample selection bias and presence-only distribution models: implications for background and pseudo-absence data. *Ecological applications.*, *19*, 181–197. https://doi.org/10.1890/07-2153.1 PMID: 19323182.

Phipps, W. L., Diekmann, M., MacTavish, L. M., Mendelsohn, J. M., Naidoo, V., Wolter, K. & Yarnell, R. W. (2017). Due South: A first assessment of the potential impacts of climate change on Cape vulture occurrence. *Biological Conservation*, *210*, 16–25.

Plaza, P. I., Martínez-López, E. & Lambertucci, S. A. (2019). The perfect threat: Pesticides and vultures. *Science of The Total Environment*, *687*, 1207-1218. https://doi.org/10.1016/j.scitotenv.2019.06.160.

Poulos, H., Chernoff, B., Fuller, P. & Butman, D. (2012). Ensemble forecasting of potential habitat for three invasive fishes. *Aquatic Invasions*, *7*(1), pp. 59–72. doi:10.3391/ai.2012.7.1.007.

Prakash, V., Bishwakarma, M. C., Chaudhary, A., Cuthbert, R., Dave, R., Kulkarni, M., Kumar, S., Paudel, K., Ranade, S., Shringarpure, R. & Green, R. E. (2012). The Population Decline of *Gyps* Vultures in India and Nepal Has Slowed since Veterinary Use of Diclofenac was Banned. *PLoS ONE*, *7*(11), e49118. https://doi.org/10.1371/journal.pone.0049118.

Prockow, M., Prockow, J., Blazej, P. & Mackiewicz, P. (2018). The influence of habitat preferences on shell morphology in ecophenotypes of *Trochulus hispidus* complex. *Science of the Total Environment, 630*, 1036-1043.

Purohit, A. & Saran, R. (2013). Population Status and Feeding Behavior of Cinereous Vulture (*Aegypus monachus*): Dynamics and Implications for the Species Conservation in and Around Jodhpur. *World Journal of Zoology, 8*(3), 312-318. DOI: 10.5829/idosi.wjz.2013.8.3.74148.

Ramesh, T., Sankar, K. & Qureshi, Q. (2011). Status of vultures in Mudumalai Tiger Reserve, Western Ghats, India. *Forktail*, *27*, 96-97.

Ramesh, V., Gopalakrishna, T., Barve, S. & Melnick, D. J. (2017). IUCN greatly underestimates threat levels of endemic birds in the Western Ghats. *Biological Conservation*, *210*, 205–221.

Rana, D. B., Chaudhary, I. P., Dangaura, H. L., Joshi, A. B. & Bhusal, K. P. (2019). Monitoring of Nest and Breeding Status of White-rumped Vulture *Gyps bengalensis* in Nepal. Bird Conservation Nepal 2019. *Gidha Sandesh*, *8*, 28-31.

Rew, J., Cho, Y., Moon J. & Hwang, E. (2020). Habitat Suitability Estimation Using a Two-Stage Ensemble Approach. *Remote Sensing*, 2020, *12*, 1475; doi:10.3390/rs12091475.

Rogers, N. (2016). Museum drawers go digital. *Science*, *352*, 762-765.

Roura-Pascual, N., Brotons, L., Peterson, A. T. & Thuiller, W. (2008). Consensual predictions of potential distributional areas for invasive species: a case study of Argentine ants in the Iberian Peninsula. *Biological Invasions*, *11*(4), 1017–1031. doi:10.1007/s10530-008-9313-3

Sala, O. E., Chapin, F. S., Armesto, J. J., Berlow, E., Bloomfield, J., Dirzo, R., Huber-Sanwald, E., Huenneke, L. F., Jackson, R. B., Kinzig, A., Leemans, R., Lodge, D. M., Mooney, H. A., Oesterheld, M., Poff, N. L., Sykes, M. T., Walker, B. H., Walker, M. & Wall, D. H. (2000). Global biodiversity scenarios for the year 2100. *Science*, *287*(5459), 1770-1774. https://doi: 10.1126/science.287.5459.1770.

Salas, E. A. L., Valdez, R., Michel, S. & Boykin, K. G. (2018). Habitat assessment of Marco Polo sheep (*Ovis ammon polii*) in eastern Tajikistan: modeling the effects of climate change. *Ecology and Evolution*, *8* (10), 5124–5138.

Samson, A. & Ramakrishnan, B. (2020). The Critically Endangered White-rumped Vulture in Sigur Plateau, Western Ghats, India. *Population, breeding ecology, and threats*, *12*(13), 16752-16763. https://doi.org/10.11609/ jott.3004.12. 13.16752-16763.

Samson, A., Ramakrishnan, B., Kannan, G., Veeramani, A. & Ramasubramanian, S. (2018). Conservation Threats on Critically Endangered *Gyps* Vultures in the Tamil Nadu Part of the Nilgiri Biosphere Reserve, Southern India. In: C. Sivaperuman, K. Venkataraman (eds.), *Indian Hotspots*. Springer Nature Singapore Pte Ltd. 2018. https://doi.org/10.1007/978-981-10-6983-3_14.

Sanjerehei, M. M. & Rundel, P. W. (2017). The impact of climate change on habitat suitability for *Artemisia sieberi* and *Artemisia aucheri* (Asteraceae)—a modeling approach. *Polish Journal of Ecology*, *65*, 97–109.

Santangeli, A., Pakanen, V. M., Bridgeford, P., Boorman, M., Kolberg, H. & Sanz-Aguilar, A. (2020). The relative contribution of camera trap technology and citizen science for estimating survival of an endangered African vulture. *Biological Conservation*, *246,* (2020), 108593.

Sehgal, A., Kumar, K., Rajan, R. & Hore, U. (2020). Nest records, Nest Site selection of *Gyps bengalensis* White-rumped Vulture and the role of Feeding Station in Kangra, Himachal Pradesh. *bioRxiv - Ecology* Pub Date: 2020-07-29. DOI: 10.1101/2020.07.29.218362.

Sharma, H. P., Rimal, B., Zhang, M., Sharma, S., Poudyal, L.P., Maharjan, S., Kunwar, R., Kaspal, P., Bhandari, N., Baral, L., Dhakal, S., Tripathi, A., Karki, N., Khadki, B., Thapa, P., Acharya, B. K., Acharya, S., Baral, K. & Katuwal, H. B. (2020). Potential Distribution of the Critically Endangered Chinese Pangolin (*Manis pentadactyla*) in Different Land Covers of Nepal: Implications for Conservation. *Sustainability,* 2020, *12*, 1282; doi:10.3390/su12031282.

Smith, A., Schoeman, M. C., Keith, M., Erasmus, B. F. N., Monadjem, A., Moilanen, A. & Di Minin, E. (2016). Synergistic effects of climate and land-use change on representation of African bats in priority conservation areas. *Ecological Indicators*, *69*, 276–283. https://doi.org/10.1016/j.ecolind.2016.04.039.

Songer, M., Delion, M., Biggs, A. & Huang, Q. (2012). Modeling Impacts of Climate Change on Giant Panda Habitat. *International Journal of Ecology*, 2012, Article ID 108752, 12 pages. doi:10.1155/2012/108752.

Stockwell, D. & Peters, D. (1999). The GARP modelling system: Problems and solutions to automated spatial prediction. *International Journal of Geographic Information Science*, *13*(2), 143–158.

Stockwell, D. R. B. & Peterson, A. T. (2002). Effects of sample size on accuracy of species distribution models. *Ecological Modelling*, *148*, 1–13. https://doi.org/10.1016/s0304-3800(01)00388-x.

Stohlgren, T. J., Ma, P., Kumar, S., Rocca, M., Morisette, J. T., Jarnevich, C. S. & Benson, N. (2010). Ensemble habitat mapping of invasive plant species. *Risk Analysis: An Official Publication of the Society for Risk Analysis*, *30*(2), 224–35. doi:10.1111/j.1539-6924.2009.01343.x.

Subedi, P. (2009). Monitoring of Gyps species vulture in Nawalparasi district, Nepal. *Banko Janakari*, *18*(2), 35-43.

Sullivan, B. L., Wood, C. L., Iliff, M. J., Bonney, R. E., Fink, D. & Kelling, S. (2009). eBird: a citizen-based bird observation network in the biological sciences. *Biological Conservation*, *142*, 2282-2292.

Summers, D. M., Bryan, B. A., Crossman, N. D. & Myer, W. S. (2012). Species vulnerability to climate change: Impacts on spatial conservation priorities and species representation. *Global Change Biology*, *18*, 2335–2348.

Swets, J. A. (1988). Measuring the accuracy of diagnostic systems. *Science*, *240*, 1285–1293.

Thakur, M. L. & Kataria, R. C. (2012). Breeding Records and Nest Site Preference of Indian White-Backed Vulture in Kangra Valley of Himachal Pradesh, India. *International Journal of Science and Nature*, *3*(2), 350-353.

Thakur, M. L. & Narang, S. K. (2012). Population status and habitat-use pattern of Indian white-backed vulture (*Gyps bengalensis*) in Himachal Pradesh, India. *Journal of Ecology and the Natural Environment*, *4*(7), 173–180.

Thapa, S., Baral, S., Hu, Y., Huang, Z., Yue, Y., Dhakal, M., Jnawali, S. R., Chettri, N., Racey, P. A., Yu, W. & Wu, Y. (2021). Will climate change impact distribution of bats in Nepal Himalayas?, *Global Ecology and Conservation*, https://doi.org/10.1016/j.gecco.2021.e01483.

Tulloch, A. I. T., Possingham, H. P., Joseph, L. N., Szabo, J. & Martin, T. G. (2013). Realising the full potential of citizen science monitoring programs. *Biological Conservation*, *165*, 128–138.

Tyrberg, T. (2010). Avifaunal responses to warm climate: The message from last interglacial. *Records of the Australian Museum*, *62*, 193–205.

Tytar, V. M. & Nekrasova, O. D. (2016). Species Distribution Modelling of the Caspian Whipsnake *Dolichophis caspius* (Squamata: Serpentes): A Tool for ranking conservation priorities in the western Pontic Steppe. *Bulletin of St. Petersburg State University*. Ser. 3. Biology. Issue. 3 pp. 143–148. DOI: 10.21638/11701/spbu03.2016.324.

Van Asselen, S. & Verburg, P. H. (2013). Land cover change or land-use intensification: Simulating land system change with a global-scale land change model. *Global Change Biology*, *19*, 3648– 3667. https://doi.org/10.1111/gcb.12331.

Wheeler, P., Ward, A. I., Smith, G. C., Petrovan, S. O. & Croft, S. (2019). Careful considerations are required when analysing mammal citizen science data - a response to Massimino et al. *Biological Conservation*, *232*, 274–275. https://doi.org/10.1016/j.biocon.2019.01.021

Wisz, M. S., Hijmans, R. J., Li, J., Peterson, A. T., Graham, C. H., Guisan, A. & NCEAS Predicting Species Distributions Working Group. (2008). Effects of sample size on the performance of species distribution models. *Diversity and Distributions*, *14*, 763–773.

Zeng, Q., Zhang, Y., Sun, G., Duo, H., Wen, L. & Lei, G. (2015). Using Species Distribution Model to Estimate the Wintering Population Size of the Endangered Scaly-Sided Merganser in China. *PLoS ONE*, *10*(2), e0117307. doi:10.1371/journal. pone.0117307.

Zhang, J., Jiang, F., Li, G., Qin, W., Li, S., Gao, H., Cai, Z., Lin, G. & Zhang T. (2019°). Maxent modeling for predicting the spatial distribution of three raptors in the Sanjiangyuan National Park, China. *Ecology and Evolution*, *9*, 6643–6654. DOI: 10.1002/ece3.5243.

Zhang, K., Zhang, Y. & Tao, J. (2019b). Predicting the Potential Distribution of *Paeonia veitchii* (Paeoniaceae) in China by Incorporating Climate Change into a Maxent Model. *Forests, 10*, 190; doi:10.3390/f10020190.

EDITORS' CONTACT INFORMATION

Kaushalendra Kumar Jha
Professor,
Tropical Forest Ecosystem and Environment Management Unit,
Technical Forestry Faculty Area,
Indian Institute of Forest Management, Bhopal, India
Department of Forest, Environment and Climate Change,
Uttar Pradesh, Lucknow, India
Jhakk1959@gmail.com

Michael O'Neal Campbell
Lecturer,
Simon Fraser University,
British Columbia, Canada
ecol55@hotmail.com

INDEX

C

D

E

F

G

H

I

J

L

M

N

O

P

Q

R

S

X

Z